AF443696

Semiconductor Nanomaterials for Flexible Technologies

Semiconductor Nanomaterials for Flexible Technologies

From Photovoltaics and Electronics to Sensors and Energy Storage/Harvesting Devices

Edited by

Yugang Sun and John A Rogers

AMSTERDAM • BOSTON • HEIDELBERG • LONDON
NEW YORK • OXFORD • PARIS • SAN DIEGO
SAN FRANCISCO • SYDNEY • TOKYO
William Andrew is an imprint of Elsevier

William Andrew is an imprint of Elsevier
The Boulevard, Langford Lane, Kidlington, Oxford OX5 1GB, UK
30 Corporate Drive, Suite 400, Burlington, MA 01803, USA

First edition 2010

British Library Cataloguing in Publication Data
A catalogue record for this book is available from the British Library

Library of Congress Cataloging-in-Publication Data
A catalog record for this book is available from the Library of Congress

ISBN–13: 978-1-43-777823-6

For information on all William Andrew publications
visit our website at books.elsevier.com

Printed and bound in USA

10 11 12 13 14   10 9 8 7 6 5 4 3 2

Working together to grow
libraries in developing countries

www.elsevier.com | www.bookaid.org | www.sabre.org

ELSEVIER   **BOOK AID** International   **Sabre Foundation**

Contents

CHAPTER 7 Zinc Oxide Nanowire Arrays on Flexible Substrates: Wet Chemical Growth and Applications in Energy Conversion . . . 197

Sheng Xu, Benjamin Weintraub and Zhong Lin Wang

CHAPTER 8 Flexible Energy Storage Devices Using Nanomaterials 227

Victor L. Pushparaj, Subbalakshmi Sreekala, Omkaram Nalamasu and Pulickel M. Ajayan

CHAPTER 9 Flexible Chemical Sensors 247

H. Hau Wang

CHAPTER 10 Mechanics of Stiff Thin Films of Controlled Wavy Geometry on Compliant Substrates for Stretchable Electronics 275

Jianliang Xiao, Hanqing Jiang, Yonggang Huang and John A. Rogers

Contact Information

[2]IBM, TJ Watson Research Center, Yorktown Heights,
New York, NY 10598, USA
E-mail: qcao@us.ibm.com

CHAPTER 5 **Flexible Field Emitters Based on Carbon Nanotubes and Other Materials**

Soo-Hwan Jeong[1] and Kun-Hong Lee[2]
[1]Department of Chemical Engineering, Kyungpook National
University, Daegu, Korea
[2]Department of Chemical Engineering, Pohang University
of Science and Technology, Pohang, Korea
E-mail: ce20047@postech.ac.kr

CHAPTER 6 **Flexible Solar Cells Made of Nanowires/Microwires**

Jongseung Yoon[1], Yugang Sun[2] and John A. Rogers[3]
[1]Department of Materials Science and Engineering, University
of Illinois at Urbana–Champaign, Urbana, IL 61801, USA
[2]Center for Nanoscale Materials, Argonne National
Laboratory, Argonne, IL 60439, USA
[3]Department of Materials Science and Engineering, Frederick
Seitz Materials Research Laboratory, University of Illinois at
Urbana–Champaign, Urbana, Illinois, USA
E-mail: ygsun@anl.gov; jrogers@uiuc.edu

CHAPTER 7 **Zinc Oxide Nanowire Arrays on Flexible Substrates: Wet Chemical Growth and Applications in Energy Conversion**

Sheng Xu, Benjamin Weintraub and Zhong Lin Wang
School of Materials Science and Engineering, Georgia Institute
of Technology, Atlanta, GA 30332-0245, USA
E-mail: zhong.wang@mse.gatech.edu

CHAPTER 8 **Flexible Energy Storage Devices Using Nanomaterials**

*Victor L. Pushparaj[1], Subbalakshmi Sreekala[1], Omkaram
Nalamasu[1] and Pulickel M. Ajayan[2]*
[1]Materials Science and Engineering Department, Rensselaer
Polytechnic Institute, Troy, New York, USA
[2]Department of Mechanical Engineering & Materials Science,
Rice University, Houston, Texas, USA
E-mail: victorpushparaj@gmail.com

Preface

Mechanically flexible systems for photovoltaics, electronics, sensing and energy storage/harvesting provide capabilities that cannot be achieved with conventional technologies. Examples include light weight, rugged form factors, wearable designs and rollable layouts. Most work in these areas involves the use of amorphous and/or polycrystalline thin films of organic or inorganic active materials that yield mechanically flexible systems when integrated on bendable substrates, such as plastic sheets or stainless steel foils. The main disadvantage of such systems is that the morphologies of the active films typically limit the performance to levels that are much lower than those obtainable with single-crystalline materials. A number of books and comprehensive reviews have been published on such approaches. This book summarizes newer work that exploits nanoscale, monocrystalline semiconductor building blocks to achieve simultaneously mechanical flexibility and high performance. Chapter 1 presents an overview of the synthesis of organic semiconductor microcrystals and nanowires with single crystallinity as well as flexible thin-film transistors (TFTs) that incorporate them. Inorganic nanowires, nanoribbons and nanomembranes synthesized via both bottom–up chemical synthesis and top–down lithographic fabrication are reviewed in Chapters 2 and 3, respectively, together with demonstrations of flexible electronic devices and circuits based on the resulting nanomaterials. Performance in these cases is similar or even superior to those of comparably designed devices on traditional rigid substrates. Carbon nanotubes can also be used as active materials for high performance of TFTs and circuits (Chapter 4) and field emitters (Chapter 5). Nanotubes offer potential advantages in mechanical and electrical properties over inorganic nanomaterials.

In addition to electronics, this book covers flexible devices for energy harvesting and conversion, energy storage and chemical sensing. Chapter 6 reviews flexible photoelectrochemical cells made of highly ordered arrays of nanowires and nanotubes for converting solar energy into electricity. Flexible solar cell modules based on arrays of monocrystalline silicon microcells are also extensively discussed in this chapter. Converting mechanical energy (vibration, motion, etc.) into electricity based on piezoelectric properties of

ZnO nanowires is highlighted in Chapter 7, with details covering the growth of ZnO nanowires and device fabrication and characterization. Some examples of energy storage devices with mechanical flexibility are presented in Chapter 8. By taking advantage of the high surface areas and chemical reactivities associated with nanomaterials, flexible chemical sensors with high sensitivities can be achieved. Such designs have potential for sensing widely ranging classes of analytes including biological molecules, pollutants and explosives (Chapter 9). Through clever mechanical designs, devices that offer not only flexibility but also fully reversible responses to large strain deformation, i.e. stretchability, are possible. This type of mechanical response can be used to advance electronic eye imagers, bending actuators, synthetic sensitive skins, structural health monitors and other devices that demand more than simple flexibility. The last chapter reviews the mechanics of nanoribbons with controlled wavy geometries on rubber substrates as one approach to stretchable devices.

Finally, we would like to thank all of the contributing authors for their time and effort in preparing the manuscripts presented in this book. Their work and that of others in this emerging area are essential to the vibrancy and fast pace of progress in this field. The community includes research sectors in academia, national laboratories and industry, spanning disciplines from chemistry and materials science to mechanical engineering, device physics and electrical engineering. We hope that this book can serve as a useful reference for those new to this field of research or already engaged in it, from graduate students to postdoctoral fellows and practicing researchers, from scientists to engineers.

Yugang Sun
John A. Rogers

Acknowledgment

Reference in this work to any specific commercial product is to facilitate understanding and does not necessarily imply endorsement by the U.S. Department of Energy.

CHAPTER 1

Flexible Organic Single-Crystal Field-Effect Transistors

Alejandro L. Briseno

Polymer Science and Engineering, University of Massachusetts, Amherst, USA

INTRODUCTION

The majority of research on flexible electronics has been focused on poly-crystalline organic thin films [1, 2], carbon nanotubes [3, 4] and inorganic structures [5]. Much less work has been accomplished in determining the limits of flexibility of organic single-crystal transistors [6, 7]. Organic single crystals have been limited to fundamental charge-transport studies and for determining the performance limitation of organic semiconductors [8–10]. Their high mobilities and outstanding electrical characteristics would make them promising candidates for electronic applications such as drivers for active matrix displays and sensor arrays. However, poor mechanical properties of bulk crystals and low throughput in device fabrication have prevented their use in flexible electronics and other wide-ranging applications. Thus, there is a strong need for the development of mechanically flexible, non-destructive, single-crystal devices with prospective applications in flexible electronics while maintaining the intrinsic properties and characteristics of organic single crystals. Our interest lies in both the fundamental aspect of device mechanics [11] and the practical applications of flexible organic single crystals.

In this chapter strategies for fabricating unconventional organic single crystals on mechanically flexible substrates are reviewed. Figure 1.1 shows a schematic of a flexible transistor test structure. The design shows a three-terminal device fabricated from a flexible substrate [either Kapton or transparent polyethylene terephthalate (PET)], a backside gate evaporated from gold, a dielectric insulating layer spun from poly-4-vinylphenol (PVP), top-contact source-drain

CONTENTS

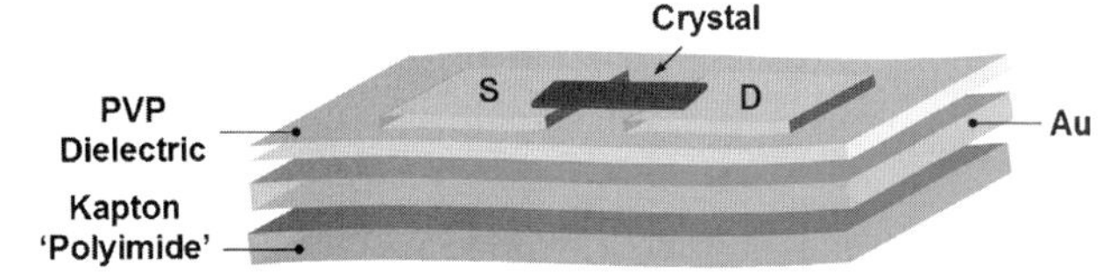

FIGURE 1.1 *Schematic of a bottom-contact flexible organic single-crystal field-effect transistor. S: source; D: drain.*

electrodes evaporated from gold and the semiconductor single-crystal active layer. The distance between the source and drain electrode is referred to as the channel length (L), while the perpendicular distance is referred to as the channel width (W). Applying a voltage to the gate electrode will form a conductive region at the insulator–semiconductor interface. A second voltage applied to the drain electrode will collect the mobile charges and enable the device to generate current across the source–drain. The effective mobility of the semiconductor can be calculated from the saturation regime using the equation $I_{DS} = (W \cdot C \cdot \mu/2 \cdot L)(V_G - V_T)$ [2], where L is the channel length, W is the channel width, C is the capacitance per unit area of the insulating layer, V_T is the threshold voltage, V_G is the gate voltage, and μ is the field-effect mobility.

This chapter will specifically focus on three classes of flexible organic single-crystal structures: (1) ultrathin macroscopic single crystals, (2) micropatterned single crystals, and (3) single-crystal nanowires. Figure 1.2 shows optical photographs of three representative single-crystal transistors fabricated from the three respective crystal forms. First, the use of field-effect transistors fabricated from ultrathin and conformable macroscopic organic single crystals is described [7, 13]. The use of ultrathin single crystals serves as a proof-of-concept for 'flexible' organic single-crystal field-effect transistors with performances exceeding those of previously reported organic thin-film flexible devices. In the second class of single-crystal structures, large-area patterning of organic single crystals onto plastic substrates is reviewed [6, 14–16]. This method is materials-general and yields single-crystal mobilities approaching

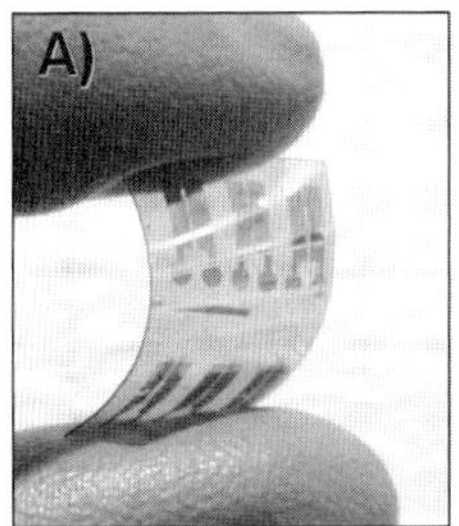

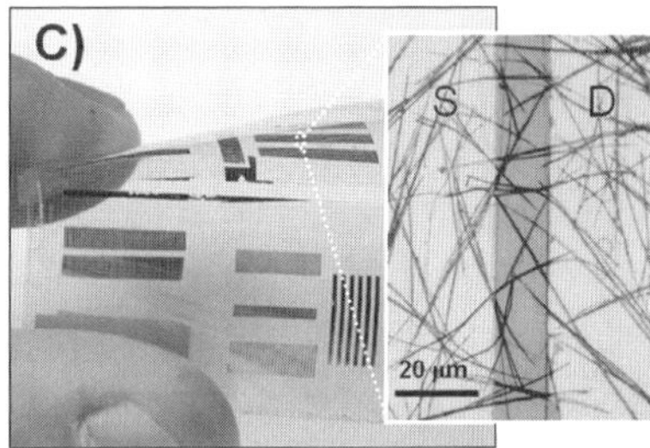

FIGURE 1.2 *Representative photographs of (A) ultrathin, (B) micropatterned, and (C) single-crystal nanowire flexible transistors.*
(Reprinted with permission from [7] © 2006 Wiley-VCH, [6] © 2006 NPG, and [12] © 2007 American Chemical Society.)

those of amorphous silicon (Si). Finally, organic single-crystal nanowires are discussed and their use in flexible transistors is demonstrated [12]. The chapter concludes with some perspectives on future research opportunities in the area of flexible organic single-crystal transistors.

MACROSCOPIC ORGANIC SINGLE CRYSTALS

Growth via physical vapor transport

The current benchmark for studying charge transport in organic semiconductors is by field-effect transistor experiments with ultrapure organic molecular crystals [8–10]. This is because one can investigate charge transport in the bulk crystal and not be concerned about the static disorder or grain boundaries that typically plague thin films produced via solution-processed methods (i.e. polymers) or thermal evaporation (organic thin films). Unfortunately, disorder in thin films affects the electrical properties of organic semiconductors and, as a result, they are not suitable for determining the performance limits, and more importantly, the intrinsic transport properties of organic semiconductors (Figure 1.3). Although several decades of intensive research have been conducted with single crystals, it was Podzorov et al. who revived their use in field-effect transistors with the high-performance organic semiconductor, rubrene [17]. Over the past five years, a tremendous amount of work has been published with regard to high-mobility single-crystal transistors. Several reviews document their advances, the current status and future outlook [9, 10]. However, one of the major impediments or drawbacks of their use is the fragility and difficulty of mass-producing devices over large areas. Because the fabrication efficiency is slow and tedious, only basic measurements and intrinsic transport phenomena can be conducted with single-crystal transistors. The

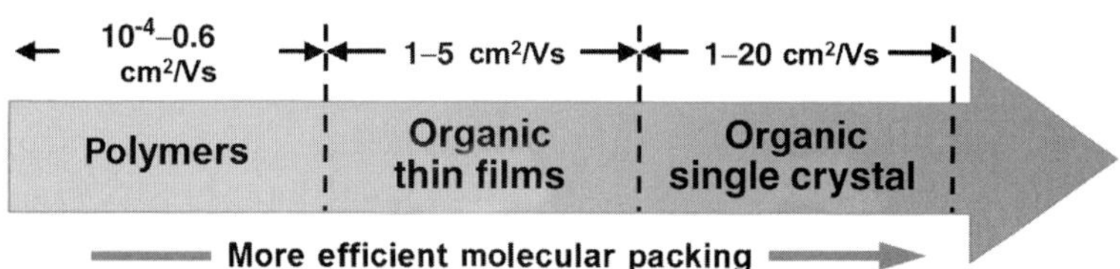

FIGURE 1.3   *Current status of organic field-effect transistors. Among the three classes of materials (polymers, thin films and single crystals), organic single crystals dominate the performance of field-effect transistors due to molecular perfection in the crystal lattice. Recent developments in each of the respective areas have produced significant breakthroughs for overcoming the drawbacks in each of the categories (i.e. new high-mobility polymers, solution-processable small molecules and patterning techniques).*

highest quality organic single crystals are grown via physical vapor transport, a technique that has been employed for well over a century [18]. Figure 1.4 shows a schematic of a furnace tube used to grow high-quality organic single crystals [19]. The same figure shows representative organic single crystal grown via physical transport. Such an instrument can be easily constructed with low-cost parts [18]. Crystals grown from the vapor phase have well-defined faces and smooth crystal surfaces. Slow growth, with a minimum of growth-nucleation sites on the inner walls of the glass cylinders, is a suitable choice to grow thick crystals at growth times of 6–24 h and sublimation temperatures of ~280–300°C [7]. Typical flow rates used are 50 ml/min. A comprehensive analysis on crystal growth conditions can be found in a previous literature paper [18].

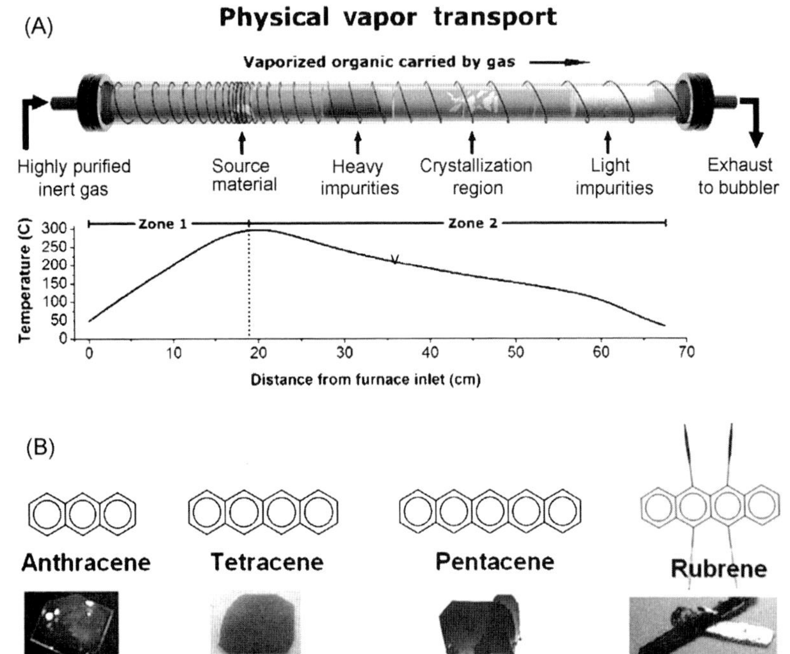

FIGURE 1.4 *(A) Physical vapor transport method for growth of single crystals [19]. The source material is placed in the hottest zone of the furnace tube where it sublimes and is carried down a temperature gradient by a stream of inert carrier gas, such as argon. The material resublimes in the cooler zone of the furnace to form large single crystals that can subsequently be hand-picked or patterned onto substrates for fabrication of single-crystal transistors. (B) Examples of oligoacene crystals grown via physical vapor transport. (Reprinted with permission from [19] © 2007 Elsevier, [9] © 2004 Wiley-VCH.)*

Growth of ultrathin single crystals

Ultrathin organic single crystals of rubrene are grown by horizontal physical vapor transport growth. Rapid crystal growth results in very thin, large and flat crystal flakes with a transparent appearance (20 min to 1 h growth time). For these results, an initial sublimation temperature of ~280–300°C is administered. At the moment nucleation is first observed on the inner glass cylinders, the temperature is increased to ~330°C while the argon flow rate is increased from 50 to 100 ml/min. It should be noted that specific parameters may vary from instrument to instrument, but the principle of rapid crystal growth remains the same. These growth conditions enable rubrene single crystals to grow as thin as 100 nm and as large as 1×1 cm in size (this is at least 500 times thinner than the standard-grown rubrene crystals). A close analysis of the thin, conformable single crystals shows a nearly defect-free surface morphology. Figure 1.5(A) shows an atom force microscope (AFM) image of the surface morphology of a thin rubrene single crystal. Thin single crystals (i.e. 100 nm to 1 μm) have smaller and less frequent surface steps than thicker crystals. A surface scan over a 2 μm^2 area at random locations on different individual samples yielded only one surface step. The surface roughness of a 300 nm thin rubrene single crystal is 0.23 nm, indicating that the surface of the crystal is smooth and free of grain boundaries. A surface island is observed with a monolayer step height of ~15 Å as shown in the inset, consistent with reported observations [20]. However, a scan of thicker crystals (> 3 μm) showed more frequent and much larger surface steps with a surface roughness

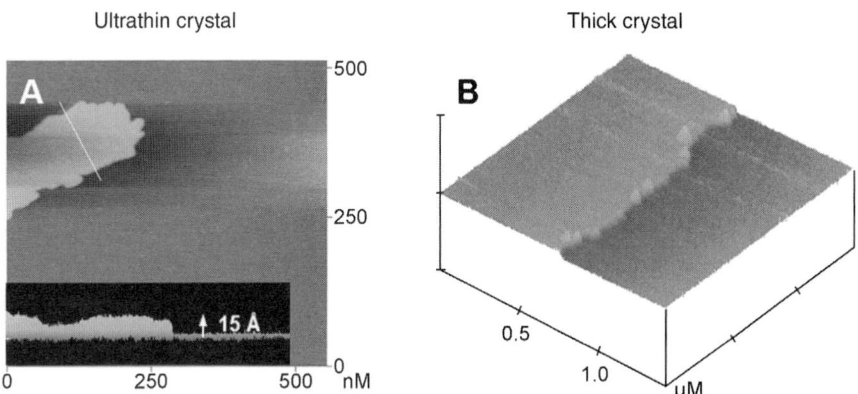

FIGURE 1.5 *Surface analysis of rubrene single crystals. (A) AFM image of a ~300 nm thin rubrene crystal (inset shows a topological side view in the direction from the indicated arrow); (B) a thicker rubrene crystal (> 3 μm). Larger and more frequent surface steps can be observed in thicker crystals.*
(Reprinted with permission from [7] © 2006 Wiley-VCH.)

of at least twice that of thin crystals (Figure 1.5B). Figure 1.6(A) shows scanning electron microscope (SEM) images of ultrathin rubrene single crystals electrostatically adhered to a device test structure. A high degree of interfacial adhesion and surface conformity is observed with thin crystals compared to thicker crystals (Figure 1.6B). The thick crystal in Figure 1.6(B) shows poor interfacial adhesion and surface brittleness. Field-effect experiments were carried out to monitor the effective mobility at various crystal thicknesses. Results from Figure 1.7 show that mobility gradually drops as a function of

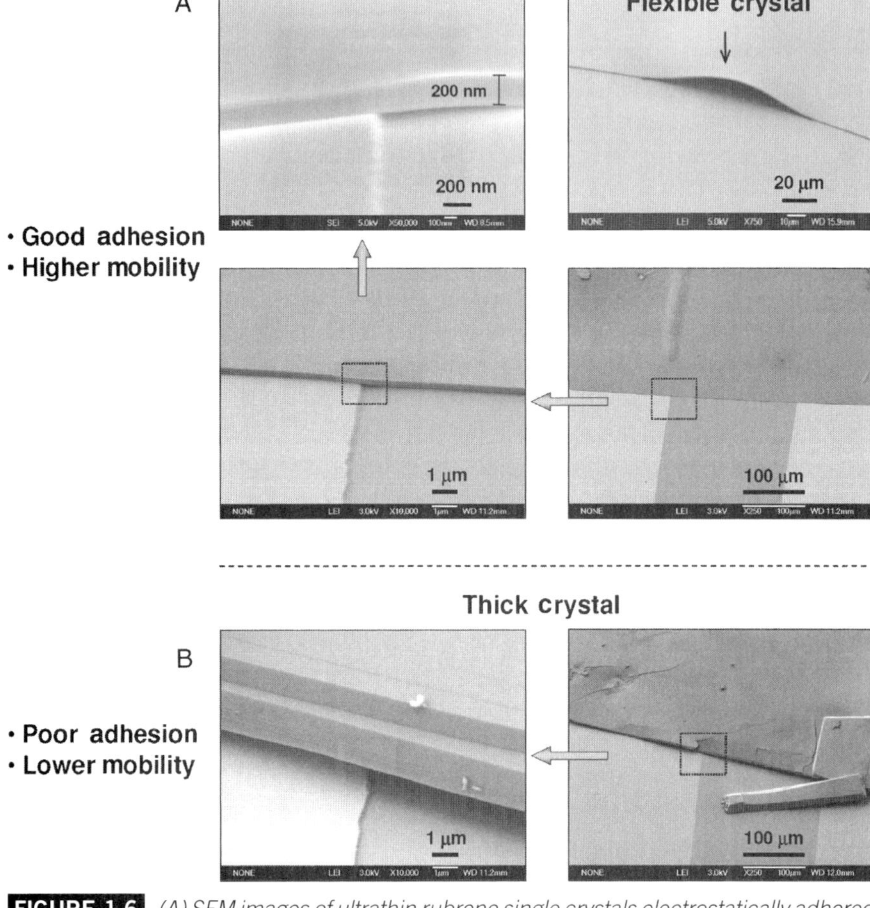

FIGURE 1.6 *(A) SEM images of ultrathin rubrene single crystals electrostatically adhered to a substrate. (B) SEM images of thick rubrene single crystals. Notice the large cracks and poor adhesion of the thick crystals compared to the good interfacial adhesion of the ultrathin single crystals from (A).*

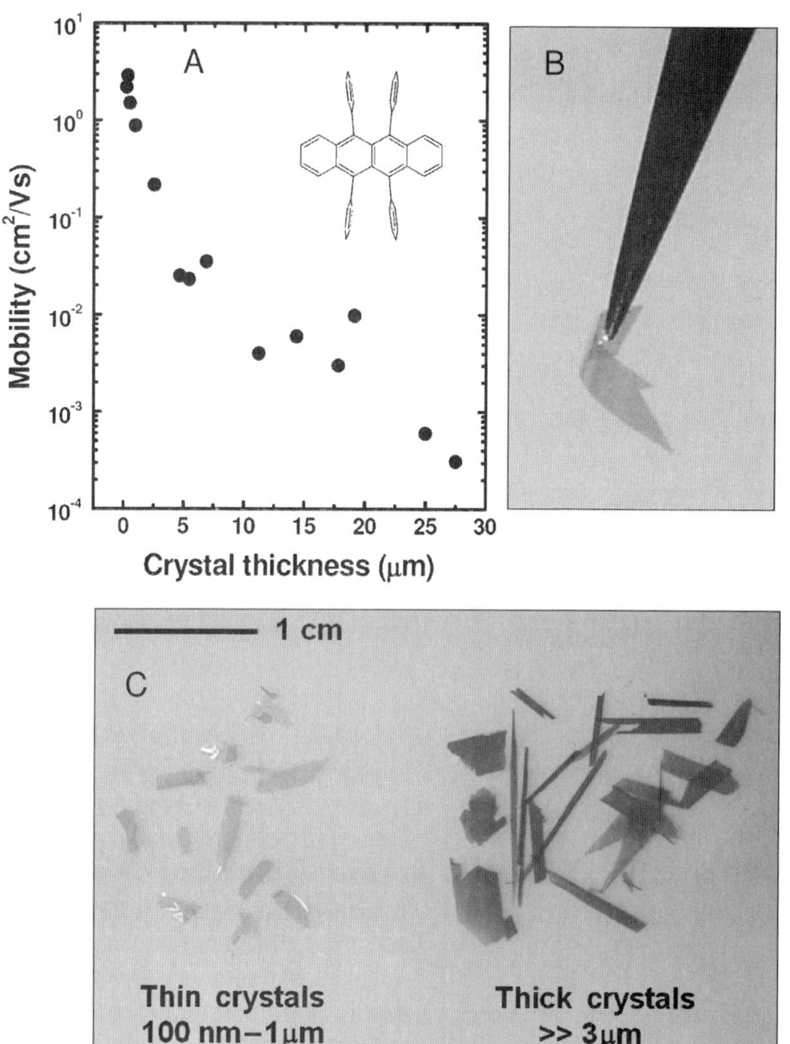

FIGURE 1.7 *Charge carrier mobility of rubrene single-crystal transistors plotted as a function of crystal thickness. (A) The plot shows the decrease in measured mobility as crystal thickness increases (0.2 ∼ 27 μm). (B) An ultrathin rubrene crystal 'bending' at the tip of a tweezer. This particular attribute enables the high-quality single crystals to conform to curvaceous substrates thus allowing bending experiments to be conducted. (C) Digital photograph comparing thin and thick rubrene single crystals. The thicker rubrene crystals (dark in color) are not capable of mechanical bending and, as a result of their fragility, applications have been limited to charge-transport studies. Both crystals are grown from the same starting material, but under different growth conditions.*
(Reprinted with permission from [7] © 2006 Wiley-VCH.)

crystal thickness. For crystals thicker than 5 μm, the mobility is less than 10^{-1} cm^2/V·s. Excellent adhesion and conformability of thin crystals to the dielectric surface is likely the reason for higher carrier mobility as opposed to the poor crystal–dielectric interface with thicker crystals (as observed in Figure 1.6). Figure 1.7 also shows a comparison of thin, nearly transparent rubrene single crystals and standard thick crystals ($>> 3$ μm). The natural flexibility of a thin rubrene single crystal as it bends on the tip of a tweezer is also shown in the same figure. The quality of single crystals was confirmed from the large birefringence observed under cross-polarized light and from the narrow peak with a full-width at half-maximum of 0.02 degree corresponding to the (002) Bragg diffraction.

Flexible single-crystal transistors

For flexible transistors, a dielectric layer of PVP was spin-coated (~2000 rpm) on 140 μm-thick Kapton® (polyimide) sheets covered with a 100 nm gold coating used for the gate electrode (Astral Technology Unlimited). The dielectric solution was prepared from a 22 wt% PVP, and 8 wt% poly(melamine-*co*-formaldehyde)methylated (Mw = 511). The substrates were heated at 100°C for 10 min then at 180°C for 30 min to initiate cross-linking. A capacitance of 1.9 nF/cm^2 was used for 1.5 μm-thick PVP (Mw = 20 000) dielectric. The source–drain electrodes were fabricated by thermal evaporation of Cr (1.5 nm) and Au (50 nm) for flexible substrates. Subthreshold slopes were multiplied by the capacitance of PVP for determining normalized subthreshold swings. The subthreshold slopes were calculated from the equation $S = dV_{\mathrm{G}}/d(\log I_{\mathrm{DS}})$.

Bending measurements are often performed by attaching the flexible devices onto curved cylinders of different radii. Device structures are tested by applying a strain on the flexible transistor device to determine the performance characteristics. The strain (ε) produced in the channel of the transistor by force bending was calculated from the following expression:

$$\varepsilon = \left(\frac{d_{\mathrm{f}} + d_{\mathrm{s}}}{2R}\right) \frac{(1 + 2\eta + \chi\eta^2)}{(1 + \eta)(1 + \chi\eta)}$$

where R is the bending radius, $\chi = Y_{\mathrm{f}}/Y_{\mathrm{s}}$ (Y_{f} and Y_{s} are the Young's moduli of the gate dielectric layer and substrate, respectively), and $\eta = d_{\mathrm{f}}/d_{\mathrm{s}}$ (d_{f} = thickness of the gate dielectric film and d_{s} = thickness of the substrate) [21]. For example, recalling that the thicknesses of the polyimide substrate and the PVP dielectric are 140 μm and 1.5 μm, respectively, and using values of ~3.2 GPa and 3.3 GPa for the moduli of polyimide and PVP, a strain of 1.23% is imposed at a radius of 5.75 mm. In most of the experiments, an outward/tensile bending was applied because theory predicts that layers under tensile strain are most prone to failure such as cracking/channeling and debonding/delamination [11, 21]. Devices were first measured at a planar geometry followed by additional measurements by decreasing the bending radius. Plastic substrates were held in bent geometries for 1 h at each of the respective radii.

Figure 1.8(A, B) shows the output and transfer characteristics of a flexible rubrene single-crystal transistor. The output curves show up to 280 μA at a gate voltage of −60 V. The flexible device exhibited a saturated field-effect mobility of 4.6 cm^2/V·s, a current on/off ratio of ~10^6, switch-on voltage of −2.1 V, and a normalized subthreshold swing of 0.9 V · nF/decade · cm^2. Note that mobilities measured here do not consider anisotropy effects and may very well deviate by a factor of 2–5 compared to reported values [8–10]. Mechanically flexible rubrene single-crystal transistors were demonstrated by bending substrates to various radii as illustrated from the transfer curve overlays in Figure 1.8(C). Figure 1.8(D) shows a plot of the field-effect mobility as a function of bending radius and strain. The flexible device can be bent to a radius of less than 1 cm (~9.4 mm; 0.74% strain) without any significant loss in performance. When the substrate is bent from a radius of 7.4 to 5.9 mm, the mobility drops significantly by more than two orders of magnitude to 0.078 and 0.0065 cm^2/V·s. Nevertheless, on/off ratios of ~10^5 and ~10^3 are still observed at such strenuous bending. The severe bending probably induces a large interfacial strain on the crystal and the dielectric, which results in a decrease in mobility. The mobility is restored to 91.3% of the original performance after releasing the flexible substrate from a bending radius of 5.9 mm (1.18% strain) to the original 'non-bent' position. This indicates that there was no permanent damage to the crystal as a result of the severe bending experiments. The current on/off ratio is also restored back to an original value of ~10^6. Figure 1.8(E) shows an optical photograph of a bent test structure containing rubrene single-crystal transistors.

Flexible single-crystal transistors have also been fabricated from tetramethylpentacene (TMPC) [13]. The pentacene derivative is grown under similar conditions to those for rubrene. Ultrathin single-crystal TMPC transistors exhibit mobilities as large as 1.0 cm^2/V·s when fabricated on Si/SiO$_2$ substrates and used in complementary inverters [13]. Figure 1.9(A) shows the chemical structure of TMPC. Figure 1.9(B, C) shows an ultrathin TMPC single crystal on a flexible device. As in the case with rubrene, TMPC also shows excellent adhesion and conformability to the underlying substrates. Flexible single-crystal transistors on polyimide substrates can be bent repeatedly with almost no change in performance. Figure 1.9(D, E) shows the output and transfer characteristics of the flexible TMPC device with a mobility of

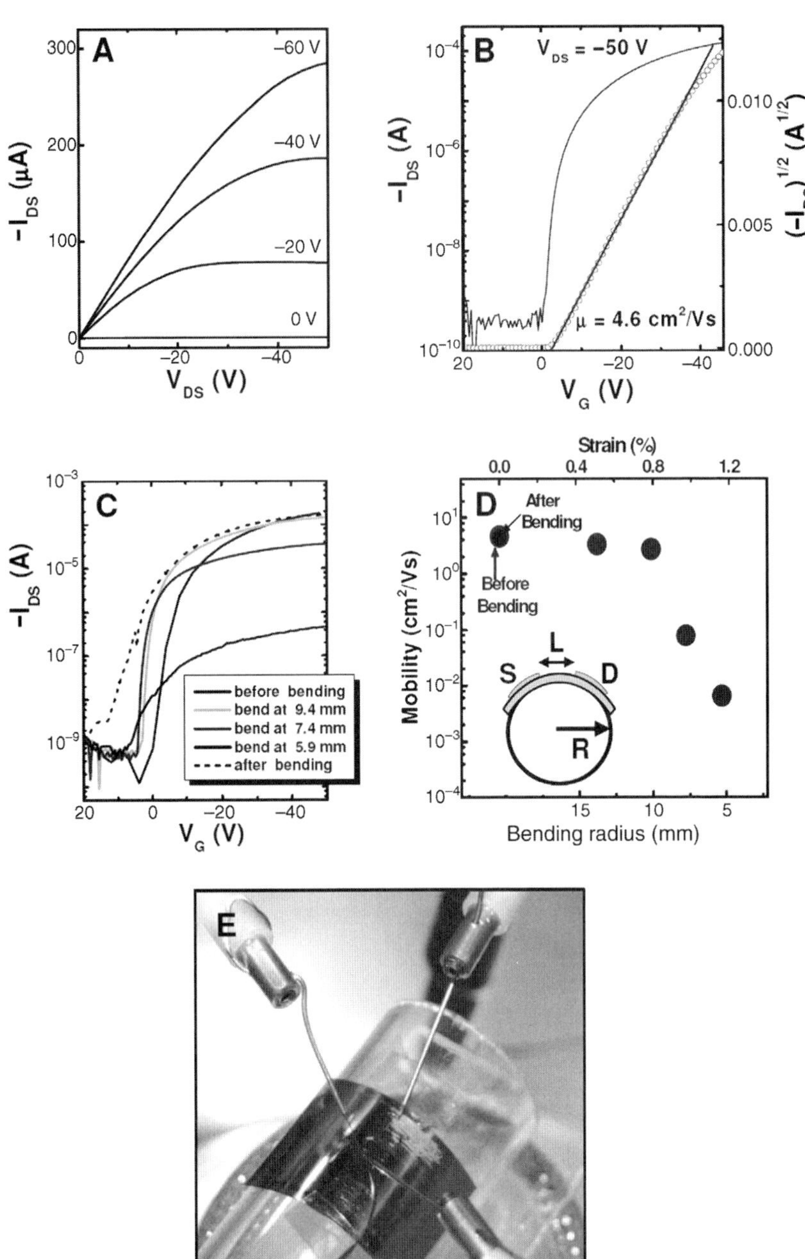

FIGURE 1.8 *Organic single-crystal field-effect transistors on flexible substrates. (A) The output and (B) transfer characteristics of ultrathin rubrene transistors fabricated from a flexible substrate. (C) An overlay of transfer curves ($V_{DS} = -50$ V) at different bending radii. Note that the magnitude of the current is restored to nearly its original current after releasing the plastic substrate from a bent radius of 5.9 mm. (D) The field-effect mobility as a function of bending radius (bottom axis) and strain (top axis). All bending measurements were performed on substrates bent across the channel length (L) as shown in the inset. (E) A photograph of an experimental flexible single-crystal device as it is measured on a curved cylinder. (Reprinted with permission from [7] © 2006 Wiley-VCH.)*

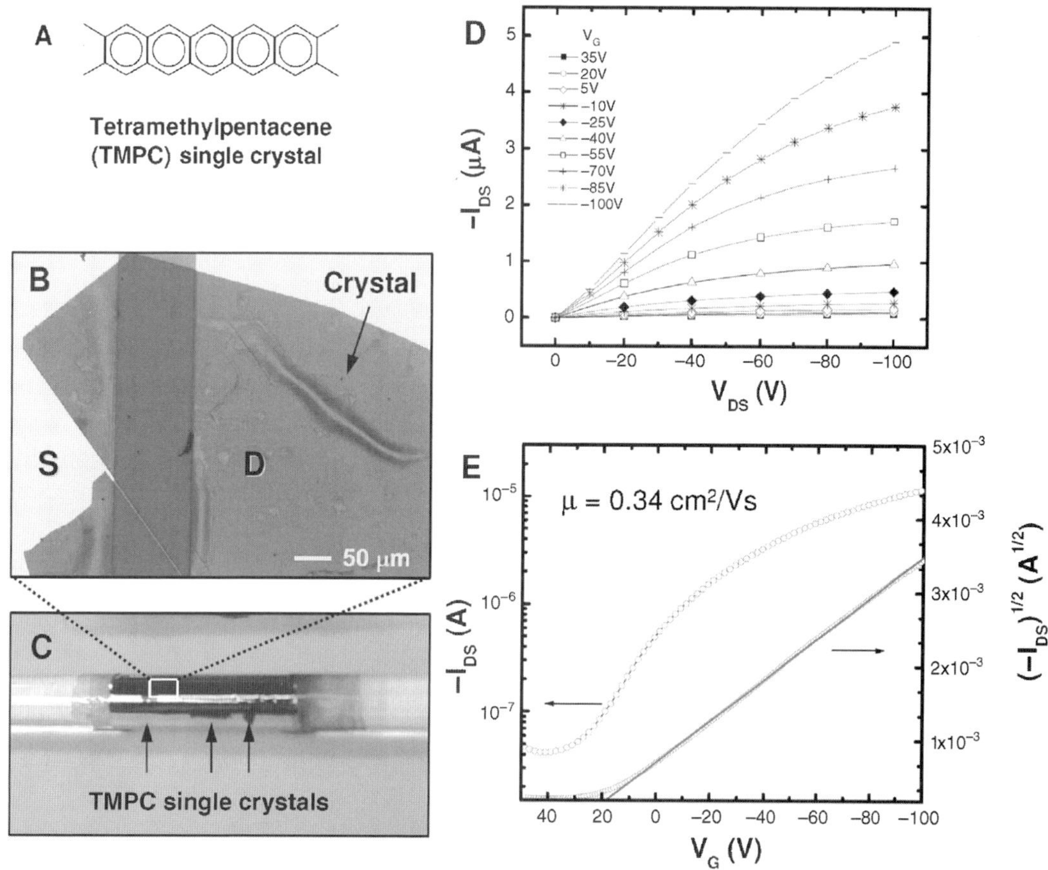

FIGURE 1.9 *(A) Chemical structure of tetramethylpentacene (TMPC); (B) an optical photograph of an ultrathin TMPC crystal adhered across source-drain electrodes; and (C) a flexible transistor wrapped around a cylinder. (D) Output and (E) transfer characteristics of the flexible single-crystal transistor.*

$0.34\ \mathrm{cm^2/V \cdot s}$. The mobility is lower than that of devices fabricated on Si/SiO_2 substrates and this is likely due to the impurity of the material as it was only recrystallized once. Small-molecule organic semiconductors are often recrystallized several times before they are grown as single crystals for use in electronic devices [9, 10, 18].

Ultrathin and flexible organic single-crystals evidently demonstrate the durability to flexing and imply that transistors fabricated from these materials may have potential use in applications where ruggedness and mechanical flexibility are a requirement. While these flexible transistors endure mild flexing, they are unlikely designed for severe treatment like those used in artificial skin applications where a bending radius of 2 mm has been achieved [2].

MICROPATTERNED ORGANIC SINGLE CRYSTALS

Control of crystal growth is an important criterion for the fabrication of high-performance materials such as organic semiconductors for use in field-effect transistors. The ability to control the sizes, orientations and nucleation sites of growing crystals remains a challenge. In theory, the nucleation and growth of crystals can be controlled by molecular interactions (or molecular specificity/recognition) at either the solution–surface interface or the air–surface interface [6, 22]. Organic single crystals are important for the fundamental understanding of charge transport in field-effect transistors. Although polycrystalline thin film devices are somewhat easier to fabricate than traditional free-standing organic single crystal devices, their performance are limited by structural imperfections. Mobilities as high as $20\ cm^2/V \cdot s$ have been reported for single crystal rubrene transistors [8]. Despite the high mobilities reported for single crystal devices, there are many factors limiting their applications. In addition to the difficulties of fabricating good electrical contacts to single crystals, the most challenging undertaking is handling the fragile crystals. Currently, single crystals are handpicked and made into an individual device, but this method is impractical for fabricating a high density of devices over a large area.

This section discusses the recently reported method of patterning large-area arrays of organic single-crystals onto microcontact-printed domains of octadecyltriethoxysilane (OTS) films (Figure 1.10) [6, 15]. This patterning method enables the growth of various organic semiconductor single crystals directly onto transistor source-drain electrodes, yielding large arrays of high-performance organic single-crystal field-effect transistors with mobilities as high as $2.4\ cm^2/V \cdot s$ and on/off ratios greater than 10^7. Furthermore, arrays of individual single crystals can be obtained by controlling the size of the printed OTS domains. Single-crystal transistors fabricated on flexible substrates were bent to a radius of 6 mm without a significant loss in performance. The described approach represents a significant step towards the realization of large arrays of high-performance organic field-effect transistors that may find applications in consumer electronics.

Substrate patterning and crystal growth

OTS printing solutions were prepared in reagent-grade toluene with concentrations ranging from 0.1–1.0 M. Cotton applicators were used to ink the polydimethylsiloxane (PDMS) stamps with the OTS solutions. PDMS stamps were brought into contact with plastic substrates for approximately 5 min and no rinsing or heat treatment was administered to the

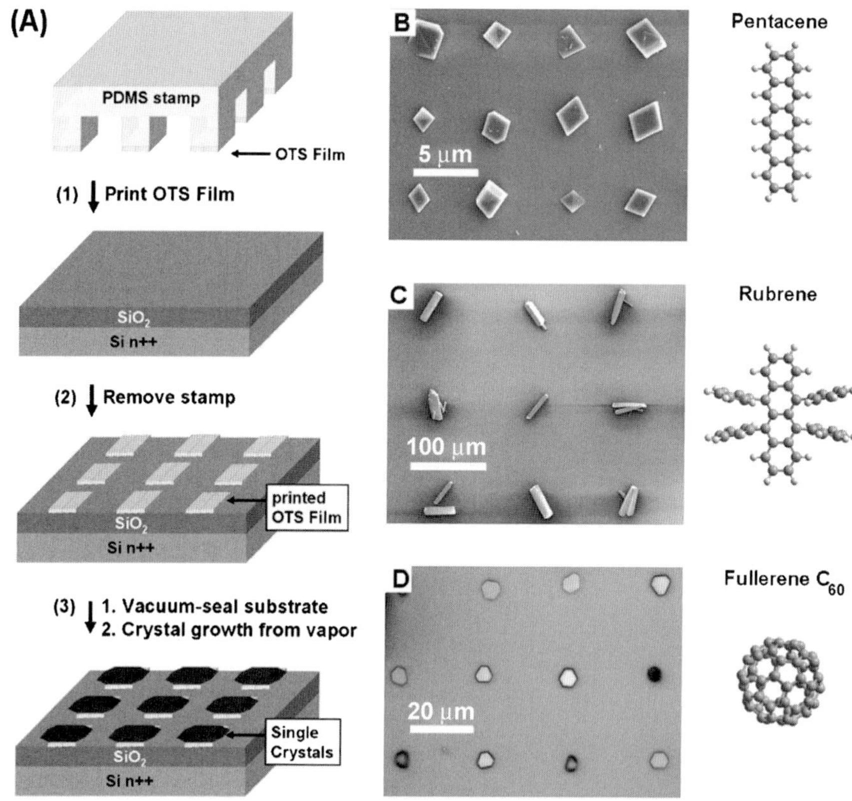

FIGURE 1.10   *Schematic procedure for patterning organic single crystals and images of patterned single crystal arrays. (A) Polydimethylsiloxane (PDMS) stamps containing relief features are inked with a thick octadecyltriethoxysilane (OTS) film and printed onto substrates. The patterned substrate is placed in a vacuum-sealed tube and placed in a temperature gradient furnace tube with the organic source material for growth of patterned single crystals. (B–D) Patterned single crystal arrays of different organic semiconductor materials.*

(Reprinted with permission from [6] © 2006 NPG.)

OTS-patterned substrates. Substrates are directly placed in a 16 mm diameter glass tube containing semiconductor material at one end of a flame-sealed tube (~10 mg of source material utilized) while the substrate is placed onto an NiCr wire clip and inserted through the opposite end of the opened tube. The patterned substrate is placed several centimeters away from the source material and flame-sealed under a vacuum pressure of ~0.38 mmHg using a standard laboratory vacuum pump. Figure 1.11 shows a sketch of the single-crystal patterning device set-up and a photograph of a sealed tube with a rubrene-patterned substrate. In this example,

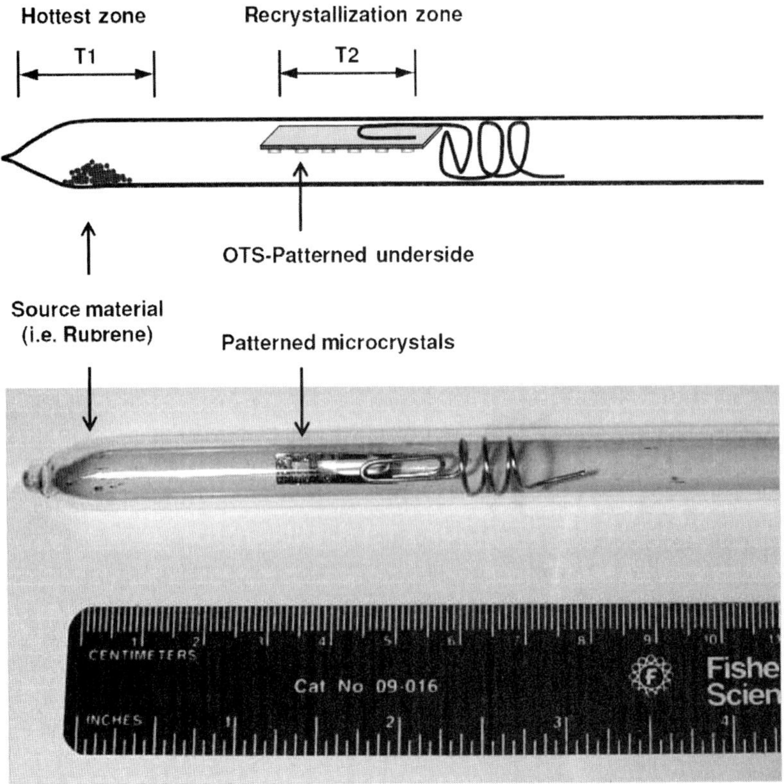

FIGURE 1.11 *Illustration of a vacuum-sealed tube used for growing patterned arrays of single crystals. The bottom photograph shows an actual tube containing a rubrene-patterned substrate.*

the rubrene source temperature is maintained at ~270°C and the closest edge of the patterned substrate is positioned 3–4 cm away from the 'hottest zone'. The recrystallization zone of rubrene crystals extends to the far edge of the sample 5 cm away from the source. A vacuum-sealed tube is placed into a temperature gradient sublimation reactor similar to the system shown in Figure 1.4. Increasing the source temperature increases the rate of crystal growth, but also increases the distance from the source to the crystal-forming region. Calibration of materials under controlled parameters was conducted to determine the exact distance of crystal growth in respect to the source zone. Larger diameter glass tubes were used to fabricate larger-area patterning. Organic source materials are often purified by temperature gradient sublimation, but in most cases, materials are often utilized without any purification and used as received.

The vapor transport method used here for crystal growth is general to a broad range of materials including high-mobility p-type materials, such as rubrene, pentacene and tetracene, and n-type materials such as C_{60}, fluorinated copper phthalocyanine ($F_{16}CuPc$) and tetracyanoquinodimethane (TCNQ). Sublimation and crystal growth occur in as little as 5 min for pentacene and as much as 2 h for C_{60}. It should be noted that high-temperature melting materials such as C_{60} and $F_{16}CuPc$ cannot be patterned onto polyimide substrates. The rough morphology of the OTS-stamped domains and the high substrate temperature are also important in achieving selective crystal growth. Because the substrate temperature is approximately $20°C$ lower than that of the evaporation source, there is a high thermal redesorption rate which results in a high energetic barrier to formation of a stable nucleus in the non-stamped regions of the substrates. Consequently, the nucleation of crystals is effectively suppressed in these substrate regions. Furthermore, according to classical nucleation theory, the barrier to heterogeneous nucleation can be significantly lowered in proximity of surface irregularities such as indentions, step edges or protrusions than on a flat surface, owing to higher surface energy at these sites [23]. Therefore, the stamped rough OTS domains act as primary centers for single-crystal nucleation. The modern phase field theory which describes nucleation processes in terms of hard-sphere fluids also predicts heterogeneous nucleation to occur favorably on rough surfaces [24].

Flexible devices

Figure 1.12(A) illustrates the schematic of the flexible single-crystal devices fabricated on a polyimide substrate. The left optical photograph in Figure 1.12 (A) shows patterned rubrene single-crystal devices attached to a vial with a diameter of 11.7 mm, while the right-most figure is an SEM of the patterned rubrene crystals. The output and transfer characteristics of a plastic rubrene transistor before and after bending (bending direction across the channel) are presented in Figure 1.12(B). A mobility of 0.9 $cm^2/V·s$, a current on/off ratio of 10^4 and a threshold voltage ~1.5 V were measured in this flexible rubrene transistor, while the best mobility for a flexible patterned pentacene transistor was 0.1 $cm^2/V·s$ with on/off ratios of $\sim10^3$ (Figure 1.13). No significant loss in performance was observed when the devices were bent to a radius as small as 6 mm. The bending experiments were performed on four randomly chosen devices, all exhibiting similar behavior. In addition to rubrene, high vapor pressure materials such as anthracene and tetracene were successfully patterned on flexible and transparent PET substrates without degradation of the plastic.

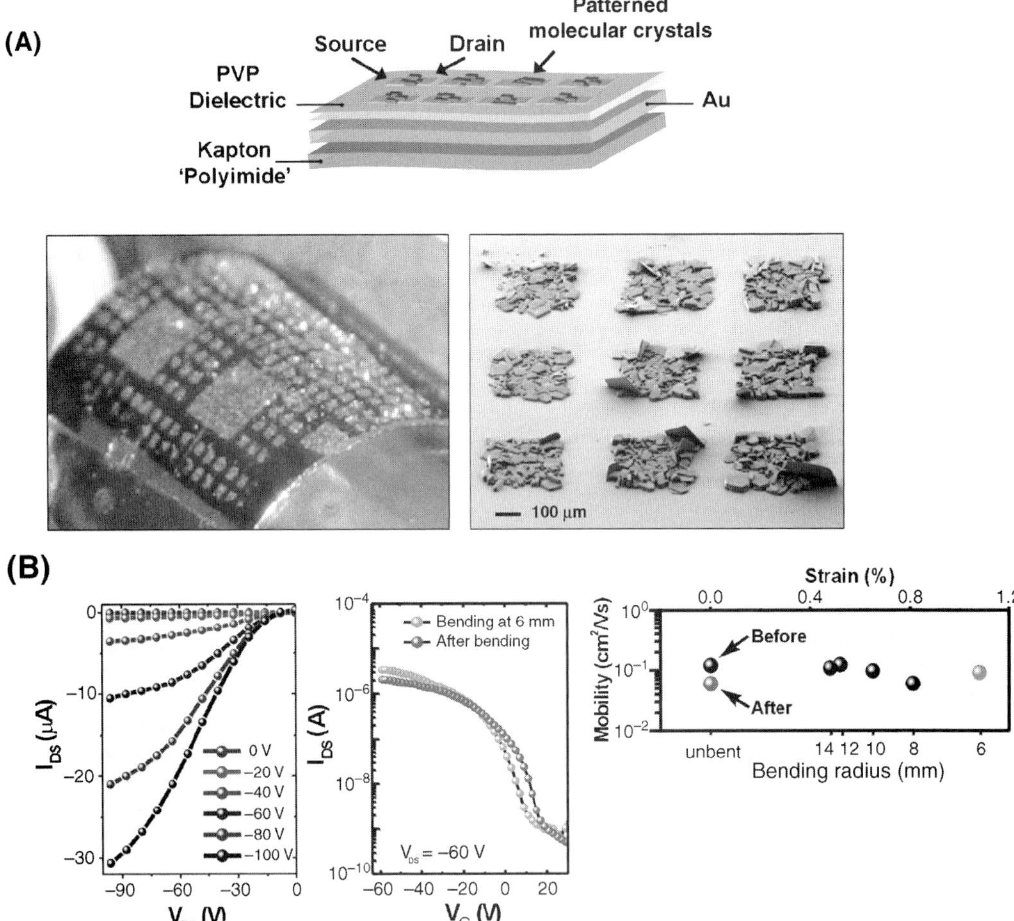

FIGURE 1.12 *Single-crystal arrays on plastic substrates and their electrical characteristics. (A) Schematic structure (top) and optical micrographs: a rubrene transistor array attached to a glass vial of diameter 11.7 mm (left), SEM of patterned rubrene crystals (right). Patterned organic crystals were grown on flexible Kapton devices that were taped onto precut glass microscope slides with high-temperature polyimide tape. (B) Left: the electrical characteristics of a flexible rubrene single-crystal transistor. Middle: a transfer curve overlay of a flexible device bent at 6 mm and after bending (planar geometry). Right: a plot of mobility as function of the bending radius (bottom axis) and strain percent (top axis). Mobilities as high as 0.9 cm^2/V·s and on/off ratios of 10^4 have been measured in the best flexible rubrene transistors.*
(Reprinted with permission from [6] © 2006 NPG.)

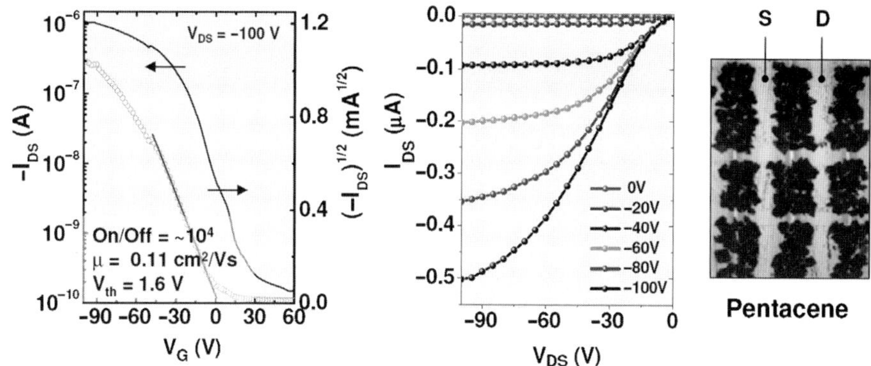

FIGURE 1.13 *Output and transfer characteristics of a patterned flexible pentacene single-crystal transistor.*
(Reprinted with permission from [6] © 2006 NPG.)

ORGANIC SINGLE-CRYSTAL NANOWIRES

Organic nanowires self-assembled from small-molecule semiconductors have attracted an enormous amount of interest for use in organic field-effect transistors [25]. This new class of materials offers solution processability, a potential for elucidating transport mechanisms, structure–property relationships and the realization of fabricating high-performance transistors on plastic that rival the performance of amorphous Si. This final section discusses the self-assembly of one-dimensional (1D) single-crystalline organic nanowires, shows the structures of commonly employed organic semiconductors, and reviews some of the advances on flexible transistors.

There is great interest in planar aromatic molecules that exhibit face-to-face π-stacking since this packing is theorized to result in high mobilities in devices such as field-effect transistors [26]. The molecules in 1D nanostructures (i.e. nanowires) predominantly self-assemble along the π–π stacking direction, which gives a high charge-carrier mobility along the long axis of the nanostructure as a result of the strong intermolecular coupling between the packed molecules. It is known that the tendency to form face-to-face stacked structures can be enhanced by adding peripheral substituents [27] as well as with an increase in π-surface-to-circumference ratios [28]. The face-to-face π-stacking motif is believed to be more efficient for charge transport than in most edge-to-face 'herringbone-packing' structures [29, 30]. Figure 1.14 shows examples of organic semiconductors that self-assemble into single-crystal nanowires through strong π–π interactions. These semiconductors have also been shown to successfully function as organic nanowire transistors.

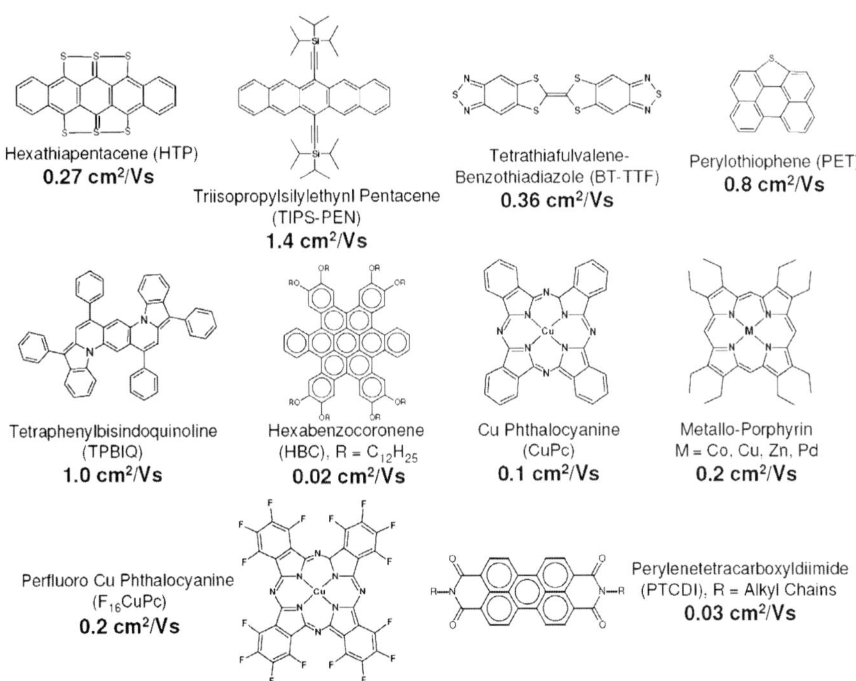

FIGURE 1.14 *Recent examples of organic semiconductors that stack face-to-face and which have been successfully used as organic nanowire field-effect transistors. Their respective mobilities are indicated below the abbreviation of the material. (Reprinted with permission from [25] © 2007 Elsevier.)*

Single-crystal nanowire transistors have shown mobilities comparable to their thin-film transistor counterparts [25]. The ability to process single-crystal nanowires from solution-phase methods and fabricate devices in a facile and low-cost manner should enable their use in flexible transistor applications.

Flexible nanowire devices

A real-world challenge in addressing the manufacturing goals for flexible electronics and displays is the ability to fabricate and process organic transistors in a facile and low-cost manner, over large areas, and on mechanically flexible substrates. With this objective in mind, researchers have investigated solution-phase routes for preparing organic single-crystal nanowires to fabricate arrays of transistors [12, 25, 31]. As an example for π-stacked small organic semiconductors, hexathiapentacene (HTP), which adopts face-to-face packing in the solid state [27], is illustrated. HTP is of interest as it combines

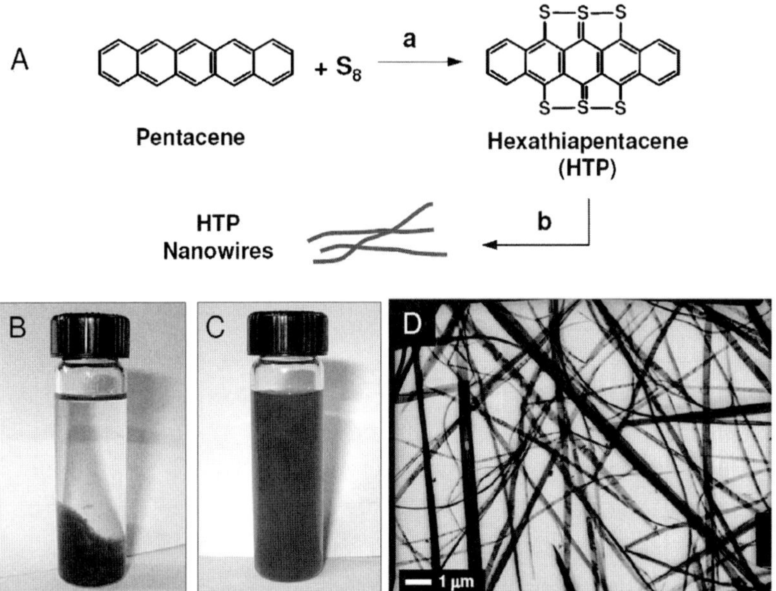

FIGURE 1.15 *(A) Synthesis of hexathiapentacene (HTP); (B) solution of settled HTP nanowires; and (C) a dispersion of the same nanowires. (D) TEM micrograph of HTP nanowires.*
(Reprinted with permission from [12] © 2007 American Chemical Society.)

the elements of pentacene and peripherally rich sulfur atoms that favor π–π stacking. The synthetic route produces bulk quantities of insoluble yet highly dispersible single-crystal HTP nanowires. HTP is synthesized from pentacene and elemental sulfur (Figure 1.15A). Since HTP is insoluble in nearly all organic solvents at room temperature, it can be readily dissolved in and recrystallized from hot organic solvents with high boiling points such as benzonitrile, nitrobenzene and *o*-dichlorobenzene. Recrystallized nanowires can be dispersed in organic solvents such as chloroform, methylene chloride or ethanol. The nanowires settle out from the solution after several hours but could be easily redispersed by tapping the vial or simple agitation (Figure 1.15B, C). Typical lengths of the nanowires ranged from several dozen microns to hundreds of micrometers, while the nanowire heights were in the range of 70–470 nm. Figure 1.15(D) shows a typical transmission electron microscope (TEM) image of a network of HTP nanowires. It is worth emphasizing that unlike pentacene and other reported derivatives [29, 30], HTP nanowires are remarkably stable [12] in most organic solvents (including chlorinated solvents) and do not exhibit any bleaching even after being exposed to ambient conditions for more than two years. The chemical stability of HTP may be attributed to the relatively high ionization potential, which makes it more

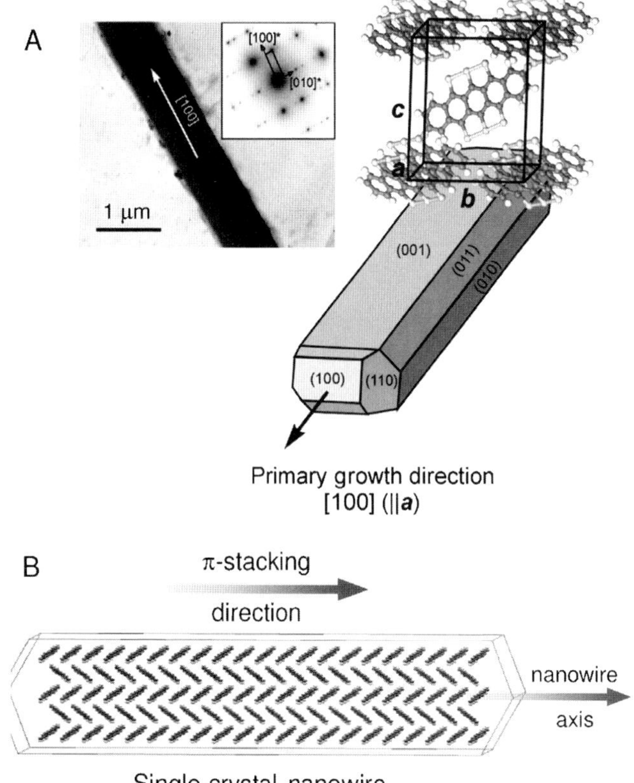

FIGURE 1.16 *(A) A TEM and corresponding electron diffraction of an individual hexathiapentacene (HTP) nanowire and the theoretically predicted growth morphology of a single-crystal nanowire. (B) Molecular packing of HTP along the a-axis of the unit cell. The illustration shows the face-to-face π-stacking direction. (Reprinted with permission from [12] © 2007 American Chemical Society.)*

difficult to be oxidized than pentacene ($HOMO_{HTP} = -5.35$ eV; $HOMO_{pentacene} = -5.0$ eV) [27]. This particular attribute is very important for employing stable solvent-processable organic semiconductors.

Figure 1.16(A) shows a TEM image of a single HTP nanowire and the corresponding electron diffraction pattern. HTP molecules stack face-to-face along the [100] crystallographic *a*-axis with π-to-π distances of 3.54 Å [27]. The π-stacking is attributed to the chalcogen atoms located on the peripheral of pentacene's backbone, which by steric hindrance prevents the edge-to-face geometry that leads to the herringbone structure of pentacene. A distance of ~3.37 Å, which is smaller than twice the van der Waals radius of S atom (3.70 Å), is observed between the central S atom and the outermost S atom of a neighboring HTP molecule [27]. The overlap of the molecular π-orbitals, crucial for the charge transport in organic semiconductor materials, is significant only along the nanowire direction (Figure 1.16B). This type of stacking is also well known for PTCDI derivatives [31], and underscores the applicability of nanowires for high-performance transistor devices.

The HTP nanowire transistors were fabricated on mechanically flexible substrates to evaluate their potential for applications in flexible electronics. This approach to measuring the devices on plastic substrates is similar to the one described in previous reports [6, 7]. To demonstrate the simple processability of the solvent-dispersible nanowires, the material from a suspension (0.1 mg/ml) of HTP nanowires was spray-coated in a mixture of 1:1 chloroform/ethanol (Figure 1.17A). The spray-coating method allows one to control the density of nanowires across source–drain electrodes. Figure 1.17(B) shows the transfer characteristics of an HTP nanowire network device on a flexible substrate. A mobility of 0.032 cm^2/V·s, a

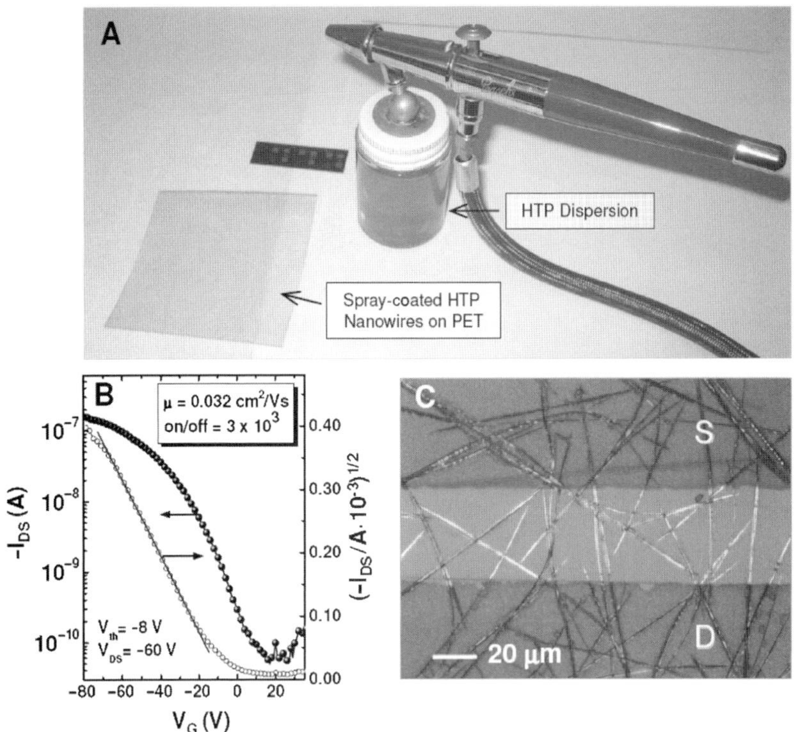

FIGURE 1.17 *(A) An airbrush used to spray-coat dispersions of hexathiapentacene (HTP) nanowires onto transparent PET and conventional substrates. (B) Transfer characteristics of an HTP nanowire network transistor fabricated on a flexible substrate. (C) Photograph of a bottom-contact network nanowire device showing the single-crystal nanowires bridging the source–drain electrodes.*
(Reprinted with permission from [12] © 2007 American Chemical Society.)

current on/off ratio of $\sim 10^3$ and a threshold voltage of -8 V were obtained from a flexible device similar to that pictured in Figure 1.17(C). Since it is difficult to estimate the active area of the nanowires, the mobilities reported here may very well be overestimated. The average mobility extracted from the HTP-network flexible devices is 0.017 ± 0.01 cm²/V·s. In addition, discrete nanowire transistors were fabricated on arrays of flexible substrates (Figure 1.18A) to determine the limits of flexibility of an individual single-crystal nanowire. Figure 1.18(B) illustrates the device configuration of a nanowire across source–drain electrodes on a curved substrate. Dispersions of HTP nanowires in ethanol were drop-cast onto plastic devices and allowed to dry in a vacuum chamber for several hours. Devices containing only one discrete nanowire were tested. The devices

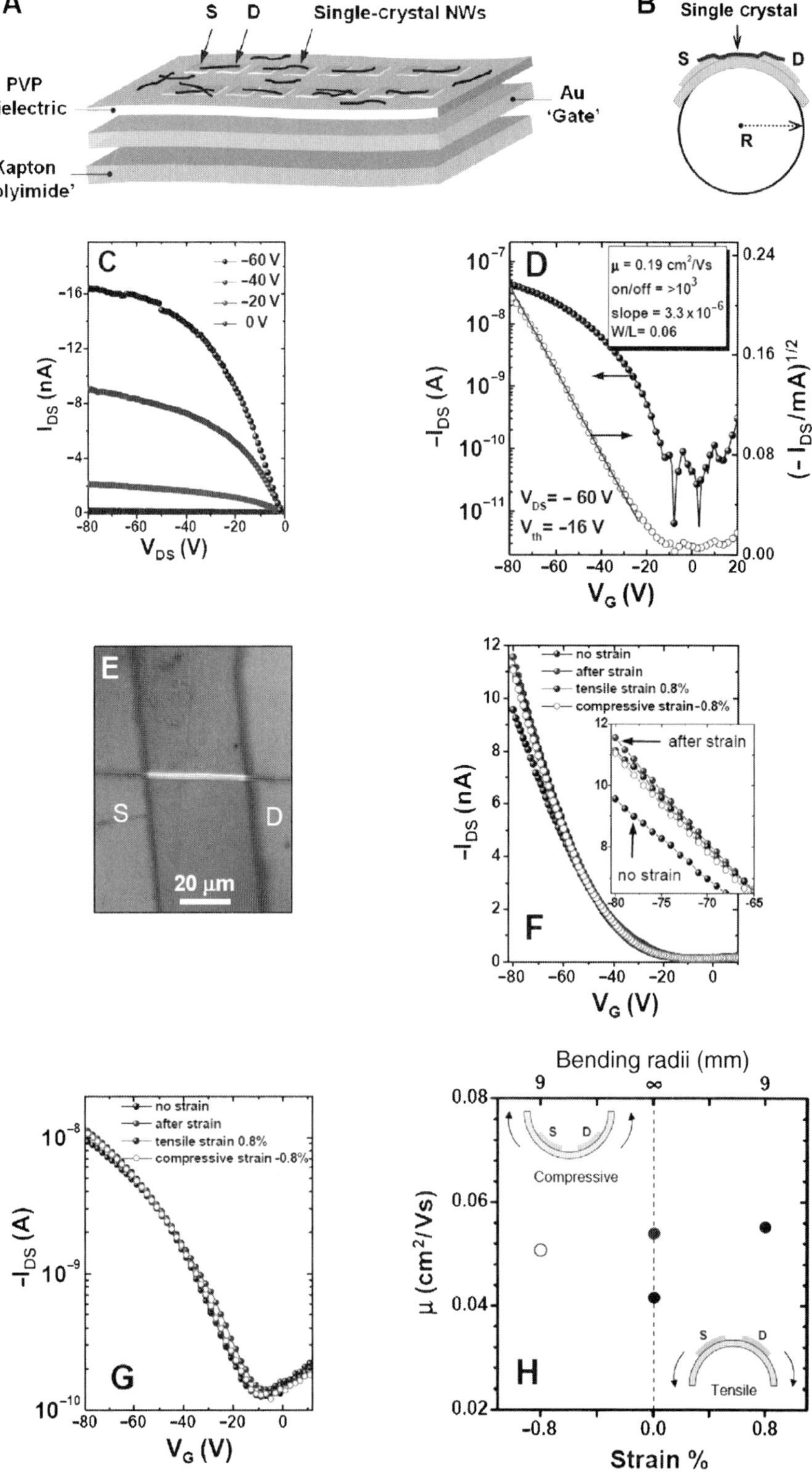

FIGURE 1.18 *(A) The configuration of a bottom-contact single-crystal device on a flexible substrate. NWs: nanowires; PVP: poly-4-vinylphenol. (B) Illustration of a single nanowire across source-drain electrodes on a curved substrate. (C) Output characteristics and (D) transfer characteristics of a hexathiapentacene (HTP) single-nanowire transistor fabricated on a flexible Kapton substrate. (E) An optical photograph of a single-crystal HTP nanowire transistor. (F) An overlay of source-drain currents (I_{DS}), and (G) log of source-drain currents under different geometrical strain conditions. All scans conducted at $V_{DS} = -60$ V. (H) A plot of mobility versus bending radii (strain %) for a single-wire transistor. (Reprinted with permission from [12] © 2007 American Chemical Society.)*

show classical field-effect behavior with excellent current modulation and well-resolved output characteristics. The highest mobility attained for such a device yielded a mobility of $0.19\ cm^2/V{\cdot}s$ and a current on/off ratio to $> 10^3$ (Figure 1.18C, D). The average mobility over six devices was $0.08 \pm 0.06\ cm^2/V{\cdot}s$.

Having demonstrated that HTP nanowire transistors successfully function on plastic, the mechanical flexibility of HTP nanowires was investigated by conducting basic experiments to determine the effects of mechanical strain [32] on the performance of such a device. Hu and co-workers demonstrated that $CuPc/F_{16}CuPc$ nanoribbons possessed excellent flexibility [33]. However, they did not report the effects of mechanical deformation on device performance. Strain experiments were performed on single nanowire devices only (Figure 1.18E), rather than on network nanowire devices. This was done to determine exclusively the performance of an individual single-crystal on a strained substrate; a network device would likely introduce inconclusive results as the devices contain overlapping nanowires. First, the flexibility of the nanowires was examined by applying a tensile (outward bending) and compressive strain (inward bending) to the device by attaching it around an 18 mm diameter (radius curvature = 9 mm) glass cylinder and ensuring the device securely conformed to the cylindrical surface (see insets in Figure 1.18H). All experiments were conducted on substrates bent across the channel length (L). The devices were first measured in the planar geometry mode followed by compressive strain, tensile strain and finally back to the original planar geometry. No significant change in mobility was found from before, during or after the applied strains. This trend has been routinely observed in at least six separate devices with the only difference being in the effective mobility among the devices. Figure 1.18(F, G) shows the I_{DS} and log of I_{DS} transfer overlays of a single-wire device subjected to various mechanical deformation geometries. Finally, the effect of mechanical stress on the gate-leakage current was measured and it was found that the gate-leakage current remained in the order of picoamps throughout the gate voltage sweep. The unaffected electrical performance under the effect of mechanical stress can be explained by the 'compliant' [21, 32, 34] substrate, which significantly takes up the stress and reduces the stress on the single-crystal. The conclusion from these experiments is that the mobility is nearly unaffected as mechanical strain is applied to the flexible substrates. The results demonstrate the durability of the test structures to mechanical stress and imply that flexible transistors fabricated from 1D nanowires may have a use in applications that require non-breakable or highly bendable substrates.

CONCLUSION

Flexible devices fabricated from organic semiconductors are among the most heavily researched topics in the area of organic electronics. This area of research is unique as it combines fundamental principles from areas of engineering, chemistry and physics. Nearly all examples on flexible transistors have used polycrystalline thin films of organic small molecules or solution-processed polymer semiconductor layers. However, flexible organic single-crystal transistors have recently debuted as next-generation devices that may very well open directions for new research opportunities. Fabrication of high-performance flexible single-crystal transistors on lightweight, flexible plastic substrates is expected to have a profound impact in the area of microelectronics. Historically, research in the area of single-crystal transistors has focused on the intrinsic properties of organic semiconductors and on fundamental issues associated with charge transport. Bearing in mind the typical perception about these crystals being extremely fragile, these results are unprecedented in showing that reducing the crystal dimensions makes it possible to achieve considerable flexibility without notable device degradation. In principle, through integration with microfabricated circuits on flexible substrates, this new area may result in a viable technology for flexible electronics.

Considerable efforts are still needed to convert organic single-crystal transistors into a practical semiconductor technology. With respect to ultrathin single crystals, it seems as if large-area device fabrication is still in the distant future. A focus on fundamental properties and device mechanics will likely pave the way in this area. Investigating the 'wavy' nature and stretchable properties of ultrathin organic single crystals is another avenue to dig deep into. With regard to single-crystal patterning, control of crystal size and azimuthal orientation will require further attention especially because high-mobility materials such as rubrene exhibit anisotropy. The quality of the electrical contacts between the crystal and the electrodes is another issue that will require further investigation. Patterning of both p- and n-type organic semiconductor single crystals on a common device structure is the ultimate challenge. Another suggestion is to use ambipolar organic semiconductors that will allow the transport of both electrons and holes. Finally, organic single-crystal nanowires will require a method of patterning, otherwise they will serve little purpose, in spite of the fact that bulk-scale quantities can be synthesized from the solution phase. Alignment of the nanowires and good electrical contacts to electrodes will be the main issues to address if they are ever to have a place in the microelectronics arena. If the aforementioned issues can be successfully addressed, the described approaches may be anticipated to represent a significant step towards the realization of high-performance flexible organic field-effect transistors for use in real-world applications.

ACKNOWLEDGMENTS

A. L. B. acknowledges all the coauthors and facilities that participated in the work described in this book chapter. Special thanks to Zhenan Bao, Younan Xia, Samson A. Jenekhe, Stefan C. B. Mannsfeld, Yang Yang and Fred Wudl, for helpful discussions and collaboration on various projects.

REFERENCES

[1] Rogers JA, Bao Z, Baldwin K, Dodabalapur A, Crone B, Raju VR, et al. Paper-like electronic displays:large-are rubber-stamped plastic sheets of electronics and microencapsulated electrophoretic inks. Proc Natl Acad Sci USA 2001;98: 4835–40.

[2] Someya T, Kato Y, Sekitani T, Iba S, Noguchi Y, Murase Y, et al. Conformable, flexible, large-area networks of pressure and thermal sensors with organic transistor active matrixes. Proc Natl Acad Sci USA 2005;102:12321–5.

[3] Cao Q, Hur S-H, Zhu Z-T, Yugang S, Wang C, Meitl MA, et al. Highly bendable, transparent thin-film transistors that use carbon-nanotube-based conductors and semiconductors with elastomeric dielectrics. Adv Mater 2006;18:304–9.

[4] LeMieux MC, Roberts M, Barman S, Jin YW, Kim JM, Bao Z. Self-sorted, aligned nanotube networks for thin-film transistors. Science 2008;321:101–4.

[5] Sun Y, Rogers JA. Fabricating semiconductor nano/microwires and transfer printing ordered arrays of them onto plastic substrates. Nano Lett 2004;4:1953–9.

[6] Briseno AL, Mannsfeld SCB, Ling M, Liu S, Tseng RJ, Reese C, et al. Patterning organic single-crystal transistor arrays. Nature 2006;444:913–7.

[7] Briseno AL, Tseng RJ, Ling M-M, Falcao EHL, Yang Y, Wudl F, Bao Z. High-performance organic single-crystal transistors on flexible substrates. Adv Mater 2006;18:2320–4.

[8] Sundar VC, Zaumseil J, Podzorov V, Menard E, Willett RL, Someya T, et al. Elastomeric transistor stamps: reversible probing of charge transport in organic crystals. Science 2004;303:1644–716.

[9] de Boer RWI, Gershenson ME, Morpurgo AF, Podzorov V. Organic single-crystal field-effect transistors. Phys Stat Sol A 2004;201:1302–31.

[10] Gershenson ME, Podzorov V, Morpurgo AF. Electronic transport in single-crystal organic transistors. Rev Mod Phys 2006;78:973–89.

[11] Lewis J. Material challenge for flexible organic devices. Mater Today 2006;9(4): 38–45.

[12] Briseno AL, Mannsfeld SCB, Lu X, Xiong Y, Jenekhe SA, Bao Z, Xia Y. Fabrication of field-effect transistors from hexathiapentacene single-crystal nanowires. Nano Lett 2007;7:668–75.

[13] Briseno AL, Tseng RJ, Li SH, Chu CW, Yang Y, Falcao EHL, et al. Organic single-crystal complementary inverter. Appl Phys Lett 2006;89:222111.

[14] Liu S, Briseno AL, Mannsfeld SCB, You W, Locklin J, Lee HW, et al. Selective crystallization of organic semiconductors on patterned templates of carbon nanotubes. Adv Funct Mater 2007;17:2891–6.

[15] Mannsfeld SCB, Briseno AL, Liu S, Reese C, Roberts ME, Bao Z. Selective nucleation of organic single crystals from vapor phase on nanoscopically rough surfaces. Adv Funct Mater 2007;17:3545–53.

[16] Mannsfeld SCB, Sharei A, Liu S, Roberts ME, McCulloch I, Heeney M, Bao Z. Highly efficient patterning of organic single-crystal transistors from the solution phase. Adv Mater 2008;20:4044–8.

[17] Podzorov V, Sysoev SE, Loginova E, Pudalov VM, Gershenson ME. Single-crystal organic field effect transistors with the hole mobility $\sim$8 cm^2/V·s. Appl Phys Lett 2003;83:3504–6.

[18] Kloc Ch, Simpkins PG, Siegrist T, Laudise RA. Physical vapor growth of centimeter-sized crystals of α-hexathiophene. J Cryst Growth 1997;182:416–27.

[19] Reese C, Bao Z. Organic single-crystal field-effect transistors. Mater Today 2007;10(3):20–7.

[20] Menard E, Podzorov V, Hur S-H, Gaur A, Gershenson ME, Rogers JE. High-performance n- and p-type single-crystal organic transistors with free-space gate dielectrics. Adv Mater 2004;16:2097–101.

[21] Gleskova H, Wagner S, Suo Z. Failure resistance of amorphous silicon transistors under extreme in-plane strain. Appl Phys Lett 1999;75:3011–3.

[22] Briseno AL, Aizenberg J, Han Y-J, Penkala RA, Moon H, Lovinger AJ, et al. Patterned growth of large oriented organic semiconductor single crystals on self-assembled monolayer templates. J Am Chem Soc 2005;127:12164–5.

[23] Venables JA, Spiller GDT, Hanbucken M. Nucleation and growth of thin films. Rep Progr Phys 1984;47:399–459.

[24] Gránásy L, Börzsönyi T, Pusztai T. Nucleation and bulk crystallization in binary phase field theory. Phys Rev Lett 2002;88:206105.

[25] Briseno AL, Mannsfeld SBC, Jenekhe SA, Bao Z, Xia Y. Introducing organic nanowire transistors. Mater Today 2008;11(4):38–47.

[26] Curtis MD, Cao J, Kampf JW. Solid-state packing of conjugated oligomers: from π-stacks to the herringbone structure. Am Chem Soc 2004;126:4318–28.

[27] Briseno AL, Miao Q, Ling MM, Reese C, Meng H, Bao Z, Wudl F. Hexathiapentacene: structure, molecular packing, and thin-film transistors. J Am Chem Soc 2006;128:15576–7.

[28] Cho DM, Parkin SR, Watson MD. Partial fluorination overcomes herringbone crystal packing in small polycyclic aromatics. Org Lett 2005;7:1067–8.

[29] Bendikov M, Wudl F, Perepichka DF. Tetrathiafulvalenes, oligoacenenes, and their buckminsterfullerene derivatives: the brick and mortar of organic electronics. Chem Rev 2004;104:4891–945.

[30] Anthony JE. The larger acenes: versatile organic semiconductors. Angew Chem Int Ed 2008;47:452–83.

[31] Briseno AL, Mannsfeld SCB, Reese C, Hancock JM, Xiong Y, Jenekhe SA, Bao Z, Xia Y. Perylenediimide nanowires and their use in fabricating field-effect transistors and complementary inverters. Nano Lett 2007;7:2847–53.

[32] Suo Z, Ma EY, Gleskova H, Wagner S. Mechanics of rollable and foldable film-on-foil electronics. Appl Phys Lett 1999;74:1177–9.

[33] Tang Q, Li H, He M, Hu W, Liu C, Chen K, et al. Low threshold voltage transistors based on individual single-crystalline submicrometer-sized ribbons of copper phthalocyanine. Adv Mater 2006;18:65–8.

[34] Gleskova H, Wagner S, Suo Z. Failure resistance of amorphous silicon transistors under extreme in-plane stain. Appl Phys Lett 1999;75:3011–3.

CHAPTER 2

Chemically Synthesized Semiconductor Nanowires for High-Performance Electronics and Optoelectronics

Xiangfeng Duan[1] and Yu Huang[2]

[1]*Department of Chemistry and Biochemistry, University of California, Los Angeles, California, USA*
[2]*Department of Materials Science and Engineering, University of California, Los Angeles, California, USA*

INTRODUCTION

Semiconductor electronics have experienced enormous growth and have become increasingly pervasive in many aspects of our daily lives over the past several decades. The advancement of semiconductor electronics is rapidly approaching two extremes in terms of physical scale. On the one hand, continued miniaturization of microelectronics according to Moore's law has led to remarkable increases in computing power while at the same time enabling cost reductions [1, 2]. In parallel, extraordinary progress has also been made in the opposite extreme, i.e. large-area electronics or macroelectronics, where electronic components are distributed and integrated over large-area substrates with sizes measured in square meters [3–7]. In microelectronics (or nanoelectronics), the smaller device dimensions are actively sought after for ever faster device switching speed, higher information storage density and larger scale integration for increasingly diverse applications in computer processors, memory and other miniaturized electronic gadgets. In macroelectronics, the large-area substrate is required because the system must be physically large and the active components must be distributed over a large area for desired functions and applications including flat-panel displays, solar cells, image sensor arrays, digital X-ray imagers, electronic paper and smart clothing.

To date, the rapid advancements in electronics have been made possible by many scientific and technological innovations associated with 'top–down'

CONTENTS

manufacturing, in which small features are patterned in bulk semiconductor materials by lithography, vacuum deposition and other processing to form functional devices. However, despite the enormous progress, substantial challenges lie ahead as the semiconductor industry quickly advances towards these physical extremes. On the one hand, as the microelectronics advances towards deep nanometer regimes, the conventional 'top–down' manufacturing approach is quickly approaching its fundamental or economic limits for further downscaling. On the other hand, conventional single-crystal silicon-based electronic manufacture cannot be readily applied to large-area electronics owing to the limited substrate size and prohibitively high cost. Current macro-electronics primarily rely on amorphous silicon (a-Si) or polycrystalline silicon (poly-Si) thin-film transistors (TFTs) on glass technology [5, 6], which is often limited by high process temperature or low carrier mobility.

To go beyond the fundamental and economic limitations of conventional top–down technologies will require fundamentally new strategies [8–10]. Revolutionary breakthroughs rather than current evolutionary progress could enable electronic systems to go far beyond these limits and fuel the expected demands of society in the future. A bottom–up approach, in which functional electronic structures are assembled from chemically synthesized, well-defined nanoscale building blocks, much like the way nature uses proteins and other macromolecules to construct complex biological systems, represents a flexible alternative to conventional top–down methods. In this approach, high-quality nanoscale building blocks, such as nanocrystals, nanotubes or nanowires, are first synthesized using various chemical approaches with precisely controlled chemical composition and physical dimension [11–13]. These nanoscale building blocks typically have nearly perfect crystalline structure and can exhibit unique electronic and optic properties. Owing to their molecular scale dimension, these presynthesized building blocks can be treated like macromolecules and processed in solution phase. Through a certain low-temperature solution assembly process, these nanoscale building blocks are organized into ordered arrays on various supporting substrates, from which the unique function and properties of nanostructures are explored for various applications in electronic and optoelectronic systems. This basic principle readily underpins several combined advantages not readily achievable in traditional technologies, including: (1) high-quality single-crystal building blocks ensure high device performance; (2) low-temperature, low-cost solution assembly processing enables large-area application on flexible substrates such as plastics; and (3) a flexible bottom–up assembly approach allows integration of different materials without the concern of chemical or structure incompatibility and therefore enables multimaterial and multifunction integration.

This chapter reviews the recent advances in bottom–up solution assembly of functional electronic and optoelectronic systems from high-quality nanoscale building blocks, with an emphasis on chemically synthesized semiconductor nanowires as a building block. First, some basic aspects of semiconductor nanowires are discussed, including the rational design of synthetic strategies, parallel and scalable assembly approaches, fundamental electronics and optical properties. Next, a series of nanoscale electronic and optoelectronic devices assembled from these building blocks is discussed, and lastly, turning towards application, a new concept of nanowire thin film-based large-area electronics is reviewed. The chapter concludes with a brief summary and perspective on future opportunities.

SEMICONDUCTOR NANOWIRE BUILDING BLOCKS

Central to the bottom–up approach are nanoscale building blocks. One-dimensional (1D) nanostructures represent the smallest dimension structure that can efficiently transport electrical carriers, and thus are ideally suited to the critical and ubiquitous task of moving and routing charges, which constitute information, in electronics. In addition, 1D nanostructures can exhibit critical device function, and thus can be exploited as both the wiring and device elements in future architectures for functional nanosystems. In this regard, two material classes, semiconductor nanowires [14, 15] and carbon nanotubes [16], have shown particular promise. Single-walled nanotubes have been used to fabricate field-effect transistors (FETs) [17, 18], diodes [19, 20] and logic circuits [21, 22]. However, the inability to control whether nanotube building blocks are semiconducting or metallic and difficulties in manipulating individual nanotubes have made device fabrication largely a random event, and thus pose a significant barrier to achieving highly integrated nanocircuits.

Semiconductor nanowires represent another important and highly versatile nanometer-scale wire structure (Figure 2.1) [23, 24]. In contrast to nanotubes, semiconductor nanowires can be rationally and predictably synthesized in single-crystal form with all key parameters, including chemical composition, diameter and length, and doping/electronic properties, controlled [25–27]. This precise control over key variables has correspondingly enabled a wide range of devices and integration strategies to be pursued. For example, semiconductor nanowires have been assembled into a series of electronic devices including crossed nanowire p–n diodes [28–31], crossed nanowire field-effect transistors (cNW-FETs) [30], nanoscale logic gates and computation circuits [30], as well as optoelectronic devices including nanoscale light-emitting diodes (LEDs) [31] and lasers [32]. In contrast to nanotubes,

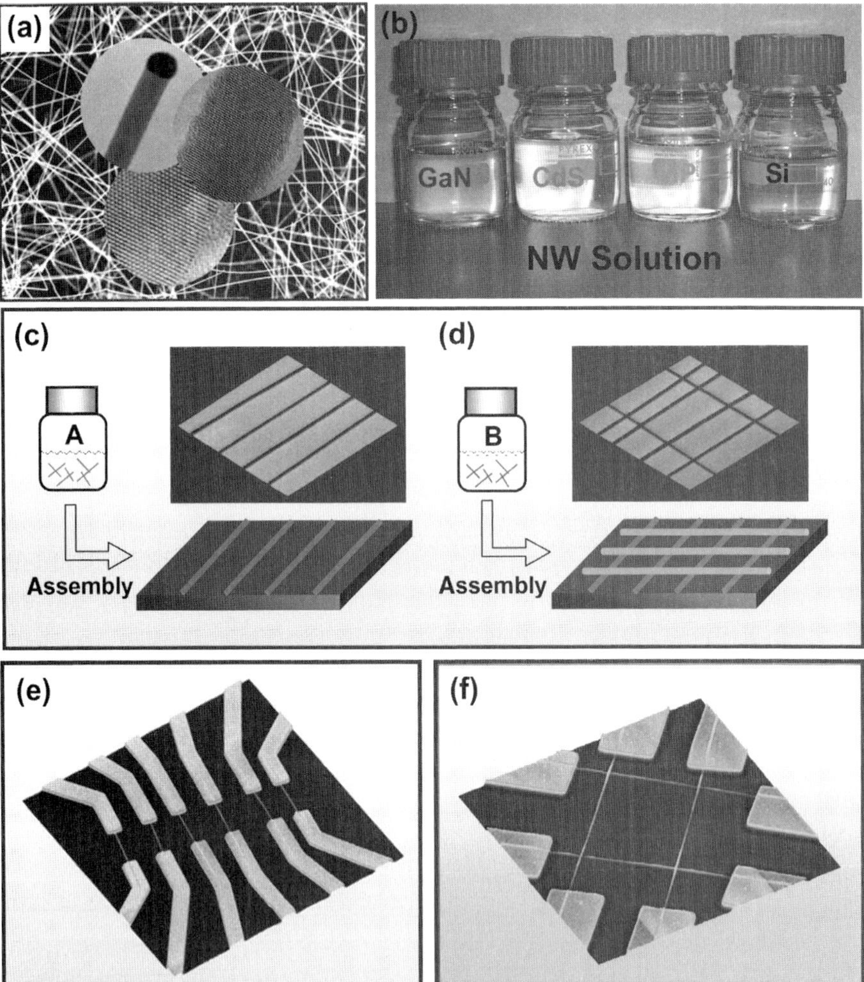

FIGURE 2.1 *Semiconductor nanowire (NW) building blocks. (a) Scanning electron microscope (SEM) image of semiconductor nanowires. The circular insets show transmission electron microscope (TEM) image of nanowires. (b) Nanowire suspension in solutions. (c, d) Schematic illustrating solution assembly process to obtain parallel or crossed nanowire array. (e, f) Images of parallel or crossed nanowire device array with lithography defined metal electrodes connecting to macroscopic systems.*

nanowire devices can be assembled in a rational and predictable manner because the size, interfacial properties and electronic properties of the nanowires can be precisely controlled during synthesis, and moreover, reliable methods have been developed for their parallel assembly [28, 33]. In addition, it is possible to combine distinct nanowire building blocks in ways not possible in conventional electronics and to leverage the knowledge base that exists for the chemical modification of inorganic surfaces [34, 35] to produce semiconductor nanowire devices that achieve new function and correspondingly lead to unexpected device and system concepts.

Synthesis of semiconductor nanowires

Rational design and reproducible synthesis of high-quality nanoscale building blocks is critical to work directed towards bottom–up assembly of functional system. In general, the growth of 1D materials requires that two dimensions are kept in the nanometer regime while the third dimension extends to macroscopic dimensions. This overall requirement is considerably more difficult than the corresponding constraints needed for successful growth of 0D and 2D structures [36, 37]. For example, many important semiconductor materials adopt a cubic zinc blende (ZB) structure, and thus when growth is stopped at an early stage, the resulting nanoscale structures are three-dimensional (3D) polyhedron nanocrystals and not 1D nanowires. To achieve 1D growth in systems where atomic bonding is relatively isotropic requires that the growth symmetry must be broken rather than simply arresting growth at an early stage.

In this regard, the authors have developed a general approach to controlled synthesis of a broad range of semiconductor nanowires via a metal nanocluster-catalyzed vapor–liquid–solid (VLS) growth mechanism [25, 38–41]. Here the catalytic nanoclusters function as an energy-favored site for localized chemical reaction and crystal nucleation, confining/directing the crystal growth in 1D, and defining the diameter of the resulted nanowire (Figure 2.2a). The growth is typically carried out in a chemical vapor deposition system at a certain specific temperature in a tube furnace (Figure 2.2b).

This synthetic concept is especially important since it readily provides the

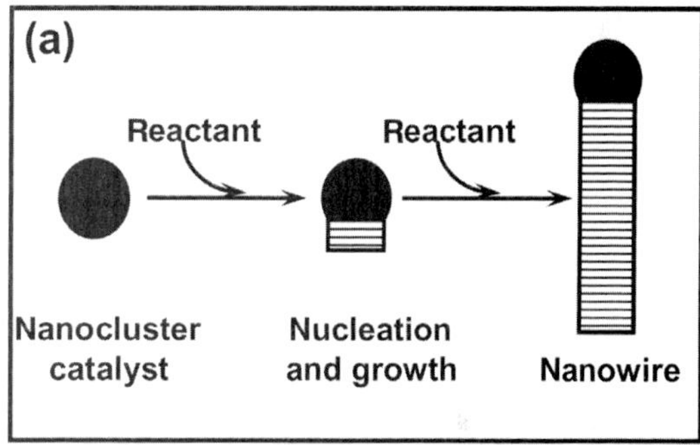

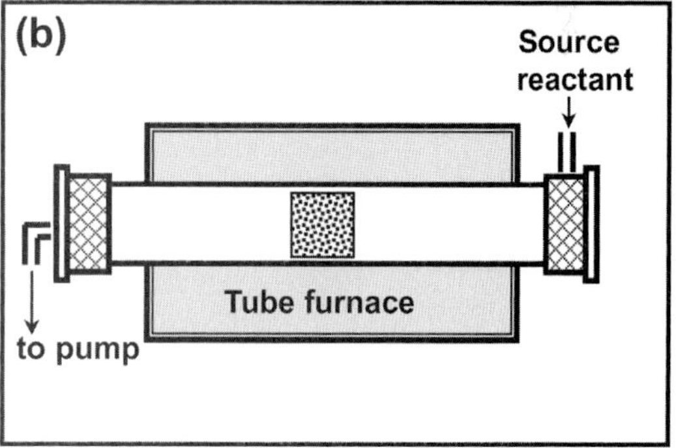

FIGURE 2.2 *Catalytic growth of nanowires. (a) Schematic illustrating the underlying concept for catalytic growth of nanowires. Liquid catalytic clusters act as the energetically favored site for localized chemical reaction, absorption of vapor phase reactant and crystallization of crystalline nanowires. (b) Schematic of a nanowire growth apparatus.*

intellectual underpinning for the specification of the catalyst and growth conditions required for predictable nanowire growth. First, equilibrium phase diagrams can be used to determine catalyst materials that form a liquid alloy with the nanowire material of interest. The phase diagram can also be used to choose a specific growth temperature such that there is a coexistence of a liquid alloy and solid nanowire phases. The liquid catalyst alloy cluster serves as the preferential site for absorption of reactant since the sticking coefficient is much higher on liquid than on solid surfaces and, when supersaturated, the nucleation site for crystallization. Preferential 1D growth occurs in the presence of reactant as long as the catalyst nanodroplet remains in the liquid state. Within this framework it is straightforward to synthesize nanowires with all materials parameters, including chemical composition, physical dimension and electronic optical properties, precisely controlled. The chemical composition (stoichiometry or doping) is controlled by gaseous reactant introduced during the nanowire growth process. The diameter is defined by the size of the catalytic nanoclusters, and the length is controlled by the growth duration. The electronic and optical properties are controlled by the chemical composition and physical dimension.

Taking GaAs nanowire as an example, the pseudobinary phase diagram of Au–GaAs (Figure 2.3a) shows that Au–Ga–As liquid and GaAs solid are the principal phases above 630°C in the GaAs-rich region of the phase diagram [42]. This information implies that Au can serve as a catalyst to grow GaAs nanowires if the growth temperature is chosen to be within this region of the phase diagram.

Indeed, reaction using Au as the catalyst produces samples consisting primarily of a wire-like structure with diameters in the order of 10 nm, and lengths extending up to tens of micrometers (Figure 2.3b). Detailed transmission electron microscope (TEM) studies along with electron diffraction (ED) and energy dispersive X-ray (EDX) analysis also revealed important features of the structure and chemical compositions of the produced GaAs nanowires. First, diffraction contrast images of individual nanowires (Figure 2.3c) indicate that nanowires are single crystal (uniform contrast) and uniform in diameter. Second, the Ga:As composition determined by EDX is consistent with stoichiometric GaAs within limits of instrument sensitivity. Third, the ED pattern recorded perpendicular to the long axis of this nanowire (inset, Figure 2.3c) can be indexed for the <112> zone axis of the ZB GaAs structure, and thus shows that growth occurs along the [111] direction. Extensive studies of individual GaAs nanowires show that growth occurs along the <111> directions in all cases. This direction and the single-crystal structure have been further confirmed by lattice-resolved TEM images (e.g. Figure 2.3d) that show clearly the (111) lattice planes perpendicular to the

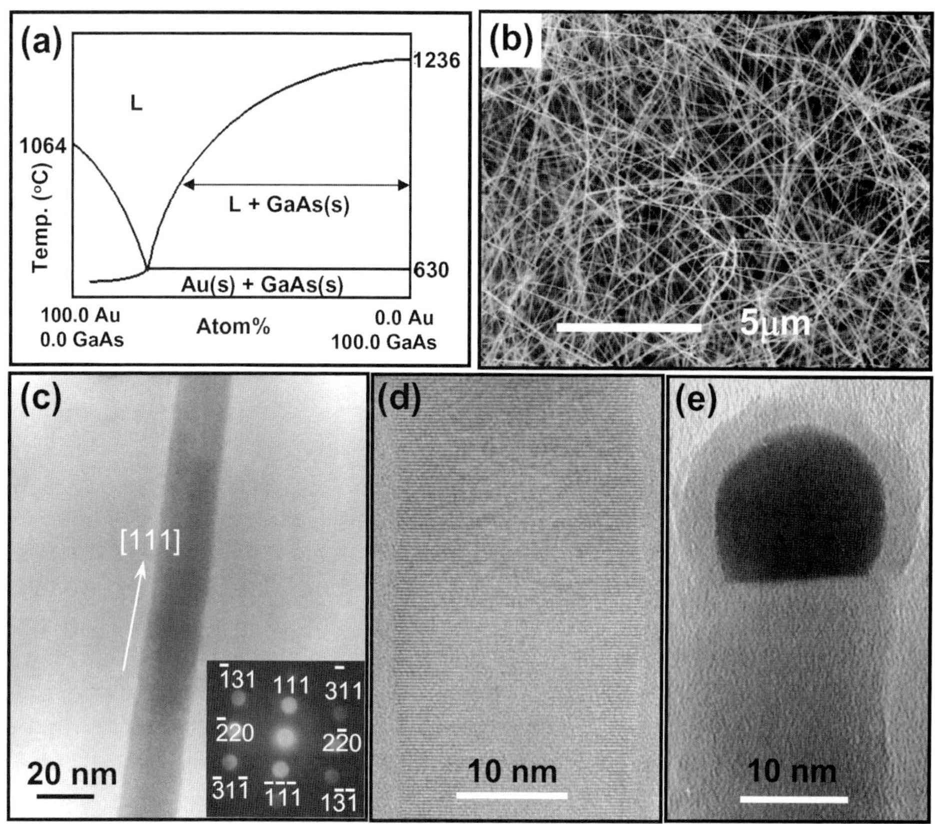

FIGURE 2.3 *Catalytic growth of GaAs nanowires. (a) Pseudobinary phase diagram of the GaAs–Au system. (b) SEM image of GaAs nanowires produced by LCG. (c) Diffraction contrast TEM image of a GaAs nanowire indicating single-crystal structure. The inset shows a convergent beam electron diffraction (ED) pattern recorded along <112> zone axis, and demonstrates wire axis (and growth) is along the [111] direction. (d) High-resolution TEM image of the nanowire. The (111) lattice planes (lattice spacing 0.32 ± 0.01 nm) are clearly seen and perpendicular to the wire axis, further confirming the [111] growth direction. (e) High-resolution TEM image of a nanowire end showing a catalyst nanocluster. Energy dispersive X-ray analysis indicates that the nanoparticle is composed primarily of Au with trace amounts of Ga and As. (Adapted from [25].)*

wire axis. Lastly, the TEM studies also revealed that most nanowires terminate at one end with a nanoparticle of higher contrast (Figure 2.3e). EDX analysis indicates that the nanoparticles are composed largely of Au. The presence of Au nanoparticles at the ends of the nanowires is consistent with the pseudobinary phase diagram, and represents strong evidence for a VLS growth mechanism.

The metal cluster-mediated catalytic growth approach represents a general synthetic approach for a broad range of nanowire materials [25, 38–41]. In a number of cases, phase diagrams could be used to define clearly catalyst, composition and growth conditions required for successful nanowire growth, although appropriate phase diagrams (e.g. pseudobinary and more complex) are not readily available for many compound semiconductor materials and catalytic metals of interest. To extend the synthetic approach to the broadest range of materials, it was recognized that catalysts for laser-assisted catalytic growth (LCG) can be chosen in the absence of detailed phase diagram data by identifying metals in which the nanowire component elements are soluble in the liquid phase but that do not form solid compounds more stable than the desired nanowire phase; that is, the ideal metal catalyst should be 'physically active' but 'chemically stable'. This basic guiding principle suggests that the noble metal Au should represent a good starting point for many semiconductor materials. Indeed, a broad range of nanowire materials, with different structure types and chemical properties has been synthesized as high-quality single crystals using Au as the catalyst (Table 2.1).

Using this approach, a broad range of semiconductor nanowires [38–41], typically with diameters in the order of 10 nm, and lengths extending up to tens of micrometers, can be rationally and predictably synthesized in single-crystal form with all key parameters controlled [26, 43, 44]. Furthermore, additional complexity can be introduced into these nanowire structures by varying the growth conditions or through subsequent processing. For example, composition modulation along the axial dimension can be obtained by changing the reactant precursor during a nanowire growth process [45–48], or through subsequent solid-state reaction diffusion [49] (Figure 2.4a, b). Structure/composition modulation along the radial direction can also be obtained through solid-state reactions or layer-by-layer deposition processes [50, 51] (Figure 2.4c, d).

Table 2.1 Summary of nanowire materials grown by the laser-assisted catalytic growth approach

IV–IV group	III–V group binary	III–V group ternary	II–VI group binary	IV–VI group binary
Si	GaN	$Ga(As_{(1-x)}P_x)$	ZnS	PbSe
Ge	GaP	$In(As_{(1-x)}P_x)$	ZnSe	PbTe
$Si_{(1-x)}Ge_x$	GaAs	$(Ga_{(1-x)}In_x)P$	CdS	
	InP	$(Ga_{(1-x)}In_x)As$	CdSe	
	InAs	$(Ga_{(1-x)}In_x)(As_{(1-x)}P_x)$	$Cd(S_{(1-x)}Se_x)$	

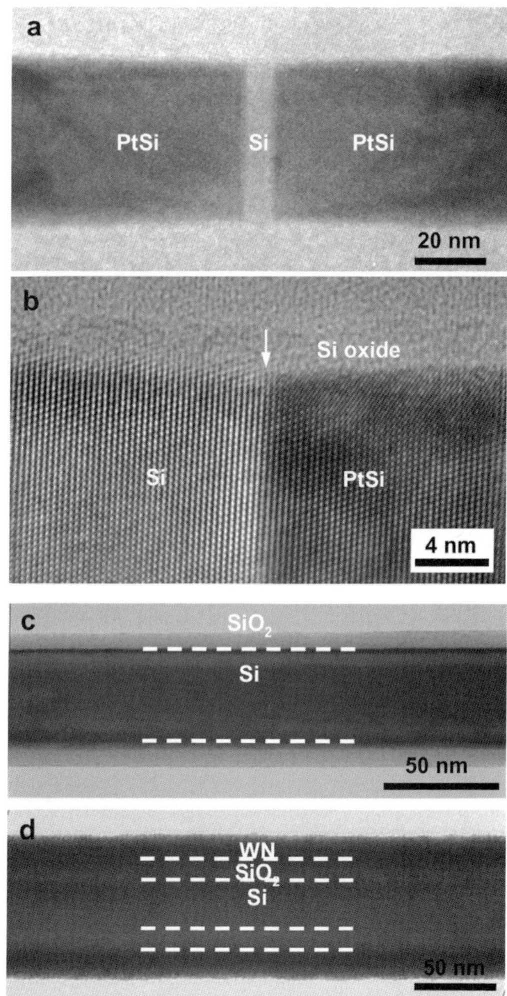

FIGURE 2.4   *Axial and radial nanowire heterostructure. (a) TEM image of a PtSi/Si/PtSi heterostructure obtained by solid-state diffusion of Pt into a silicon nanowire. (b) High-resolution TEM image of PtSi/Si heterostructure showing clean and sharp interfaces between PtSi and Si. The arrow highlights the Si/PtSi interface. (c) TEM image of Si/SiO₂ core–shell nanowire where the SiO₂ shell is obtained through high-temperature oxidation of silicon nanowires. (d) TEM image Si/SiO₂/WN (tungsten nitride) core–shell–shell nanowires where the SiO₂ shell was grown using thermal oxidation, and the WN outer shell was deposited using atomic-layer deposition process.*
(Adapted from [49, 51].)

Assembly of nanowires

The availability of a broad range of semiconductor nanowires with controlled dimension and properties opens many exciting opportunities in fundamental sciences as well as technological applications. To explore fully the potential of these nanowire building blocks for integrated nanosystems will require the development and implementation of efficient and scalable strategies for assembling them into increasingly complex architectures. First, methods are needed to assemble nanowires into highly integrated arrays with controlled orientation and spatial position. Second, approaches must be devised to assemble nanowires on multiple length scales and to make interconnects between nanoscopic, microscopic and macroscopic worlds. To address these critical next levels of organization, several complementary strategies have been explored for the hierarchical assembly of nanowires on surfaces.

Electrical field-directed assembly

Owing to their highly anisotropic morphologies and large polarizabilities, nanowires can be readily polarized and aligned within electric fields (E-fields) (Figure 2.5) [28]. This underlying idea of E-field-directed assembly can be readily seen in images of nanowires aligned between parallel electrodes (Figure 2.5b), which demonstrate that virtually all of the nanowires aligned

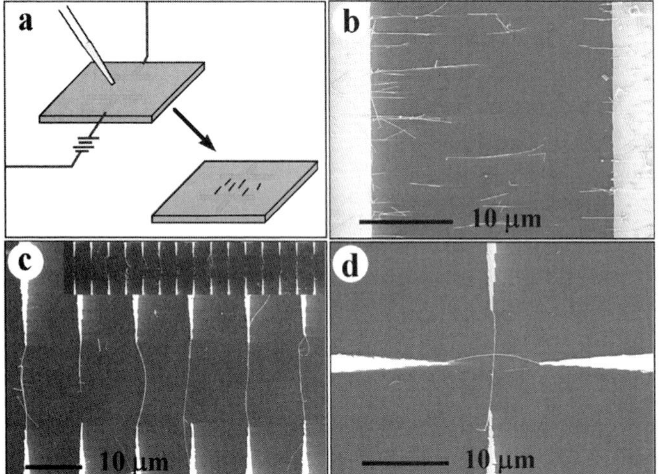

FIGURE 2.5 E-field directed assembly of nanowires. (a) Schematic view of E-field alignment. (b) Parallel array of nanowires aligned between two parallel electrodes. (c) Spatially positioned parallel array of nanowires obtained following E-field assembly. The top inset shows 15 pairs of parallel electrodes with individual nanowires bridging each diametrically opposed electrode pair. (d) Crossed nanowire junction obtained using layer-by-layer alignment with the E-field applied in orthogonal directions in the two assembly steps.
(Adapted from [28].)

in parallel along the E-field direction. E-field-directed assembly can also be used to position individual nanowires at specific positions with controlled directionality. For example, E-field assembly of nanowires between an array of electrodes (Figure 2.5c) clearly shows that individual nanowires can be positioned to bridge pairs of diametrically opposed electrodes and form a parallel array. In addition, by changing the E-field direction with sequential nanowire solutions, the alignment can be carried out in a layer-by-layer fashion to produce crossed nanowire junctions (Figure 2.5d). These results demonstrate clearly that E-field-directed assembly can be used to align and position individual nanowires into parallel and crossed arrays, which correspond to two basic geometries for integration, and thus provide one robust approach for rational and parallel assembly of nanoscale device arrays.

Fluidic flow-directed assembly

Exploiting the linear shear flow near a flat surface, fluidic flow can also be explored to assemble nanowires with well-controlled orientation [33]. In this method, nanowires can be aligned by passing a suspension of nanowires through microfluidic channel structures, for example, formed between a poly-dimethylsiloxane (PDMS) mold [52] and a flat substrate (Figure 2.6a). Images of nanowires assembled on substrate surfaces (Figure 2.6b) within microfluidic flows demonstrate that virtually all of the nanowires are aligned along the flow direction. This alignment readily extends over large-area substrate, and is limited only by the size of the fluidic channels used. The alignment of nanowires within the channel flow can be understood within the framework of shear flow [53, 54]. Specifically, the channel flow near the substrate surface resembles a shear flow, and linear shear force aligns the nanowires in the flow direction before they are immobilized on the substrate.

The fluidic flow assembly approach can be used to organize nanowires into more complex crossed nanowire structures, which are critical for building high-density nanodevice arrays, using a layer-by-layer deposition process (Figure 2.6c). The formation of crossed and more complex structures requires that the nanostructure–substrate interaction is sufficiently strong that sequential flow steps do not affect preceding ones: this condition is readily achieved by modifying the substrate surface with proper functional chemical groups. For example, alternating the flow in orthogonal directions in a two-step assembly process yields crossbar structures in high yield (Figure 2.6d). Experiments have demonstrated that crossbars extending over hundreds of micrometers on a substrate with only hundreds of nanometers' separation between individual cross-points are obtained through a very straightforward, parallel, low-cost and fast process.

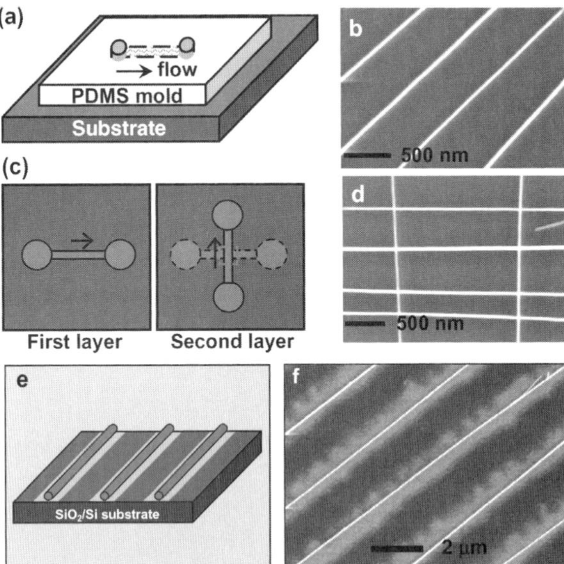

FIGURE 2.6 *Fluidic flow-directed assembly of nanowires. (a) Schematic and (b) SEM image of parallel nanowire arrays obtained by passing a nanowire solution through a channel on a substrate. (c) Schematic and (d) SEM image of crossed nanowire matrix obtained by orthogonally changing the flow direction in a sequential flow alignment process. (e) Schematic and (f) SEM image of regular nanowire arrays obtained by flowing nanowire solution over a chemically patterned surface. PDMS: polydimethylsiloxane. (Adapted from [33].)*

Fluidic flow-directed assembly of multiple crossed nanowire arrays offers significant advantages over previous efforts. First, it is intrinsically very parallel and scalable with the alignment readily extending over very large length scales. Second, this approach is general for virtually any elongated nanostructure including carbon nanotubes and DNA molecules. Third, it allows for the directed assembly of geometrically complex structures by simply controlling the angles between flow directions in sequential assembly steps. For example, equilateral triangles have been assembled in a three-layer deposition sequence using 60 degree angles between the three flow directions [33]. The method of flow alignment thus provides a flexible way to meet the requirements of many device configurations in the future. An important feature of this sequential assembly scheme is that each layer is independent, and thus a variety of homojunction and heterojunction configurations can be obtained simply by changing the composition of the nanowire suspension used for each flow step.

To enable further control of spatial location and periodicity of assembled nanowire arrays, complementary chemical interactions between chemically patterned substrates and nanowires were explored (Figure 2.6e). Substrates for alignment are first patterned with two different functional groups, with one of the functional groups designed to have a strong attractive interaction with the nanowire surface, and then following flow alignment, regular, parallel nanowire arrays with lateral periods the same as those of the surface patterns are produced (Figure 2.6f). The experiment demonstrates that the nanowires are preferentially assembled at positions defined by the chemical pattern, and moreover, show that the periodic patterns can organize the nanowires into regular superstructures. It is important to recognize that the patterned surface alone does not provide good control of the 1D nanostructure organization. Assembly of nanotubes [55, 56] and nanowires on patterned substrates shows that 1D nanostructures align with bridging and looping structures over the patterned areas and show little directional control. The use of fluidic flows avoids these significant problems and enables controlled assembly in one or more directions. By combining this approach with other surface patterning methods, such as phase separation in diblock copolymers [57] and spontaneous ordering of molecules [58], it should be possible to generate well-ordered nanowire arrays without the limitations of conventional lithography.

Langmuir–Blodgett technique assisted nanowire assembly

The Langmuir–Blodgett (LB) technique has been widely used to prepare monolayers of fatty acids and amphiphilic molecules and later was extended to prepare molecular electronic devices based on a monolayer of molecules or nanocrystals [59, 60]. Recently, LB has been exploited for controlled assembly of nanowires over large scale (Figure 2.7) [61, 62]: LB was applied to compress uniaxially a nanowire monolayer floating on a water surface to produce aligned nanowire arrays parallel to the direction of the trough with controlled spacing (Figure 2.7a). The aligned nanowire monolayer was then transferred to any desired planar substrate, such as Si, plastic or glass, yielding densely packed nanowire arrays on the substrate surface

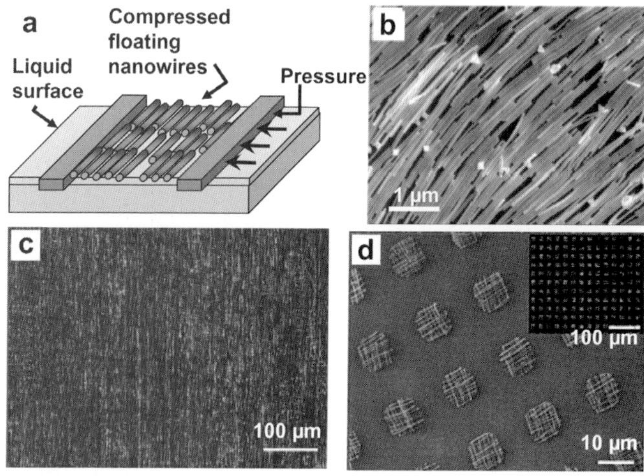

FIGURE 2.7 *Langmuir–Blodgett (LB) technique assisted nanowire assembly. (a) Schematic of LB-guided assembly. Nanowires are floated on a water surface and compressed through computer-guided pressure. (b) SEM image of silver nanowires deposited on Si wafer. (c) Large-area image of parallel nanowire array deposited uniformly on a substrate. (d) SEM image of patterned crossed nanowire arrays; inset: large-area dark-field optical micrograph of the patterned crossed nanowire arrays. (Adapted from [61, 62].)*

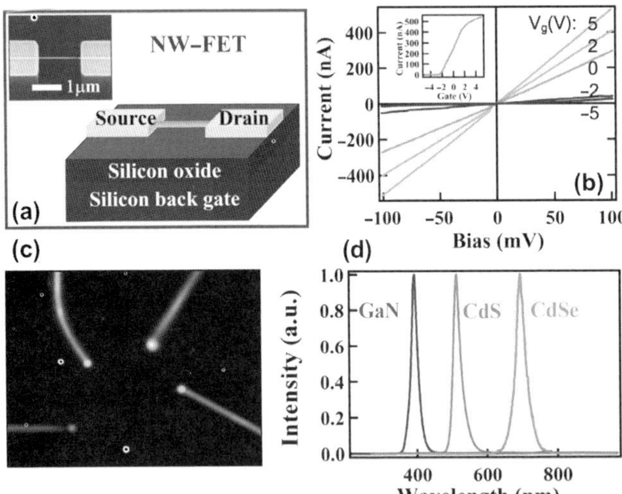

FIGURE 2.8 *Nanowire electronic and optical properties. (a) Schematic of a nanowire field-effect transistor (NW-FET) used to characterize electrical transport properties of individual nanowires. Inset: SEM image of an NW-FET; two metal electrodes, which correspond to source and drain, are visible at the left and right sides of the image. (b) Current vs voltage for an n-type InP NW-FET. The numbers inside the plot indicate the corresponding gate voltages (V_g). Inset: current vs V_g for V_{sd} of 0.1 V. (c) Real-color photoluminescence image of various nanowires showing different color emissions. (d) Spectra of individual nanowire photoluminescence. All nanowire materials show a clean band-edge emission spectrum with narrow FWHM around 20 nm.*

(Figure 2.7b, c). The LB technique can be applied to large areas of nanowire assemblies, limited only by the numbers of nanowires dispersed on water surface. Similar to fluidic flow-directed assembly, this approach also offers the flexibility of assembling various nanowires in a layer-by-layer fashion and produces both parallel and cross-packed nanowire arrays (Figure 2.7d). However, to do this usually requires coating the nanowire surface with a molecular layer to render a hydrophobic surface, which needs to be removed before further processing.

Properties of nanowires

The basic electronic properties of nanowires are characterized using electrical transport studies in a single nanowire field-effect transistor (NW-FET) configuration (Figure 2.8a, b). The NW-FETs are prepared by placing nanowires onto the surface of an oxidized Si substrate, where the underlying conducting Si is used as a global back gate [63, 64]. Source and drain electrodes are defined by electron-beam lithography, followed by electron-beam evaporation of metal contacts, and electrical transport measurements are taken at room temperature. Current (I_{sd}) versus source–drain voltage (V_{sd}) and I_{sd} versus gate voltage (V_g) are then recorded for a NW-FET to characterize its electrical properties. Gate sweeping measurement of the NW-FET enables elucidation of important qualitative and quantitative properties of nanowires. For example, changes in V_g produce variations in the electrostatic potential of the nanowire, and hence modulate the carrier concentration and conductance of the nanowire (Figure 2.8b). Depending on the conductance modulation, it is possible to estimate the carrier mobility in individual nanowires using standard transistor formula: $dI_{sd}/dV_g = \mu(C/L^2)V_{sd}$ and $C \cong 2\pi\varepsilon\varepsilon_0 L/\ln(2h/r)$, where μ is the carrier mobility, C is the capacitance, ε is the dielectric constant, h is the thickness of the SiO_2 dielectric, L is the length and r is the radius of the nanowire. A broad range of nanowire materials including p-type Si, n-type GaN, CdS and InP nanowires shows excellent carrier mobility comparable to bulk materials [65], which demonstrates the high quality of these materials.

Optical studies on various direct band gap semiconductor nanowires have demonstrated that these nanowire materials can also exhibit excellent photoluminescence properties with the emission color tuneable from ultraviolet (UV) to near-infrared (IR) regime by controlling either the chemical composition or physical dimension of nanowire materials (Figure 2.8c). For example, photoluminescence studies on individual GaN, CdS and CdSe nanowires showed clean luminescence with spectra maxima of ~370, ~510 and ~700 nm, respectively [65], which is consistent with near band edge emission in these materials (Figure 2.8d). In addition, a photoluminescence study on InP nanowires of various diameters showed that the emission spectra systematically shifted from 900 nm to 700 nm when the nanowire diameter was reduced from 50 nm to 10 nm. This systematic blue shift of emission with reduced nanowire diameter can be attributed to the quantum confinement effect in these molecular scale nanowires [66].

The fundamental physical properties of nanowire materials can be further improved to surpass greatly their bulk counterpart using precisely engineered nanowire heterostructures. It has been recently demonstrated that Si/Ge/Si core–shell nanowires exhibit electron mobility surpassing that of state-of-the-art technology [67]. Group III–V nitride core–shell nanowires of multiple layers of epitaxial structures with atomically sharp interfaces have also been demonstrated with well-controlled and tunable optical and electronic properties [68, 69]. Together, the studies demonstrate that semiconductor nanowires represent one of the best defined nanoscale building block classes, with well-controlled chemical composition, physical size and superior electronic/optical properties, and therefore they are ideally suited for assembly of more complex functional systems.

NANOWIRE NANOELECTRONICS

The ability rationally to synthesize nanowires with controlled electronic properties and to assemble nanowires into regular arrays readily enabled these nanowire structures to be explored for a variety of functional device arrays.

Single nanowire transistors

With the high-quality single-crystalline nature, a single semiconductor nanowire can function as a nanoscale semiconducting channel for a high-performance FET. In addition to single-crystalline nature, to achieve a high-performance transistor requires optimized source–drain contacts and

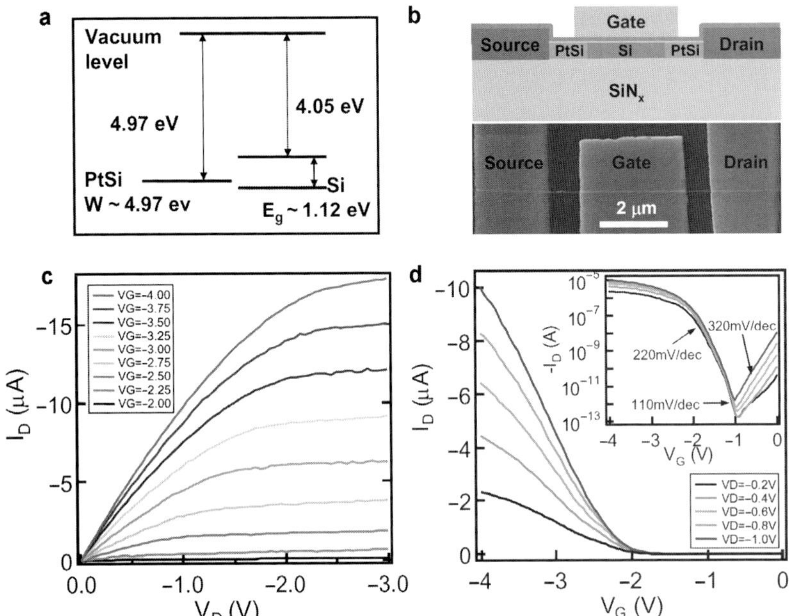

FIGURE 2.9 *Single nanowire transistor using PtSi/i–Si/PtSi nanowire heterostructures. (a) Relative energy band alignment between PtSi and Si. (b) Top: schematics of PtSi/i–Si/PtSi nanowire heterostructure device with HfO₂ as gate dielectrics; bottom: SEM images of a device (scale bar: 2 µm). (c) Drain current (I_D) vs drain–source voltage (V_D) at increasing gate negative voltages (V_{GS}) in steps of 0.25 V starting from bottom at $V_G = -2.0$ V. (d) I_D vs V_G at $V_D = -0.2, -0.4, -0.6, -0.8, -1.0$ V. The inset shows $-I_D$ vs V_G in the exponential scale, highlighting the on/off ratio $> 10^7$, and ambipolar transport with a subthreshold swing of 110–220 mV/decade for hole transport and 320 mV/decade for electron transport. (Adapted from [49].)*

high-quality gate dielectrics. Importantly, single-crystalline metal silicide can form atomically sharp junctions with Si nanowires, which represent a natural candidate for high-quality source and drain contacts [49, 70–74]. With many different silicide materials of variable work function available, one can readily select proper silicide material to make nearly ideal contact with either p-channel or n-channel transistors. For example, it has recently been demonstrated that PtSi/Si nanowire junction can be formed in through solid-state reaction between a Si nanowire and Pt pads. PtSi has a work function of 4.97 eV [75, 76], aligning well with the Si valence band with a small energy barrier of ~0.2 eV, and therefore can be used as nearly perfect ohmic contact for p-channel Si nanowire transistors (Figure 2.9a). In addition, the formation of atomically clean Si/PtSi interfaces prevents Fermi-level pinning and helps to maintain a low Schottky barrier determined by work function rather than by interface states. This has enabled the fabrication of high-performance nanowire transistors from intrinsic Si nanowires using PtSi as the source and drain contacts, and high-k dielectrics (HfO₂) as the gate dielectrics. To make the device (Figure 2.9b), a 7 nm thick layer of HfO₂ was deposited on the surface of a partially silicidized PtSi/Si/PtSi device using atomic layer deposition (ALD) to form gate dielectrics, and a top Cr/Au electrode was defined with electron-beam lithography and deposited using electron-beam evaporation. Electrical transport measurement on the device shows nearly perfect transistor characteristics. Figure 2.9(c) shows drain current (I_D) versus drain–source bias voltage (V_D) relations at various gate voltages (V_G) in steps of 0.25 V. The device shows typical p-channel enhancement mode (normally off) transistor behavior. I_D increases linearly with V_D at low V_D, and saturates at higher V_D. It is important to note that a nearly perfect linear relationship is observed in the I_D–V_D plot in the low bias region, which suggests that a satisfactory ohmic contact is achieved for hole transport in the intrinsic Si nanowire device owing to a very low Schottky barrier. The plot of I_D versus V_G (Figure 2.9d) at constant $V_D = -0.2, -0.4,$ $-0.6, -0.8$ and -1.0 shows that little current can pass through the device when the gate voltage is below a threshold voltage of approximately -2.0 V, and I_D increases nearly linearly when the gate voltage increases in the negative direction beyond the threshold voltage.

To gauge the device performance, it is important to analyze some key device parameters, including transconductance, mobility, on/off current ratio and subthreshold swing. High transconductance is a critical measure of transistor performance and determines voltage gains of transistor-based devices including amplifiers and logic circuits. The maximum transconductance of the tested device in the saturation region is $\sim$12 µS. Assuming that the effective channel width equals to the nanowire diameter (40 nm), a normalized trans-conductance of $\sim$0.3 mS/µm is obtained. Importantly, this value is significantly better than those reported previously for a chemical synthesized Si nanowire device and is nearly comparable to the state-of-the-art metal oxide semiconductor field-effect transistor (MOSFET) devices ($\sim$0.6 mS/µm) [77]. In addition, considering the present device has a relatively long channel length (2.3 µm), the transconductance can be further improved by reducing the channel length. The estimated field-effect hole-mobility of the device is 168 cm^2/V·s, which is closely comparable to p-type single-crystal Si material such as silicon-on-insulator MOSFET devices ($\sim$180 cm^2/V·s) [78].

The plot of the $-I_D$–V_G curve in the exponential scale (inset, Figure 2.9d) shows that I_D decreases exponentially below the threshold voltage, and upon reaching a minimum of $\sim$0.2 pA, it increases again, further sweeping the gate voltage towards the positive direction. This behavior suggests ambipolar transport, with the left branch corresponding to hole transport (p-type) and right branch corresponding to electron transport (n-type). The ambipolar transport is typical of metal-contacted Schottky barrier MOSFET and has been previously observed in silicide contacted Si on insulator devices and nanowire devices [79, 80]. The plot shows that the transistor has a maximum on/off current ratio greater than seven orders of magnitude. The exponential decrease in current defines a key transistor parameter, the subthreshold swing $S = -dV_G/d\lg|I_D|$. For the hole transport branch (left), the sub-threshold swing ranges from 110 mV/decade near the current minimum to 220 mV/decade near the threshold; and for the electron transport branch (right), a constant subthreshold swing of 320 mV/decade is observed. In conventional MOSFETs, subthreshold transport is dominated by thermal emission with the swing S determined by $S = (k_B T/e) \cdot \ln[(10)(1 + \alpha)]$, where T is temperature, k_B is Boltzmann's constant, e is elementary charge, and α depends on capacitances in the devices and is 0 when the gate capacitance is much larger than other capacitances such as interface trap state capacitance. The lowest theoretical limit for S is therefore $S = (k_B T/e) \cdot \ln(10)$ ~60 mV/decade at room temperature. The existence of trapping states in gate dielectrics can lead to the larger than ideal subthreshold swing. In addition, it is also often seen that Schottky barrier MOSFET has a larger subthreshold swing due to Schottky barrier limited carrier injection. Both gate dielectric (native silicon oxide and HfO_2) interface trapping states and Schottky barrier may contribute to the non-ideal subthreshold swing observed in these devices. In hole transport branch (p-branch), the transport near the current minimum (with subthreshold swing of ~110 mV/decade) is dominated by thermal emission, and the transport near the threshold (with swing ~220 mV/decade) is probably more also affected by the existence of a small Schottky barrier. Therefore, the subthreshold swing can be improved by reducing the interface trapping states with improved gate dielectrics and by further reducing the effective Schottky barrier with alternative contact materials (e.g. IrSi) [75] or dopants near the contact region [79]. The significant larger swing (~320 mV/decade) in the electron transport branch (n-branch) is due to a much larger Schottky barrier for electron transport [79]. These results clearly demonstrate that a high-performance transistor can be obtained from single-crystalline semiconductor nanowires with proper selection of source–drain contact and gate dielectrics.

Crossed nanowire devices

Direct assembly of highly integrated functional electronic circuits based on nanowires requires (1) the development of new device concepts with scalable device configuration and (2) high-yield assembly of these devices with controllable functional properties. The crossed nanowire matrix represents an ideal configuration since the critical device dimension is usually defined by the cross-point and can be readily scaled down to nanometer level, and crossed nanowire configuration itself is naturally a scalable architecture and thus enables the possibility of massive system integration. Moreover, the crossed nanowire matrix is a versatile structure and can be configured into a variety of critical device elements, such as diodes and transistors. For example, a p–n diode can be obtained by simply crossing a p- and n-type nanowire as demonstrated in the cases of p-Si/n-Si, p-InP/n-InP, p-InP/n-CdS and p-Si/n-GaN materials [28–31]. Electrical measurements of such crossed junctions show

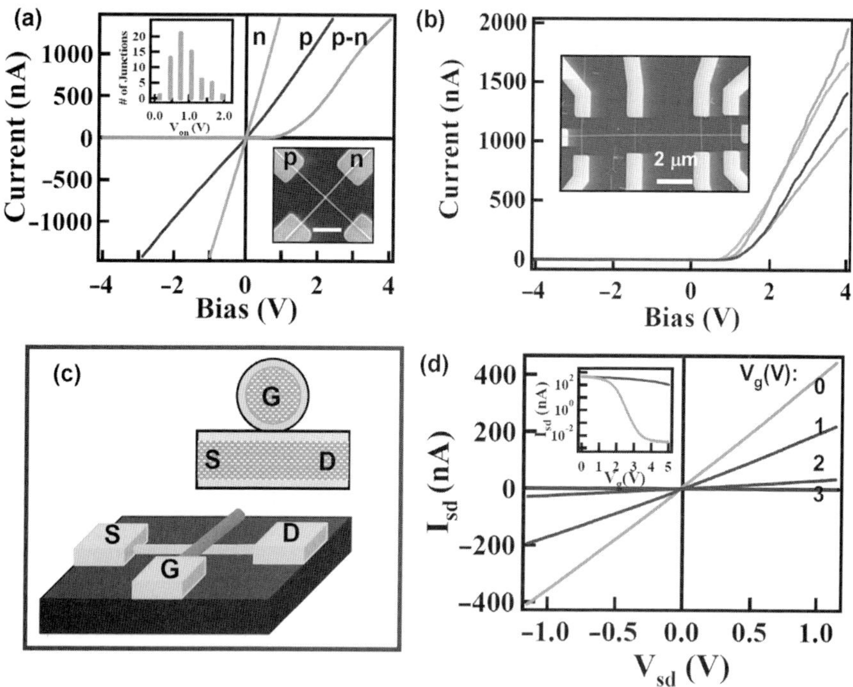

FIGURE 2.10 *Crossed nanowire devices. (a) Current–voltage (I–V) relation of the crossed p–n diode. Linear or nearly linear I–V behavior of the p- (blue) and n-type (green) nanowires indicates good contact between nanowires and metal electrodes. I–V curves across the junction (red) show clear current rectification. The top left inset shows a histogram of turn-on voltage for over 70 as-assembled junctions showing a narrow distribution around 1 V. The bottom right inset shows a typical SEM image of a crossed nanowire p–n diode. Scale bar: 1 μm. (b) I–V behavior for a 4(p) × 1(n) multiple junction array. Inset shows a SEM image of a nanowire p–n diode array. (c) Schematics and illustrating the crossed nanowire field-effect transistor (cNW-FET) concept. (d) Gate-dependent I–V characteristics of a cNW-FET formed using a p-NW as the conducting channel and n-NW as the local gate. The red and blue curves in the inset show I_{sd} vs V_{gate} for n-NW (red) and global back (blue) gates when the V_{sd} is set at 1 V. The conductance modulation (> 10^5) of the field-effect transistor is much more significant with the nanowire gate than with a global back gate (< 10). (Adapted from [30].)*

clear current rectification across the junction and linear current behavior in individual nanowires, demonstrating the formation of p–n diode at the crossing point (Figure 2.10a). To gauge the reproducibility of these assembled nanowire p–n diodes, a large number of p–n junctions assembled from p-Si nanowires and n-GaN nanowires was studied (inset, Figure 2.10a) [30]. *I–V* measurements made on over 100 crossed p-Si/n-GaN nanowire devices show that over 95% of the junctions exhibit current rectification with turn-on voltages of around 1.0 V. Reproducible assembly of crossed nanowire structures with predictable electrical properties contrasts sharply with results from nanotube-based device, and readily enabled the assembly and properties of integrated p–n diode arrays to be explored. Significantly, electrical transport measurements made on a typical 4 by 1 crossed p-Si/n-GaN junction array (Figure 2.10b) show that the four nanoscale cross-points form independently addressable p–n diodes with clear current rectification and similar turn-on voltages.

Nanoscale FETs can also be achieved in the crossed nanowire configuration using one nanowire as the active conducting channel and the other crossed nanowire as the gate electrode (Figure 2.10c). Significantly, the three critical FET device metrics are naturally defined at the nanometer scale in the assembled crossed nanowire FETs (cNW-FETs): (1) a nanoscale channel width determined by the diameter of the active nanowire ($\sim$2–20 nm depending chosen nanowire); (2) a nanoscale channel length defined by the crossed gate nanowire diameter ($\sim$10 nm); and (3) a nanoscale gate dielectric thickness determined by the nanowire surface oxide ($\sim$1 nm). These distinct nanoscale device metrics lead to greatly improved device characteristics such as high gain, high speed and low power dissipation. For example, the conductance modulation of an NW-FET is much more significant with the nanowire gate ($> 10^5$) than that with a global back gate (< 10) (inset in Figure 2.10d). Moreover, the local nanowire gate enables independently addressable FET arrays and thus enables highly integrated nanocircuits.

Nanoscale logic gates and computational circuits

High-yield assembly of crossed nanowire p–n diodes and cNW-FETs from p-Si and n-GaN materials enables more complex functional electronic circuits, such as logic gates, to be produced. Logic gates are critical blocks of hardware in current computing systems that produce a logic-1 and logic-0 output when the input logic requirements are satisfied. Diodes and transistors represent two basic device elements in logic gates [80]. Transistors are more typically used in current computing systems because they can exhibit voltage gain. Diodes do not usually exhibit voltage gain, although they may also be desirable in some cases [30]; for example, the circuit architecture and constraints on the assembly of nanoelectronics might be simplified using diodes since they are two-terminal devices, in contrast to three-terminal transistors. In addition, by combining the diodes and transistors in logic circuits, it is possible to achieve high voltage gain, while simultaneously maintaining a simplified device architecture. To demonstrate the flexibility of these nanowire device elements, both diode- and FET-based logic were investigated.

For example, a two-input logic OR gate was realized using a 2(p) by 1(n) crossed nanowire p–n diode array [30]. When either of the inputs to the p-NW is high, a high output is obtained at the n-NW as the p–n diode is forward biased; a low output is only achieved when both inputs are low; and thus realizing the same function as a conventional logic 'OR' gate (Figure 2.11a, b). A logic 'AND' gate was also assembled from two p–n diodes and one cNW-FET (Figure 2.11c, d), and a logic 'NOR' gate with gain over 5 was assembled from three cNW-FETs in series (Figure 2.11e, f). Importantly, logic OR, AND and NOR gates form a complete set of logic elements and enable the organization of virtually any logic circuit. For example, nanowire logic gates have been interconnected to form an XOR gate and a logic half adder, which was used to carry out digital computations in a way similar to conventional electronics (Figure 2.11g, h).

NANOSCALE PHOTONICS AND OPTOELECTRONICS

The availability of a wide range of nanowire materials readily allows us to choose materials with different properties to tailor device functions in a manner that is unique to the bottom–up assembly approach. In addition to nanoscale electronics, the broad range of optically active III–V and II–VI group compound semiconductor nanowire materials is attractive as building blocks for miniaturized photonic and optoelectronic devices. To this end, a variety of photonic devices, including nanoscale LEDs and diode arrays, single nanowire optical waveguides, cavities and lasers, has been demonstrated.

Nanoscale light-emitting diode and diode array

A p–n diode can be obtained by crossing a p- and n-type nanowire as described previously. In direct band gap semiconductors such as InP, the p–n diode also forms the basis for the critical optoelectronics devices, including LEDs and laser diodes. To assess whether nanoscale devices might behave similarly, the electroluminescence from crossed nanowire p–n junctions was studied. Significantly, electroluminescence can be readily observed from these nanoscale junctions in forward bias. A 3D plot of the electrolumine- scence intensity taken from a typical nanowire p–n diode at forward bias (Figure 2.12a) shows the emitted light comes from a point-like source, and moreover, comparison of electroluminescence and photoluminescence (inset, Figure 2.12a) images recorded on the same sample shows that the position of the electroluminescence maximum corresponds to the crossing point in the photoluminescence image. These data thus demonstrate that the emitted light indeed comes from the crossed nanowire p–n junction.

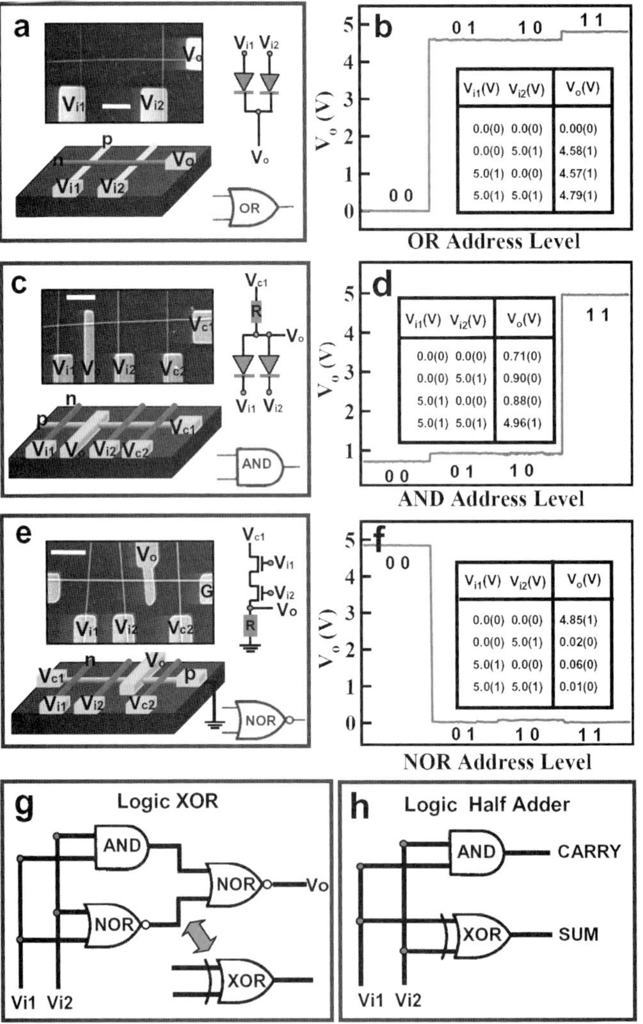

FIGURE 2.11 *Nanoscale logic gates. (a, b) Schematic and measured output vs address level relation for a logic OR gate assembled from two crossed nanowire p–n diodes. The insets in (a) show the SEM image and equivalent electronic circuit of the device. Scale bar: 1 μm. The inset in (b) shows the experimental truth table for nano-OR gate. (c, d) Schematic and measured output vs address level relation for a logic AND gate assembled from two diodes and one crossed nanowire field-effect transistor (cNW-FET). The inset in (d) shows the experimental truth table for the nano-AND gate. (e, f) Schematic and measured output vs address-level relation for a logic NOR gate assembled from three cNW-FETs. The inset in (f) shows the experimental truth table for the nano-NOR gate. (g, h) Schematics for a logic XOR gate and logic half adder realized by interconnecting individual logic gate elements. Significantly, the logic half adder assembled this way can be used to do digital computation just as conventional electronics do.*
(Adapted from [30].)

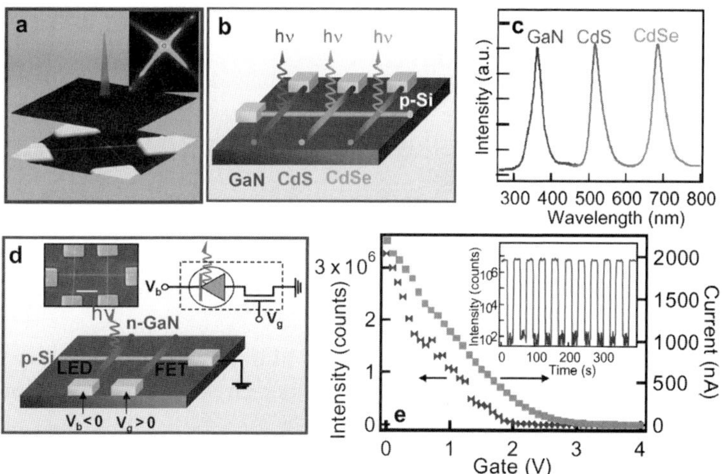

FIGURE 2.12 *Nanoscale light-emitting diode (nanoLED). (a) Crossed InP nanowire LED. Top: Three-dimensional (3D) plot of light intensity of the electroluminescence from a crossed nanowire LED. Light is only observed around the crossing region. Bottom: 3D atomic force microscope image of a crossed nanowire LED. Inset: Photoluminescence image of a crossed nanowire junction. (b) Multicolor nanoLED array. Schematic and corresponding of a tricolor nanoLED array assembled by crossing one n-GaN, n-CdS and n-CdS nanowire with a p-Si nanowire. The array was obtained by fluidic assembly and photolithography with ∼5 μm separation between nanowire emitters. (c) Normalized electroluminescence spectra obtained from the three elements. (d) Integrated nanoFET–nanoLED array. Schematic of an integrated crossed nanowire field-effect transistor (cNW-FET) and LED. Inset shows SEM image of a representative device (scale bar: 3 μm) and the equivalent circuit. (e) Plots of current and emission intensity of the nanoLED as a function of voltage applied to the nanowire gate at a fixed bias of −6 V. Inset: Electroluminescence intensity vs time relation: a voltage applied to the nanowire gate is switched between 0 and +4 V for fixed bias of −6 V. (Adapted from [31].)*

Electroluminescence spectra (peaked 820 nm) recorded from the cNW-LEDs exhibit blue shifts relative to the bulk band gap of InP (925 nm). The blue shifts are due in part to quantum confinement of excitons, although other factors may also contribute. Furthermore, photoluminescence studies have demonstrated that the photoluminescence peak can be systematically blue-shifted as the nanowire diameter is decreased [66], and thus these results provide a means for controlling the color of the LEDs in a well-defined way. Indeed, electroluminescence results recorded from p–n junctions assembled from smaller (and larger) diameter nanowires show larger (smaller) blue shifts. In addition, the emission color from nanoLEDs can be further varied by using chemically distinct semiconductor nanowires with different band gaps. Considering the wide range of group IV, III–V and II–VI semiconductor nanowire materials available [25], it is possible to assemble a variety of nanowire-based nanoLEDs for different spectral regimes [31]. Indeed, it has been recently demonstrated that crossing p-type Si nanowires with n-type GaN, CdS, CdSeS, CdSe and InP can produce nanoLEDs with emission spectra covering the whole spectrum regime from UV to near-IR [31]. In such heterostructure devices, p-Si nanowires are used as the passive hole injector, and the n-type compound nanowires are used as electron injector and active emitter, which defines the wavelength of emission.

The bottom–up assembly approaches allows flexibly combination of chemically distinct nanoscale building blocks that would otherwise be structurally and/or chemically incompatible in a sequential growth process typical of planar fabrication. This capability should enable assembly of nanostructures with functions not readily obtained by other methods and open up new opportunities. For example, the ability to form nanoLEDs with non-emissive Si nanowire hole injectors has been exploited to assemble multicolor arrays consisting of n-type GaN, CdS and CdSe nanowires crossing a single p-type Si nanowire (Figure 2.12b). Normalized emission spectra recorded from the array demonstrate three spectrally distinct peaks with maxima at 365, 510 and 690 nm (Figure 2.12c) consistent with band edge emission from GaN, CdS and CdSe, respectively. The ability to assemble and integrate different materials together seamlessly and to tune the emission from each nanoLED independently offers substantial potential producing specific wavelength sources, and demonstrates an important step towards integrated nanoscale photonic circuits. Although lithography sets the integration scale of these multicolor arrays, it should be possible to create much denser nanoLED arrays via (1) controlled growth of modulated nanowire superlattice structures [45–49] and/or (2) selective assembly of different semiconductor materials [81].

Optoelectronic circuits consisting of integrated crossed nanowire LED and FET elements can also be assembled (Figure 2.12d). Specifically, one GaN nanowire forms a p–n diode with the SiNW and a second GaN nanowire functions as a local gate as described previously. Measurements of current and emission intensity versus gate voltage show that (1) the current decreases rapidly with increasing voltage, as expected for a depletion mode FET, and (2) the intensity of emitted light also decreases with increasing gate voltage. When the gate voltage is increased from 0 to +3 V, the current is reduced from ~2200 nA to an off state, where the supply voltage is −6 V (Figure 2.12e). Advantages of this integrated approach include switching with much smaller changes in voltage (0 to 3 V vs 0 to 6 V) and the potential for much more rapid switching. The ability to use the nanoscale FET to switch the nanoLED on and off reversibly has also been demonstrated (inset, Figure 2.12e).

The potential of coupling the bottom–up assembly of nanophotonic devices together with top–down fabricated Si structures has also been investigated, since this coupling could provide a new approach for introducing efficient photonic capabilities into integrated Si electronics. To this end, a hybrid top–down/bottom–up approach was employed: (1) using lithography to pattern p-type Si wires on the surface of a silicon-on-insulator (SOI) substrate, and then (2) assembling n-type emissive nanowires on top of Si structures to form arrays consisting of p–n junctions at cross-points (Figure 2.13a). Conceptually, this hybrid structure (Figure 2.13b) is virtually the same as the crossed nanowire structures described above and should produce electroluminescence in forward bias. Notably, I–V data recorded for a hybrid p–n diode formed between the p-Si and a n-CdS nanowire show clear current rectification (Figure 2.13c) and a sharp electroluminescence spectrum peaking at 510 nm (Figure 2.13d), which is consistent with CdS band edge emission. Importantly, the photonic devices produced in this approach are highly reproducible and can be readily implemented in integrated arrays. For example, a 1×7 crossed array consisting of a single CdS nanowire over 7-fabricated p-Si wires (Figure 2.13e) exhibits well-defined emission from each of the cross-points in the array (Figure 2.13f). Similar results were also obtained for two-dimensional arrays, and demonstrate clearly that bottom–up assembly has the potential to introduce photonic function into integrated Si microelectronics.

Single nanowire waveguide, Fabry–Perot cavity and laser

In addition to LEDs, nanoscale optical waveguides, cavities and lasers are vital for the realization of highly integrated photonic circuits. Importantly, free-standing semiconductor nanowires can function as stand-alone optical waveguides, cavities and gain media to support lasing emission [32, 82, 83]. In general, a nanowire can function as a single mode optical waveguide much like a conventional optical fiber [84] when $1 \sim (\pi D/\lambda)(n_1{}^2 - n_0{}^2)^{0.5} < 2.4$, where D is the nanowire diameter, λ is the wavelength, and n_1 and n_0 are the refractive indices of the nanowire and surrounding medium, respectively (Figure 2.14a). If the ends of the nanowire are cleaved (Figure 2.14b), they can function as two reflecting mirrors that provide optical feedback and define a Fabry–Perot optical cavity with modes $m(\lambda/2n_1) = L$, where m is an integer and L is the length of the cavity. A photoluminescence image of CdS nanowires of proper diameter shows pronounced emission from both ends (Figure 2.14c), clearly demonstrating the waveguide effect along the nanowire axis. Furthermore, spectra of the emissions from nanowire ends show prominent periodic

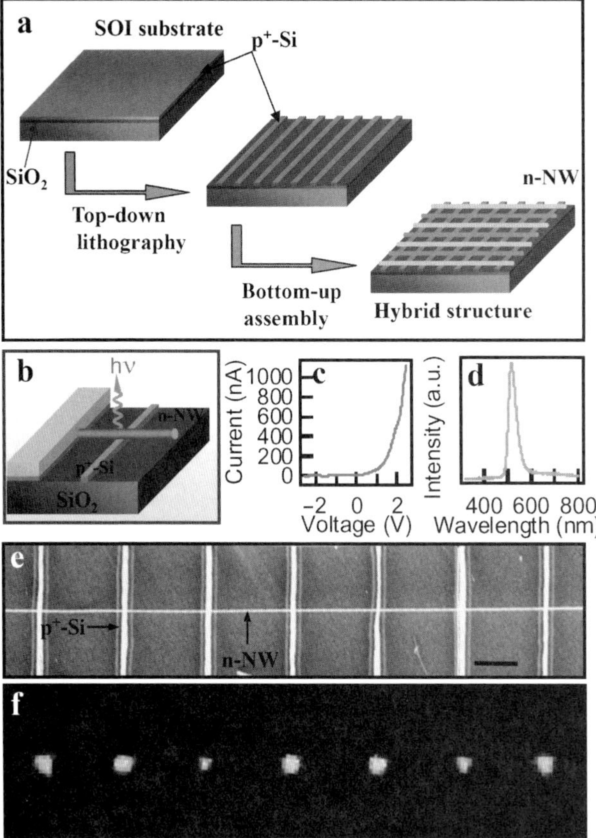

FIGURE 2.13 *Integration nanowire (NW) photonics with silicon electronics. (a) Schematic illustrating fabrication of hybrid structures. A silicon-on-insulator (SOI) substrate is patterned by standard electron-beam or photolithography followed by reactive ion etching. Emissive nanowires are then aligned on to patterned SOI substrate to form photonic sources. (b) Schematic of a single LED fabricated by the method outlined in (a). (c) I–V behavior for a crossed p–n junction formed between a fabricated p⁺-Si electrode and an n-CdS nanowire. (d) Electroluminescence spectrum from the forward biased junction. (e) SEM image of a CdS nanowire assembled over seven p⁺-Si electrodes on a SOI wafer; scale bar is 3 μm. (f) Electroluminescence image recorded from an array consisting of a CdS nanowire crossing seven p⁺-Si electrodes. The image was acquired with +5 V applied to each Si electrode while the CdS nanowire was grounded. (Adapted from [31].)*

modulation in intensity, which can be attributed to the longitudinal modes of a Fabry–Perot cavity (Figure 2.14d). For a cavity of length L, the mode spacing, $\Delta\lambda$, is given by $(\lambda^2/2L)(n_1 - \lambda(dn_1/d\lambda))^{-1}$, where $dn_1/d\lambda$ is the dispersion relation for the refractive index. This expression provides a good description of the observed spacing with the measured nanowire length. Moreover, analysis of similar data from nanowires of varying length demonstrates that the mode spacing is inversely proportional to the wire length, as expected. Together, these results demonstrate that the individual nanowire can function as a nanoscale optical waveguide and Fabry–Perot cavity. The observation of sharp modes in the uniform CdS nanowire gain medium suggests that a single nanowire can support laser emission. Indeed, optical excitation at higher powers leads to preferential gain in a single mode and the onset of lasing (Figure 2.14e).

The observation of laser emission with optical excitation further prompts the investigation of electrically pumped

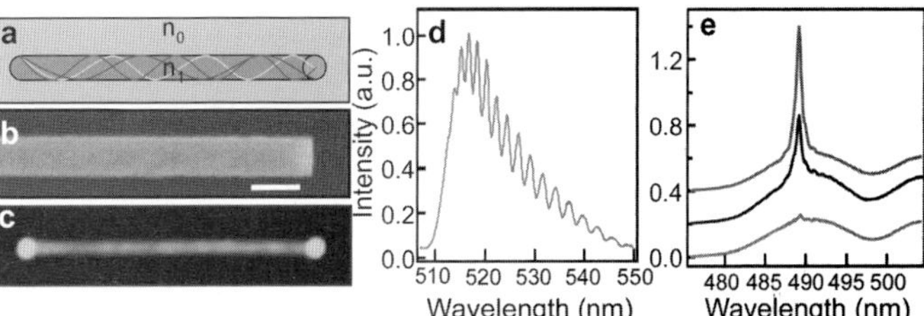

FIGURE 2.14 *Nanowire optical waveguides and cavities and laser with optical excitation. (a) Schematic showing a nanowire as an optical waveguide; with cleaved ends it defines a Fabry–Perot cavity. (b) SEM image of a cleaved CdS nanowire end. Scale bar: 100 nm. (c) Room-temperature photoluminescence image of a CdS nanowire uniformly excited with a mercury lamp. (d) Photoluminescence spectrum from the nanowire end exhibiting periodic intensity modulation which corresponds to the Fabry–Perot modes of the nanowire. (e) Emission spectra from a CdS nanowire end with a pump power of 190, 197 and 200 mW (red, blue and green) recorded at 8 K show preferential gain a single mode and demonstrate laser emission. (Adapted from [32].)*

nanolasers. To achieve electrical pumped lasers requires efficient injection of both electrons and holes into the nanowire cavity and gain medium. One approach to achieve electrical injection into CdS nanowire cavities was carried out using n-type CdS and p-type silicon (p-Si) crossed nanowire structures (Figure 2.15a, b). In forward bias, these crossed nanowire structures exhibit strong electroluminescence with several important characteristics. Images of the electroluminescence (Figure 2.15c, d) show two points of emission: one corresponding to the n-CdS/p-Si nanowire cross-point and the other to the end of the CdS nanowire. Significantly, the intensity of the end emission is at least two orders of magnitude larger than the cross-point emission. Furthermore, electroluminescence spectra recorded from the CdS nanowire end exhibit a prominent modulation in intensity, which can be assigned to the longitudinal modes of a Fabry–Perot cavity formed in the CdS nanowire in this crossed device (Figure 2.15e). These results clearly demonstrate that the CdS nanowires can function as excellent waveguides and Fabry–Perot cavity in this relatively simple device configuration, and suggest that at sufficiently high injection currents lasing should be achieved. However, the crossed nanowire devices is clearly not optimal for achieving high-density injection into the whole nanowire cavity, as high injection current invariably led to breaking down of the device at the cross-point owing to the nature of localized injection.

To enable more uniform injection a hybrid structure was implemented (Figure 2.15f, g) in which holes are injected along the length of a CdS nanowire cavity from a p-Si electrode defined in a heavily doped p-Si layer on a planar substrate [32]. Images of the room-temperature electroluminescence produced in forward bias from these structures (Figure 2.15h) show strong emission from the exposed CdS nanowire end. Low-temperature

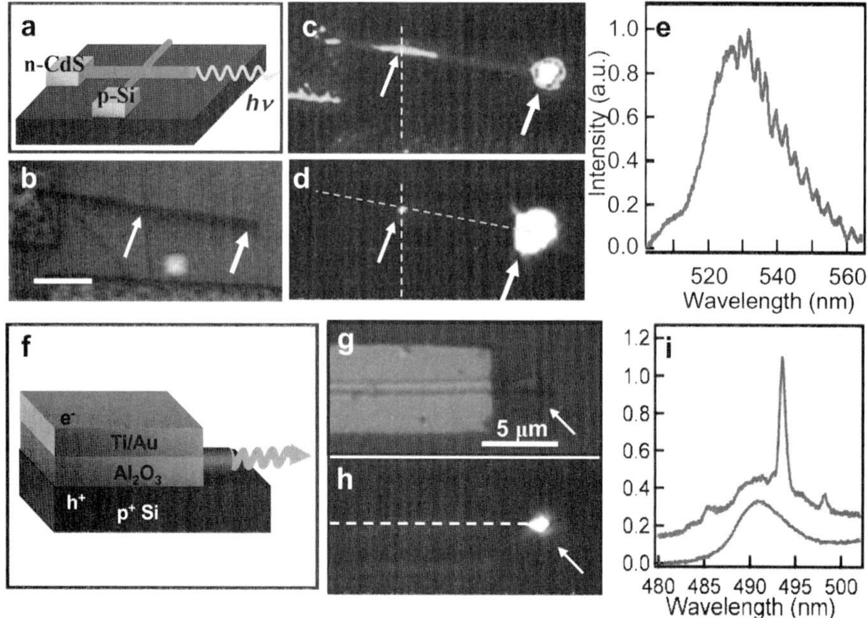

FIGURE 2.15 *Nanowire optical waveguides and cavities and laser with electrical excitation. (a) Device schematic illustrating a p–n diode formed between p-Si and n-CdS nanowires, where the CdS nanowire forms the cavity and active medium. (b) Optical image of a device with arrows highlighting the cross-point (blue) and CdS nanowire end (green). Scale bar: 5 µm. (c, d) The electroluminescence image obtained at room temperature with the device forward biased at 5 V. White-light illumination is used in image (c) to show the electrodes. The two bright spots highlighted by blue and green arrows correspond to the emission from the cross-point and end, respectively. (e) Electroluminescence spectra from the nanowire end show periodic intensity modulation, demonstrating that the nanowire can function as an optical waveguide and cavity in this injection device. (f) Schematic showing the device structure. In this structure, electrons and holes can be injected into the CdS nanowire cavity along the whole length from the top metal layer and the bottom p-Si layer, respectively. (g) Optical image of a device described in (d). The arrow highlights the exposed CdS nanowire end. Scale bar: 5 µm. (h) Electroluminescence image recorded from this device at room temperature with an injection current of ~80 µA. The arrow highlights emission from the CdS nanowire end. The dashed line highlights the nanowire position. (i) Emission spectra from a CdS nanowire device with injection currents of 200 µA (red) and 280 µA (green) recorded at 8 K. These spectra are offset by 0.10 intensity units for clarity. (Adapted from [32].)*

measurements made on the devices have shown the preferential pumping into single mode. At low injection currents, the spectrum of the end emission (Figure 2.15i) shows a broad peak with full width at half-maximum (FWHM) $\sim$5 nm. Significantly, when the injection current was increased further, the emission intensity increased abruptly, and the spectrum quickly collapsed into a limited number of sharp peaks with a dominant emission line around 493 nm (Figure 2.15i). Importantly, the dominant mode has an instrument resolution limited line width of only 0.7 nm, demonstrating that the injection lasing is achieved in this new type of device. Using individual nanowires as the laser cavity and gain medium for laser diodes represents a new and powerful approach for producing integrated electrically driven photonic devices. This basic approach, which relies on bottom–up assembly of the key laser cavity/medium in a single step, can be extended to other materials, such as GaN and InP nanowires, to produce nanoscale lasers that not only cover the UV through near-IR spectral regions but also can be integrated as single or multicolor laser source arrays in Si microelectronics and lab-on-a-chip devices.

NANOWIRE THIN-FILM ELECTRONICS

The assembly of a broad range of nanoscale electronic and optoelectronic devices using nanowire building blocks demonstrates the great potential of nanowires for powering the next generation of electronics. However, significant challenges remain to be solved before these nanoscale devices can be brought into real applications. For example, better assembly schemes are needed to organize nanowires in a precisely controlled manner and new circuit architectures must be developed to be compatible with nanowire device structures. While a totally new paradigm of nanowire nanoelectronics is farther away, more intermediate application of nanowires in current technologies may arise by adopting a more conventional device configuration.

Importantly, a new concept of nanowire thin-film transistors (NW-TFTs) has recently been proposed and demonstrated using oriented semiconductor nanowire thin films as semiconducting channels, displaying carrier mobilities approaching those of single-crystal materials and the low-temperature solution processiability of organic materials [85–87]. This new technology opens up brand new opportunities in macroelectronics that will not only greatly improve the existing technologies, but also enable an entirely new generation of flexible, wearable, disposable electronics for computing, data storage and wireless communications.

Nanowire thin-film transistors: underlying concept

The concept of NW-TFTs takes a new perspective and presents a paradigm shift in nanomaterial-enabled electronics, exploiting nanomaterials not for smaller devices, but for higher performance electronics over larger area substrates. In short, an oriented nanowire thin film was assembled to form a novel electronic substrate that can be processed using standard methods to yield high-performance NW-TFTs. Unlike a-Si or poly-Si TFTs (Figure 2.16a) in which carriers have to travel across multiple grain boundaries resulting in low mobility [6], NW-TFTs have conducting channels formed by multiple parallel single-crystal nanowire paths (Figure 2.16b). Therefore, charges travel from source to drain within single crystals, ensuring high carrier mobility.

Figure 2.16(c–f) illustrates a typical NW-TFT fabrication process. Single-crystal nanowires with controlled diameters are synthesized using metal cluster catalyzed chemical vapor deposition. The nanowires are then dispersed

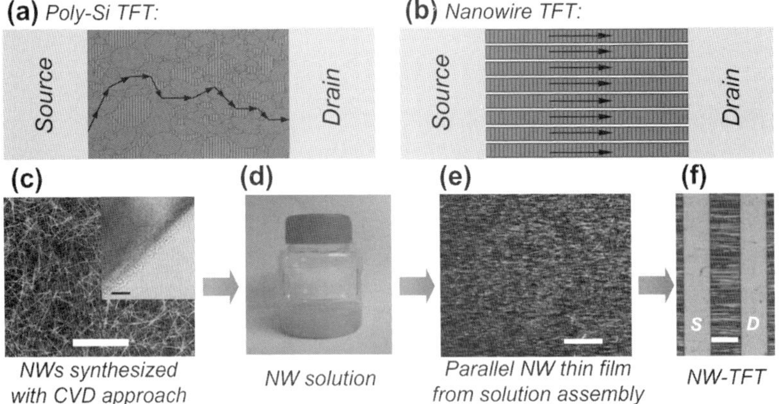

FIGURE 2.16 *Underlying concept of nanowire thin-film transistors (NW-TFTs). (a) In polycrystalline silicon TFTs, electrical carriers have to travel across multiple grain boundaries, resulting in low carrier mobility. (b) NW-TFTs have conducting channels consisting of multiple single-crystal nanowires in parallel, and thus charges travel from source to drain within single crystals, ensuring high carrier mobility. (c–f) NW-TFT fabrication process. (c) SEM image of single-crystal Si nanowires synthesized at high temperature using catalytic chemical vapor deposition (CVD) processes. Scale bar: 5 µm. Inset: TEM image of a nanowire displaying single-crystalline structure. Scale bar: 5 nm. (d) Nanowire solution obtained by dispersing nanowires into proper solvent such as ethanol. (e) Optical micrograph of nanowire thin film assembled from solution using fluidic flow-directed assembly approach. Scale bar: 100 µm. (f) Optical micrograph of an NW-TFT with the source to the drain electrodes bridged by parallel arrays of nanowires. Scale bar: 5 µm. (Adapted from [85].)*

into solution and assembled onto the surface of chosen substrate using various assembly approaches to produce a densely oriented nanowire thin film. The nanowire thin film is then processed using standard lithography and metallization processes to yield TFTs.

An important attribute of the NW-TFT technology is that the crystalline semiconductor material growth, which typically requires high temperatures to produce high-quality materials, is separated from device fabrication. This enables the assembly of TFTs at low temperatures, for compatibility with materials such as plastics, without sacrificing the ability to incorporate high-quality single-crystal semiconductor elements. It is also possible to incorporate a high-quality gate oxide and highly conductive source and drain electrodes through composition modulation along the radial or axial direction during the chemical synthesis of nanowires before the assembly of NW-TFT, and therefore ensure high-performance devices. Together, the dissociation of material synthesis from device substrate enables a general technology platform to apply a broad range of semiconductor materials onto many different substrates.

Nanowire thin-film transistors using single-crystal silicon nanowires

Figure 2.17 shows the transport characteristics of a typical NW-TFT assembled from p-type Si nanowires. The device was assembled on Si substrates

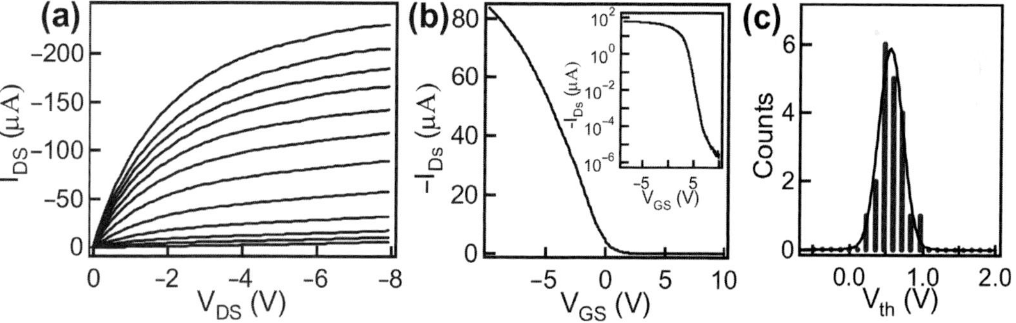

FIGURE 2.17 *Electrical transport characteristics of nanowire thin-film transistors (NW-TFTs). (a) Drain current (I_{DS}) vs drain–source voltage (V_{DS}) (I_{DS}–V_{DS}) at increasing gate voltages (V_{GS}) in steps of 1 V starting from the top at $V_{GS} = -10$ V. (b) $-I_{DS}$ vs V_{GS} at $V_{DS} = -1$ V. Inset: $-I_{DS}$ vs V_{GS} at $V_{DS} = -1$ V in the exponential scale, highlighting the on/off ratio of nearly 10^8 and subthreshold swing of ~600 mV/decade. (c) Histogram of threshold voltage (V_{th}) distribution from 20 NW-TFT devices showing a high device-to-device reproducibility and a tight distribution. Gaussian fitting shows a standard deviation of only 0.22 V.*
(Adapted from [85].)

with 100 nm Si nitride dielectrics as back gates. Electrical transport studies show typical characteristics similar to those of conventional enhancement mode p-channel TFTs with an on/off ratio of nearly eight orders of magnitude (Figure 2.17a, b). Modeling the device characteristics using standard MOSFET equations, one can deduce hole mobilities in the NW-TFT to be $\sim$119 cm^2/V·s, which is significantly larger than those of organic, a-Si ($<$ 1 cm^2/V·s) or conventional p-type poly-Si TFTs ($\sim$10–40 cm^2/V·s) [6], and comparable to that of p-type single-crystal Si material such as SOI MOS-FET ($\sim$180 cm^2/V·s) [88]. Hara et al. recently achieved hole mobility of 200 cm^2/V·s in poly-Si TFTs using laterally crystallized large-grain poly-Si, in which source and drain electrodes are bridged by single-crystal grains [89]. This represents a significant improvement in poly-Si devices. However, it requires a costly laser crystallization process, and may have device-to-device uniformity issues because of the large-grain Si used. Importantly, NW-TFTs can exhibit highly reproducible and predictable device characteristics, which are critical for practical applications and determine the viability of the technology. For example, a histogram of the threshold voltage distribution (Figure 2.17c) of 20 devices shows a standard deviation of only 0.22 V, which can further be improved upon by optimizing the device fabrication process. Together, these studies clearly demonstrate that NW-TFTs have an overall performance comparable to devices made from single-crystal materials.

In addition to single-crystalline channel, high-quality gate oxide is an important factor that can impact many critical device parameters, such as subthreshold swing, i.e. the voltage required to reduce the channel current by one order of magnitude. The plot of the $-I_{DS}$–V_{GS} curve in the exponential scale for the NW-TFT (inset, Figure 2.17b) shows a subthreshold swing $S = -dV_{GS}/d\ln|I_{DS}|$ of $\sim$ 600 mV/decade, which is significantly better than those of organic [90] or amorphous Si TFTs, which typically range from one to many volts per decade, and comparable to those of most poly-Si TFTs [6], but substantially larger than the best values in poly-Si TFTs ($\sim$200 mV/decade) and single-crystal Si devices ($\sim$70 mV/decade) [88]. These non-ideal sub-threshold swings in NW-TFTs can be largely attributed to non-ideal gate oxide quality and thickness: (1) a relatively large parasitic capacitance due to the existence of surface trapping states on the nanowire surface, and interface states between the nanowires and the substrate due to the low-quality native oxide on the nanowire surface and unprotected nanowire–substrate interface [6]; and (2) a relatively small gate capacitance due to the relatively thick gate dielectric (100 nm) used. These limitations can be overcome altogether by adopting a surrounding conformal gate. For example, using a solution conformal gate, a subthreshold swing of 70 mV/decade has been demonstrated. Furthermore, the surrounding conformal gate can also be realized in a solid-state device through the growth of core–shell–shell nano-wires, in which a single-crystalline semiconductor channel, high-quality thermal gate oxide and surrounding conformal metal gate are integrated in a single nanowire through precisely controlled chemical synthesis. Using core–shell–shell nanowires, NW-TFTs with subthreshold swings as small as 80–100 mV/decade have been achieved.

Group III–V and II–VI nanowire materials for high-performance macroelectronics

It is important to recognize that the NW-TFT fabrication process represents a general technology platform for virtually any material since nanowire synthesis is independent of the final device substrate. Other semiconductor nanowires or other similar structures such as nanoribbon/microribbon structures can also be explored for macroelectronic applications. Therefore, a broad range of materials including group III–V and II–VI compound semiconductors [25, 38–41] can be exploited as the TFT channel materials, opening up a broad range of opportunities to improve the device performance or create new functions.

This great flexibility allows us to choose materials with desired properties. For example, CdS is an excellent material for optical as well as electronic applications owing to its intrinsically low density of surface trapping states [32], and was one of the earliest materials used for TFTs [91]. However, the application of CdS in modern electronic technology is limited primarily due to the lack of suitable substrates for growing high-quality thin films. The dissociation of CdS nanoribbon growth from device substrate enables exploration of CdS for high-performance electronics. Single-crystal CdS nanoribbons (Figure 2.18a), with thickness in the order of 10–100 nm, width in the order of 1 μm, and length up to a few tens of micrometers, were grown using a vapor transport approach. These nanoribbons are particularly interesting nanostructures for assembling TFTs since they closely resemble conventional single-crystal thin films and can be flexibly manipulated in solution and assembled on a variety of substrates. Significantly, TFTs fabricated from CdS nanoribbons exhibit electron mobilities of 200–300 cm^2/V·s, comparable to the electron mobilities in single-crystal CdS materials (Figure 2.18b, c) [85]. Furthermore, CdS nanoribbon TFTs show a subthreshold swing S as small as 70 mV/decade (Figure 2.18d), approaching the theoretical limit of 60 mV/decade. Such a small subthreshold swing can be largely attributed to the intrinsically low surface state density in CdS material.

In addition, group III–V materials (e.g. GaAs, InP or InAs) of high intrinsic carrier mobility can be employed to produce TFTs with carrier mobilities well

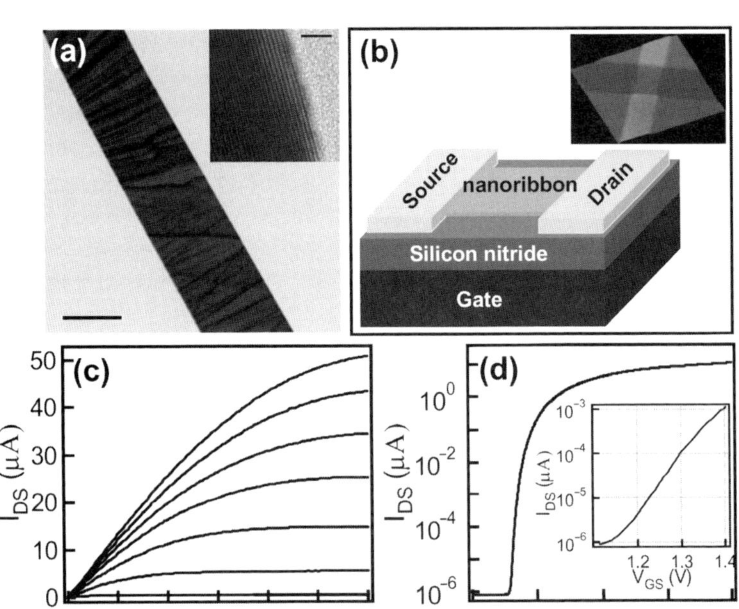

FIGURE 2.18 *Compound semiconductor materials for high-performance multifunctional macroelectronics. (a) TEM image of a CdS nanoribbon. Scale bar: 2 μm. A patterned contrast observed in the TEM image is due to strain resulting from bending/ twisting of the ribbons sitting on the TEM grid. Inset: high-resolution TEM image shows that the nanoribbon has a single-crystal structure almost free of defects and a surface oxide layer. Scale bar: 5 nm. (b) Schematic showing the device configuration of the nanoribbon thin-film transistor (TFT). Inset: 3D atomic force microscope topographic image of a nanoribbon TFT. (c) I_{DS}–V_{DS} at variable gate voltages (V_{GS}) starting from the top at $V_{GS} = +8$ V and decreasing in steps of 1 V. (d) I_{DS}–V_{GS} relation of CdS nanoribbon TFT at $V_{DS} = 1$ V in the exponential scale showing an on/off ratio greater than 10^7. The inset shows a zoomed-in plot in the subthreshold region that highlights a subthreshold swing as small as 70 mV/decade.*
(Adapted from [85].)

above 1000 cm^2/V·s. The carrier mobility may be further enhanced by fabricating novel core–shell nanowire structures and fully exploiting quantum electronic effects at reduced dimensions [92]. For example, recent work on core–shell nanowire structures has demonstrated significantly improved device performance with carrier mobilities as high as 21 000 cm^2/V·s for GaN/AlN/AlGaN [68] and 11 500 cm^2/V·s for InAs/InP core–shell nanowire [93]. Such high-mobility nanowire materials are attractive building blocks for macroelectronics and can open up a range of exciting opportunities not currently possible.

Nanowire thin-film transistors on plastics

One important aspect of the NW-TFT concept is that the entire device fabrication process is performed essentially at room temperature, except for the nanowire synthesis step. Therefore, the assembly of high-performance NW-TFTs can be readily applied to low-cost glass and plastic substrates. Indeed, it has been shown that NW-TFTs can be fabricated on plastic substrate (Figure 2.19a) with carrier mobilities comparable to those made on Si substrates, which are among the best TFTs that have ever been made on plastics. Importantly, slightly flexing of the plastic with NW-TFTs does not significantly change the device behavior (Figure 2.19b), clearly demonstrating the flexibility of NW-TFTs. The use of free-standing nanostructures for flexible electronics has also been demonstrated by various groups using nanotubes [94], nanowires [95], and nanoribbons or microribbons [96].

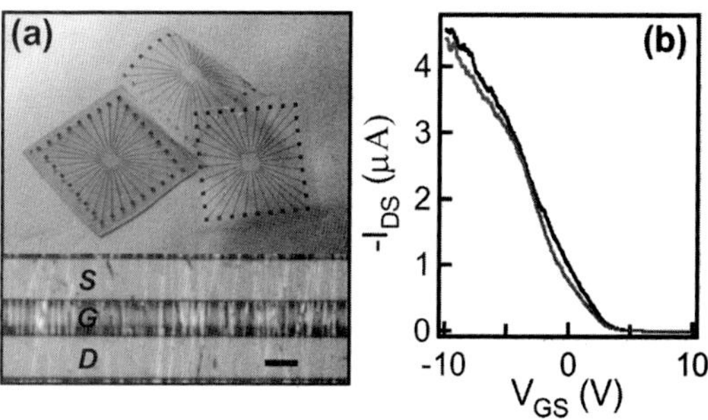

FIGURE 2.19 *Nanowire field-effect transistors (NW-TFTs) on plastic. (a) Top: a picture of plastic devices with NW-TFTs. The plastic devices show high mechanical flexibility. Bottom: optical micrograph of an NW-TFT on plastic. Scale bar: 5 μm. (b) −I_DS–V_GS relation at V_DS = −1 V. The black and gray curves show the transfer characteristics of the same device before and after flexing (and releasing) the plastic substrate (radius of curvature ~55 mm), demonstrating the mechanical flexibility of NW-TFTs on plastics. (Adapted from [85].)*

CONCLUSION

In summary, this chapter has reviewed the recent advances in exploring chemically synthesized nanowires to assemble functional electronics and optoelectronics. In this approach, high-quality single-crystal semiconductor nanowires are first synthesized through a chemical approach at high temperature to ensure nearly perfect crystalline structure. The synthesized nanowires are then dispersed in solution and organized onto a device substrate to obtain functional electronics and optoelectronics through solution assembly processes. This bottom–up approach offers several critical advantages compared to conventional top–down approach for electronics fabrication. First, precisely controlled chemical synthesis allows a well-defined nanostructure to be obtained, with critical dimension down to 1–2 nm regime, well beyond the capability of traditional lithography. Second, the low-temperature solution assembly process allows placement of high-quality semiconductor nanostructures and fabrication of high-performance semiconductor electronics on a wide range of substrates, including plastics.

Third, the bottom–up assembly approach allows flexible combination of many different semiconductors together to create new device functions, which would not be possible with traditional technology. Lastly, precise chemical synthesis and flexible solution assembly allow facile integration over multiple length scales, from nanometer-scale building blocks to micrometer-scale devices and macroscale functional systems. Further development in this bottom–up approach can open up exciting opportunities for a paradigm shift in the area of electronics: moving microelectronics from single-crystal substrates to glass and plastic substrates; integrating macroelectronics, microelectronics and potentially nanoelectronics at the device level; and integrating different semiconductor materials on a single substrate to achieve multiple functionalities in a way that is not possible with current technologies. This can impact on a broad range of existing applications from flat-panel displays to image sensor arrays, and enable a whole new range of flexible, wearable or disposable electronics for computing, storage and wireless communication.

REFERENCES

[1] Gelsinger PP, Gargini PA, Parker GH, Yu AYC. Microprocessors circa 2000. IEEE Spectrum 1989;(October):43–7.

[2] Meindl JD, Chen Q, Davis JA. Limits on silicon nanoelectronics for terascale integration. Science 2001;293:2044–9.

[3] Reuss RH, Hopper DG, Park JG. Macroelectronics. Mater Res Soc Bull 2006;31:447–50.

[4] Shur MS, Wilson P, Urban D, editors. Electronics on unconventional substrates-electrotextiles and giant-area flexible circuits. Mater Res Soc Proc 2002; Vol. 736.

[5] Uchikoga S. Low-temperature polycrystalline silicon thin-film transistor technologies for system-on-glass displays. Mater Res Soc Bull 2002;27:881–6.

[6] Street RA Technology and applications of amorphous silicon. Berlin: Springer; 2000.

[7] Rogers JA, Bao Z, Baldwin K, Dodabalapur A, Crone B, Raju VR, et al. Paper-like electronic displays: large-area rubber-stamped plastic sheets of electronics and microencapsulated electrophoretic inks. Proc Natl Acad Sci USA 2001;98: 4835–40.

[8] Lieber CM. The incredible shrinking circuit. Sci Am 2001;285:50–6.

[9] Jacoby M. Nanoscale electronics. Chem Eng News 2002(September 30):38.

[10] Heath JR, Kuekes PJ, Snider GS, Williams RS. A defect-tolerant computer architecture: opportunities for nanotechnology. Science 1998;280:1716–21.

[11] Burda C, Chen X, Narayanan R, El-Sayed MA. Chemistry and properties of nanocrystals of different shapes. Chem Rev 2005;105:1025–102.

[12] Dai H. Carbon nanotubes: synthesis, integration, and properties. Acc Chem Res 2002;35:1035–44.

[13] Duan X, Lieber CM. Semiconductor nanowires: rational synthesis. In: Schwarz JA, Contescu CI, Putyera K, editors. Dekker encyclopedia of nanoscience and nanotechnology. New York: Marcel Dekker; 2005.

[14] Duan X, Huang Y, Cui Y, Lieber CM. Nanowire nanoelectronics assembled from the bottom-up. In: Reed MA, Lee T, editors.Molecular nanoelectronics, American Scientific Publishers; 2003. p. 199–227.

[15] Lu W, Lieber CM. Nanoelectronics from the bottom up. Nat Mater 2007;6: 841–50.

[16] Collins PG, Avouris P. Nanotube for electronics. Sci Am 2000;283:62–9.

[17] Tans SJ, Verschueren ARM, Dekker C. Room-temperature transistor based on a single carbon nanotube. Nature 1998;393:49–52.

[18] Martel R, Schmidt T, Shea HR, Hertel T, Avouris Ph. Single- and multi-wall carbon nanotube field-effect transistors. Appl Phys Lett 1998;73:2447–9.

[19] Yao Z, Postma HWCh, Balents L, Dekker C. Carbon nanotube intramolecular junctions. Nature 1999;402:273–6.

[20] Fuhrer MS, Nygrad J, Shih L, Forero M, Yoon YG, Mazzoni MSC, et al. Crossed nanotube junctions. Science 2000;288:494–7.

[21] Derycke V, Martel R, Appenzeller J, Avouris P. Carbon nanotube inter- and intra-molecular logic gates. Nano Lett 2001;1:453–6.

[22] Bachtold A, Hadley P, Nakanishi T, Dekker C. Logic circuits with carbon nanotube transistors. Science 2001;294:1317–20.

[23] Lieber CM. One-dimensional nanostructure: chemistry, physics and application. Solid State Commun 1998;107:607–16.

[24] Hu J, Odom TW, Lieber CM. Chemistry and physics in one dimension: synthesis and properties of nanowires and nanotubes. Acc Chem Res 1999;32:435–4.

[25] Duan X, Lieber CM. General synthesis of compound semiconductor nanowires. Adv Mater 2001;12:298–302.

[26] Gudiksen MS, Wang J, Lieber CM. Synthetic control of the diameter and length of single crystal semiconductor nanowires. J Phys Chem B 2001;105:4062–4.

[27] Cui Y, Duan X, Hu J, Lieber CM. Doping and electrical transport in silicon nanowires. J Phys Chem B 2000;104:5213–6.

[28] Duan X, Huang Y, Cui Y, Wang J, Lieber CM. Indium phosphine nanowires as building blocks for nanoscale electronic and optoelectronic devices. Nature 2001;409:66–9.

[29] Cui Y, Lieber CM. Functional nanoscale electronic devices assembled using silicon nanowire building blocks. Science 2001;291:851–3.

[30] Huang Y, Duan X, Cui Y, Lauhon L, Kim K, Lieber CM. Logic gates and computation from assembled nanowire building blocks. Science 2001;294:1313–7.

[31] Huang Y, Duan X, Lieber CM. Nanowire for integrated multicolor nanophotonics. Small 2005;1:142–7.

[32] Duan X, Huang Y, Argarawal R, Lieber CM. Single-nanowire electrically driven lasers. Nature 2003;421:241–5.

[33] Huang Y, Duan X, Wei Q, Lieber CM. Directed assembly of one-dimensional nanostructures into functional networks. Science 2001;291:630–3.

[34] Seker F, Meeker K, Kuech TF, Ellis AB. Surface chemistry of prototypical bulk II–VI and III–V semiconductors and implications for chemical sensing. Chem Rev 2000;100:2505–36.

[35] Iler RKThe chemistry of silica. New York: Wiley; 1979.

[36] Murray CB, Norris DJ, Bawendi MG. Synthesis and characterization of nearly monodisperse CdE (E = sulfur, selenium, tellurium) semiconductor nanocrystallites. J Am Chem Soc 1993;115:8706–15.

[37] Esaki LIn: Namba S, Hamaguchi C, Ando T, editors. Science and technology for mesoscopic structures. Tokyo: Springer; 1992.

[38] Morales AM, Lieber CM. A laser ablation method for the synthesis of crystalline semiconductor nanowires. Science 1998;279:208–11.

[39] Duan X, Wang J, Lieber CM. Synthesis and optical properties of gallium arsenide nanowires. Appl Phys Lett 2000;76:1116–8.

[40] Duan X, Lieber CM. Laser-assisted catalytic growth of single crystal GaN nanowires. J Am Chem Soc 2000;122:188–9.

[41] Wei Q, Lieber CM. Synthesis of single crystal bismuth-telluride and lead-telluride nanowires for new thermoelectrical materials. Mater Res Soc Symp Proc 2000; 581:219–23.

[42] Panish MB. Ternary condensed phase systems of gallium and arsenic with group IB elements. J Electrochem Soc 1967;114:516–21.

[43] Gudiksen MS, Lieber CM. Diameter-selective synthesis of semiconductor nanowires. J Am Chem Soc 2000;122:8801–2.

[44] Cui Y, Lauhon LJ, Gudiksen MS, Wang J, Lieber CM. Diameter-controlled synthesis of single-crystal silicon nanowires. Appl Phys Lett 2001;78:2214–26.

[45] Hu J, Ouyang M, Yang P, Lieber CM. Controlled growth and electrical properties of heterojunctions of carbon nanotubes and silicon nanowires. Nature 1999;399: 48–51.

[46] Gudiksen MS, Lauhon LJ, Wang J, Smith D, Lieber CM. Growth of nanowire superlattice structures for nanoscale photonics and electronics. Nature 2002; 415:617–20.

[47] Wu Y, Fan R, Yang P. Block-by-block growth of single-crystalline Si/SiGe superlattice nanowires. Nano Lett 2002;2:83–6.

[48] Bjork MT, Ohlsson BJ, Sass T, Persson AI, Thelander C, Magnusson MH, et al. One-dimensional steeplechase for electrons realized. Nano Lett 2002;2:87–9.

[49] Lin YC, Lu KC, Wu WW, Bai J, Chen LJ, Tu KN, Huang Y. Single crystalline PtSi nanowires, PtSi/Si/PtSi nanowire heterostructures, and nanodevices. Nano Lett 2008;8:913–8.

[50] Lauhon LJ, Gudiksen MS, Wang D, Lieber CM. Epitaxial core-shell and core-multi-shell nanowire heterostructures. Nature 2002;420:57–61.

[51] Duan X. Assembled semiconductor nanowire thin films for high-performance flexible macroelectronics. Mater Res Soc Bull 2007;32:134–41.

[52] Duffy DC, McDonald JC, Schueller OJA, Whitesides GM. Rapid prototyping of microfluidic systems in poly(dimethylsiloxane). Anal Chem 1998;70:4974–8.

[53] Stover CA, Koch DL, Cohen C. Observations of fibre orientation in simple shear flow of semi-dilute suspensions. J Fluid Mech 1992;238:277–96.

[54] Koch DL, Shaqfeh ESG. The average rotation rate of a fiber in the linear flow of a semi-dilute suspension. Phys Fluids A 1990;2:2093.

[55] Liu J, Casavant MJ, Cox M, Walters DA, Boul P, Lu W, et al. Controlled deposition of individual single-walled carbon nanotubes on chemically functionalized templates. Chem Phys Lett 1999;303:125–9.

[56] Burghard M, Dueberg D, Philipp G, Muster J, Roth S. Controlled adsorption of carbon nanotubes on chemically modified electrode arrays. Adv Mater 1998;10:584–8.

[57] De Rosa C, Park C, Lotz B, Wittmann JC, Fetters LJ, Thomas EL. Control of molecular and microdomain orientation in a semicrystalline block copolymer thin film by epitaxy. Macromolecules 2000;33:4871–6.

[58] Gleiche M, Chi LF, Fuchs H. Nanoscopic channel lattices with controlled anisotropic wetting. Nature 2000;403:173–5.

[59] Yang P. Wires on water. Nature 2003;425:243–4.

[60] Ulman A. An introduction to ultrathin organic films: from Langmuir–Blodgett to self-assembly, San Diego, CA: Academic; 1991.

[61] Tao A, Kim F, Hess C, Goldberger J, He R, Sun Y, et al. Langmuir–Blodgett silver nanowires monolayers for molecular sensing using surface-enhanced Raman spectroscopy. Nano Lett 2003;3:1229–33.

[62] Whang D, Jin S, Wu Y, Lieber CM. Large-scale hierarchical organization of nanowire arrays for integrated nanosystems. Nano Lett 2003;3:1255–9.

[63] Huang Y, Duan X, Cui Y, Lieber CM. Gallium nitride nanowire nanodevices. Nano Lett 2002;2:101–4.

[64] Cui Y, Duan X, Hu J, Lieber CM. Doping and electrical transport in silicon nanowires. J Phys Chem B 2000;104:5213–6.

[65] Huang Y, Lieber CM. Integrated nanoscale electronics and optoelectronics: exploring nanoscale science and technology through semiconductor nanowires. Pure Appl Chem 2004;76:2051–68.

[66] Gudiksen MS, Wang J, Lieber CM. Size-dependent photoluminescence from single indium phosphide nanowires. J Phys Chem B 2002;106:4036–9.

[67] Xiang J, Lu W, Hu Y, Wu Y, Yan H, Lieber CM. Ge/Si nanowire heterostructures as high-performance field-effect transistors. Nature 2006;441:489–93.

[68] Li Y, Xiang J, Qian F, Gradecak S, Wu Y, Yan H, et al. Dopant-free GaN/AlN/AlGaN radial nanowire heterostructures as high electron mobility transistors. Nano Lett 2006;6:1468–73.

[69] Qian F, Gradecak S, Li Y, Wen CY, Lieber CM. Core/multishell nanowire heterostructures as multicolor, high-efficiency light-emitting diodes. Nano Lett 2005;5:2287–91.

[70] Wu Y, Xiang J, Yang C, Lu W, Lieber CM. Single-crystal metallic nanowires and metal/semiconductor nanowire heterostructures. Nature 2004;430:61–4.

[71] Lu KC, Tu KN, Wu WW, Chen LJ, Yoo BY, Myung NV. Point contact reactions between Ni and Si nanowires and reactive epitaxial growth of axial nano-NiSi/Si. Appl Phys Lett 2007;90:253111.

[72] Lu KC, Wu WW, Wu HW, Tanner CM, Chang JP, Chen LJ, Tu KN. In situ control of atomic-scale Si layer with huge strain in the nanoheterostructure NiSi/Si/NiSi through point contact reaction. Nano Lett 2007;7:2389–94.

[73] Weber WM, Geelhaar L, Graham AP, Unger E, Duesberg GS, Liebau M, et al. Silicon-nanowire transistors with intruded nickel-silicide contacts. Nano Lett 2006;6:2660–6.

[74] Byon K, Tham D, Fischer JE, Johnson AT. Synthesis and postgrowth doping of silicon nanowires. Appl Phys Lett 2005;87:193104.

[75] Bucher E, Schulz S, Lux-Steiner MCh, Munz P, Gubler U, Greuter F. Work function and barrier heights of transition metal silicides. Appl Phys A 1986;40:71–7.

[76] Freeouf JL. Silicide Schottky barriers: an elemental description. Solid State Commun 1980;33:1059–61.

[77] Chau R, Datta S, Doczy M, Doyle B, Kavalieros J, Metz M High-k/metal-gate stack and its MOSFET characteristics. IEEE Elec Dev Lett 2004;25:408–10.

[78] Mizuno T, Sugiyama N, Kurobe A, Takagi SI. Advanced SOI p-MOSFETs with strained-Si channel on SiGe-on-insulator substrate fabricated by SIMOX technology. IEEE Trans Electron Devices 2001;48:1612–8.

[79] Zhang M, Knoch J, Zhao QT, Breuer U, Mantl S. Impact of dopant segregation on fully depleted Schottky-barrier SOI-MOSFETs. Solid State Electron 2006; 50:594–600.

[80] Horowitz P, Hill WThe art of electronics. Cambridge: Cambridge University Press; 1989.

[81] Whaley SR, English DS, Hu EL, Barbara PF, Belcher AM. Selection of peptides with semiconductor binding specificity for directed nanocrystals assembly. Nature 2000;405:665–8.

[82] Huang MH, Mao S, Feick H, Yan H, Wu Y, Kind H, et al. Room-temperature ultraviolet nanowire nanolasers. Science 2001;292:1897–9.

[83] Johnson JC, Choi HJ, Knutsen KP, Schaller RD, Yang P, Saykally RJ. Single gallium nitride nanowire lasers. Nat Mater 2002;1:106–10.

[84] Chen CLElements of optoelectronics and fiber optics. Chicago, IL: Irwin; 1996.

[85] Duan X, Niu C, Sahi V, Chen J, Parce JW, Empedocles S, Goldman JL. High-performance thin-film transistors using semiconductor nanowires and nanoribbons. Nature 2003;425:274–8.

[86] Jin S, Whang D, McAlpine MC, Friedman RS, Wu Y, Lieber CM. Scalable interconnection and integration of nanowire devices without registration. Nano Lett 2004;4:915–9.

[87] Menard E, Lee KJ, Khang DY, Nuzzo RG, Rogers JA. A printable form of silicon for high performance thin film transistors on plastic substrates. Appl Phys Lett 2004;84:5398–400.

[88] Mizuno T, Sugiyama N, Kurobe A, Takagi SI. Advanced SOI p-MOSFETs with strained-Si channel on SiGe-on-insulator substrate fabricated by SIMOX technology. IEEE Trans Electron Devices 2001;48:1612–8.

[89] Hara A, Takei M, Takeuchi F, Suga K, Yoshino K, Chida M, et al. High performance low temperature polycrystalline silicon thin film transistors on nonalkaline glass produced using diode pumped solid state continuous wave laser lateral crystallization. Jpn J Appl Phys Pt 1 2004;43:1269–76.

[90] Dimitrakopoulos CD, Mascaro DJ. Organic thin-film transistors: a review of recent advances. IBM J Res Dev 2001;45:11–27.

[91] Weimer PK. Proc IEEE 1962; 56:1462.

[92] Sakaki H. Quantum wires, quantum boxes and related structures: physics, device potentials and structural requirements. Surf Sci 1992;267:623–9.

[93] Jiang X, Xiong Q, Nam S, Qian F, Li Y, Lieber CM. InAs/InP radial nanowire heterostructures as high electron mobility devices. Nano Lett 2007;7:3214–8.

[94] Snow ES, Campbell PM, Ancona MG, Novak JP. High-mobility carbon-nanotube thin-film transistors on polymeric substrate. Appl Phys Lett 2005;86:033105.

[95] McAlpine MC, Friedman RS, Jin S, Lin KH, Wang WU, Lieber CM. High-performance nanowire electronics and photonics on glass and plastic substrates. Nano Lett 2003;3:1531–5.

[96] Sun Y, Menard E, Rogers JA, Kim HS, Kim S, Chen G, et al. Gigahertz operation in flexible transistors on plastic substrates. Appl Phys Lett 2006;88:183509.

CHAPTER 3

Fast Flexible Electronics Made from Nanomembranes Derived from High-Quality Wafers

Zhenqiang Ma and Guoxuan Qin

Department of Electrical and Computer Engineering, University of Wisconsin–Madison, Madison, Wisconsin, USA

INTRODUCTION

Flexible electronics employing organic semiconductors, amorphous and poly-crystalline semiconductors have existed for many years. The organic polymer-based flexible electronics exhibit extraordinary flexibility, bendability and stretchability. Such excellent mechanical performance enables them to be used in a number of unique applications. They are particularly useful for electronic papers and electronic textiles, although the mobility values are very low in comparison to most inorganic semiconductors. Amorphous silicon (Si) and polycrystalline Si (and a few other III–V materials) have much higher carrier mobility values than organic polymers. Some of them can also be manufactured in large areas at relatively low cost. However, the mechanical properties are generally not comparable to organic polymers. For flexible elec-tronics applications, amorphous and polycrystalline materials are sometimes also limited by the temperature constraint of substrate, such as plastic sub-strate. The poor crystalline quality of these materials is a major limiting factor for the related devices to deliver high performance. Because of these reasons, flexible electronics using these traditional semiconductor materials have been limited to low-speed electronics applications, such as displays, radiofrequency (RF) tags and smart cards. On the other hand, high-speed electronics are only accessible by single-crystal based rigid chips. In many advanced electronics applications, such as WiFi portable devices and military antennae for com-munication and surveillance, high speed is required. Making the relevant devices and systems mechanically flexible offers significant advantages compared with a rigid form. The flexible form will allow devices to be easily bent and to improve the mechanical reliability, absorbing the shock energy when impacted. Large flexible systems can be folded or rolled to ease storage and launching. A flexible form can allow the system to be conformally attached on rugged and uneven surfaces to reduce protrusion and dragging force in motion.

The ideal materials for high-speed applications are high-mobility (also high saturation velocity if used for high-power devices) materials. This also holds for high-speed flexible electronics (fast flexible electronics). The material list for potential fast flexible electronics includes carbon nanotubes, graphene and a number of single crystal inorganic semiconductors (e.g. Si, Ge, GaAs, InP and GaN). Carbon nanotubes have very high intrinsic carrier mobility. However, they have been difficult to use owing to the difficult manipulation of individual tubes, the difficulty of realizing low contact resistance, some alignment issues (which seem to have been solved in some studies) and the difficulty in separating metallic and semiconductor tubes. Graphene is a relatively new material, which exhibits very high mobility values. The lack of bandgap is currently the bottleneck preventing it from being used as a useful type of semiconductor.

Single-crystal inorganic semiconductors recently emerged as a very promising semiconductor material for fast flexible electronics. Obtaining these materials in very thin forms has been the key to their practical application. Simply thinning down the wafers to the desired thickness is not economic for flexible electronics, particularly large-area electronics applications. By taking advantage of etching selectivity between certain materials and between orientations of the same materials, the selective release of some of the layers (membrane or nanomembranes) seems to be a very promising way to realize very thin semiconductors with high mobilities. The mobilities of the membranes are as high as their bulk counterparts. The thinness has enabled the desired mechanical flexibility for flexible electronics applications. After the membranes are transferred to flexible substrates, high-speed devices are expected to be realized. Many challenges remain in effectively transferring the high mobility values of the single-crystal membranes into high device speeds. In this chapter, several important experimental demonstrations of fast, flexible thin-film transistors (TFTs) employing transferable single-crystalline Si nanomembranes, which have solved these challenges, are described. The first section describes channel mobility enhancement via strain-sharing techniques. In the next section, the effective doping techniques of thin nanomembranes are briefly discussed. The following section discusses TFT design, fabrication and characterization employing the properly doped semiconductor nanomembranes. The modeling of record-speed device performance is also included in the discussion. The results show the potential of microwave amplification using flexible TFTs. Then, complementary TFTs employing hybrid oriented nanomembranes are described. In the final section, high-performance flexible passives compatible with microwave TFTs are detailed. All these demonstrations show the rise of a new RF era, where high-speed devices can be deployed on plastic substrates with simple and low-cost fabrication methods.

STRAINED SILICON NANOMEMBRANES

Single-crystal semiconductor materials transferred to flexible polymers are excellent candidates for high-performance flexible electronics. Both single-crystal Si and GaAs have recently been transferred to create TFTs on plastic [1–3]. These transfer methods usually require the use of soft stamps and special structures such as anchors or tethers [4]. In this section, a simpler transferring technique is described and it is combined with a novel membrane fabrication method to create strained Si TFTs. Strained single-crystal Si is used as the channel material for the TFTs, with its well-known enhancement of carrier mobility and thus transistor performance [5].

Device fabrication

Thin membranes are fabricated as the active layer of TFTs on silicon-on-insulator (SOI) wafers and transferred onto a plastic host substrate. For comparison, unstrained Si TFTs were also fabricated. The fabrication of both strained Si and unstrained Si TFTs is identical except for the preparation of the active layers. For the unstrained Si, the starting material is a UniBond® SOI wafer with a 200 nm Si top layer and a 200 nm buried oxide (BOX) layer. For the strained Si, the starting substrate is SOI with a 10 nm Si top layer (such a thin Si top layer can be obtained by thinning down a thicker one using oxidation or chemical mechanical polishing). A sandwich structure of Si, SiGe alloy and another Si layer is pseudomorphically grown on the thin Si layer in an ultrahigh-vacuum chemical vapor deposition (UHV-CVD) system with no intentional doping. The final structure, from top to bottom, is 45 nm Si, 150 nm $Si_{0.83}Ge_{0.17}$ and 50 nm Si. The lattice mismatch between Si and $Si_{0.83}Ge_{0.17}$ (with a larger lattice constant than that of Si) results in compressive strain in the SiGe layer, while the two Si layers are completely relaxed, as verified by X-ray diffraction (XRD). Both types of active layer (Si for the unstrained film and Si/SiGe/Si for the strained films) are first patterned into 1 mm long, 20 µm wide strips with 5 µm gaps between the strips, using optical photolithography. The regions between the strips of active layer are etched down to the BOX layer with reactive ion etching (RIE). To release the strips of the active layer, photoresist on top of the strips is first stripped and then the sample is immersed in aqueous concentrated hydrofluoric acid (HF) for 10–20 min to etch away the BOX layer selectively. During the BOX removal process, the active layer falls down and attaches to the starting Si substrate [6] by van der Waals' forces. No anchors or tethers are used in this process, while the released strips do not float around in the HF solution. This is possible when the BOX layer is thin (200 nm used). Once the BOX under the strips has been fully removed, the sample is thoroughly rinsed in distilled (DI) water and is then ready for transfer. In the Si/SiGe/Si sandwich structure, the SiGe alloy layer, which was compressively strained before the BOX removal and is thicker than the two Si layers, is able to relax. The relaxation/stretching of the SiGe layer thus induces tensile strain in the two Si layers. As a result, upon removal of the BOX layer, both Si layers of the sandwich membrane become biaxially tensile strained, creating an elastic strain-sharing status between the thicker SiGe alloy layer and two thin Si layers. The strain status in the sandwich membrane can be engineered by tuning the SiGe/Si thickness ratio and also the Ge fraction in the SiGe alloy [7]. Such a strain-sharing technique can also be used to create a compressive SiGe layer for enhanced hole mobility, by starting with SiGe-on-insulator (also commercially available) and growing a final SiGe/Si/SiGe sandwich structure. The advantage of this strain-sharing method is that no misfit dislocations are generated [8]. Hence, one can expect to fully exploit the mobility enhancement in strained Si, without concern about possible degradation by threading dislocations [9] in graded SiGe buffer layers.

Indium–tin oxide (ITO)-coated, 175 μm thick, polyethylene terephthalate (PET) is chosen as the plastic host substrate. The sheet resistance of the ITO is 50 Ω/□. Before transferring the released active layer to the PET host, the PET is thoroughly rinsed with acetone, isopropyl alcohol (IPA) and DI water. A layer of SU-8 2002 (Microchem Corp.) photoresist is spun on the polymer at 4000 rpm for 30 s after treatment in an oxygen plasma for 10 s to increase the adhesion between the polymer and the SU-8. The SU-8 2002 serves as an adhesion layer to facilitate the subsequent transfer and also as the gate dielectric layer for the TFTs. The spin-on SU-8 2002 is baked at 50°C for 1 min. Then, the Si substrate on which the released active layer settled upon BOX etching is brought *facing down* to the SU-8. After gently pressing the Si handle substrate against the plastic, then picking up the Si substrate, the active layer, which is now upside down relative to its original position on the SOI, is transferred to the plastic host. The flip-transfer method does not need the use of a soft stamp. The SU-8 2002 is then baked at 100°C for 5 min, followed by ultraviolet (UV) flood exposure from the back side. It is further baked at

115°C for 5 min after the UV exposure to cross-link the entire SU-8. Finally, source/drain metal consisting of 70 nm thick Ti is formed on the transferred unstrained or strained Si surface by optical photolithography and lift-off. A short dip in dilute HF is applied before evaporating Ti to remove the native oxide on the Si surface. No extra annealing is introduced after the metal deposition. The process flow for transferring strained Si to the polymer host by dry printing is illustrated schematically in Figure 3.1. The strain status (compressive or tensile) is indicated by arrows in Figure 3.1(d, e).

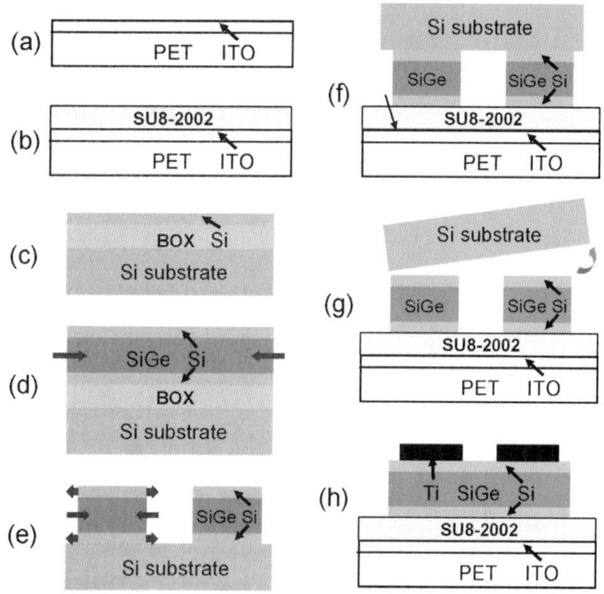

Strained silicon thin-film transistor performance

Figure 3.2(a) shows optical microscope images of the unstrained Si strips transferred onto the plastic host. A very high transfer rate is achieved with this flip-transfer technique. The inset in Figure 3.2(a) shows a zoom-in image of the transferred strips. The gaps between strips stay very uniform and unchanged across the 1 mm length of the strips. This result demonstrates that this flip-transfer technique should be very attractive for device fabrication with regard to align-

FIGURE 3.1 *Schematic process flow for transferring single-crystalline Si/SiGe/Si structure to a flexible polymer host and thin-film transistor (TFT) fabrication. (a) Begin with indium–tin oxide (ITO)-coated polyethylene terephthalate (PET) host. (b) Spin on SU-8 2002 as adhesive dielectric layer. (c) Begin with SOI substrate for active layer preparation. (d) Grow SiGe and Si layers on SOI (this step is skipped for the unstrained case). Only the SiGe alloy layer is compressively strained at this stage. (e) The active Si/SiGe/Si layer is dry etched into strips and the buried oxide (BOX) layer is selectively etched by concentrated hydrofluoric acid. The structure sits on the Si substrate and the two thin Si layers are now tensile strained due to strain sharing. (f) The starting substrate is brought upside down in contact with the adhesive layer. (g) The Si/SiGe/Si active layer is transferred to the plastic host. (h) Deposit source and drain contact metal (Ti).*

ment considerations. The transfer technique provides some advantages over the printing technique demonstrated by Menard et al. [2]. First, no stamping or holding anchors are necessary, greatly reducing the complexity and possibly enhancing the yield of the transfer process. Second, in this transfer process, the bottom side of the active layer is exposed after transfer, providing an opportunity for manufacturing double-sided devices that other transfer techniques are unable to provide. However, it is further noted that in order to apply this anchor-free method, SOI source materials with a very thin BOX layer are needed. Further, the current flip transfer cannot spread active materials to a large area, which can be easily done with a stamp-based transfer method.

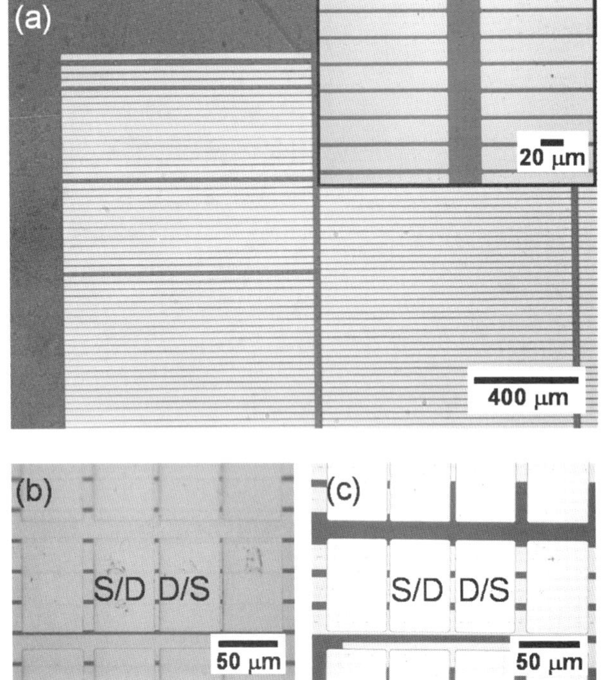

FIGURE 3.2 *Optical microscope images. (a) As-transferred unstrained Si strips. The inset shows a magnified image. (b) Finished devices on unstrained Si. The 50 × 75 μm rectangles are 70-nm-thick Ti metal pads that are patterned on the Si strips, which run horizontally. The distances between the edges of the metal pads are 10 μm, 5 μm and 3 μm, which correspond to different channel lengths. Measurements are made between neighboring metal pads. Because the metal pads cover three strips, the gate width is 60 μm. (c) Finished devices on strained Si; same conditions as in (b).*

In this section, the focus is on the strain effects on the devices' static electrical properties, by making a comparison between the unstrained and strained Si TFTs. Figure 3.2(b, c) shows images of finished devices on unstrained Si and strained Si, respectively. The distances between the edges of the rectangular metal pads shown in the figures are 3 μm, 5 μm and 10 μm, which correspond to different channel lengths. As shown in Figure 3.1, all these devices are made in the bottom gate form, although the fabrication of top-gated devices is also straightforward.

The XRD test is performed on the Si/SiGe/Si sandwich strips before and after they are transferred to the flexible polymer host and when the TFT fabrication is complete. Strain sharing has already occurred before transfer, as explained previously. With these data, the crystalline quality can be verified after transfer, and the strain status of the Si/SiGe/Si membrane determined and compared at different steps. Figure 3.3(a) shows the Si handle wafer diffraction peak from the SOI and the thin SiGe alloy along with thin Si layer peaks of the sandwich structure. Note that not all the area of the sample was patterned into strips, so the BOX is still present on some of the sample, providing an opportunity to compare diffraction peak shifts on the same sample. In the transfer step, only the fully undercut strips are transferred to the polymer host. Diffraction shows two (narrow) SiGe alloy peaks and two (broad) Si peaks (averaged lines were drawn for these two broad peaks to assist the observation), as expected. The pair at the lower angle corresponds to less compressively strained SiGe, due to strain sharing $(\theta = 34.269°)$ on the free parts of the membrane, and to fully compressively strained SiGe alloy $(\theta = 34.159°)$ (from the attached parts of the membrane). The peaks at the higher angle correspond to the very thin (hence broad) tensile strained Si films, again due to strain sharing in the free-standing parts of the membranes $(\theta = 34.697°)$ and unstrained Si $(\theta = 34.585°)$ (this peak coincides with the Si handle wafer peak). The angular shift in the Si peak is caused by the tensile strain induced by strain sharing, as known from transfers of membranes onto Si substrates [10]. The widths of the peaks reflect the thinness of the layers. The rigid $0.11°$ angular shift (indicated by arrows in Figure 3.3a) between the members of each pair reveals that all the strain relieved in the SiGe alloy layer is transferred to the Si layers without any inelastic relaxation. Figure 3.3(b) shows the XRD results on these released strained Si strips transferred onto the polymer host after complete TFT fabrication. Diffraction from unstrained Si strips transferred to a different piece of the same polymer host is also shown. As expected, only the Si thin-film peak is observed in the unstrained Si membrane, while the strained sample has a SiGe alloy peak at the lower Bragg angle and a Si thin-film peak at the higher angle. Thickness fringes on both samples can be clearly observed: they indicate that supreme crystalline quality is maintained throughout the transfer and the device fabrication. The existence of thickness fringes and the sharpness of diffraction peaks also verify that no or negligible numbers of dislocations are generated in the strain-sharing approach.

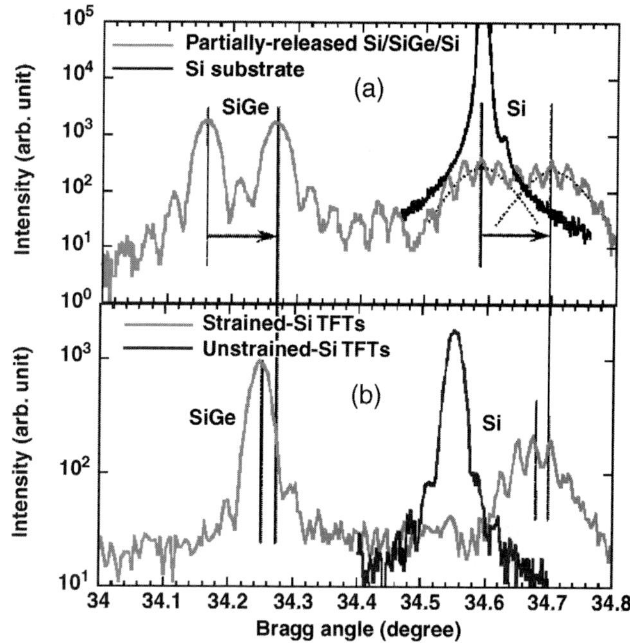

FIGURE 3.3 *X-ray diffraction line scan ($\theta/2\theta$) results. (a) An average over released and unreleased portions of an Si/SiGe/Si membrane structure on the starting silicon-on-insulator (SOI) substrate. The arrows indicate a 0.11° angular shift of both the SiGe alloy peak and the Si thin-film peak for the released portions of the membrane relative to the unreleased portions, because of strain sharing in the released portion. The Si peak for the unreleased portion is at the same angle as the bulk-Si peak, as expected. The membrane peaks are broad because the layers are thin (averaged lines are drawn to assist the observation). (b) Diffraction from a released, strain-relaxed Si/SiGe/Si membrane and an unstrained Si layer transferred onto polymer hosts, and after thin-film transistor (TFT) fabrication. A small systematic shift of the peaks relative to the relaxed, non-transferred membrane (a) is observed and is likely due to the flexibility of the polymer.*

There is no reference (such as the Si substrate Bragg peak) for the membranes transferred to the flexible polymer host. Bragg angles for membranes on the polymer may have slight, possibly random, shifts caused by the flexibility of the PET. The lines in Figure 3.3(a) extended to Figure 3.3(b) mark the positions of the Si and SiGe peaks of the strain-sharing (released) membrane. The corresponding peaks for the membranes on the polymer host are slightly, but rigidly, shifted, as indicated by the lines. The fact that the separation of the

Si and SiGe peaks is maintained when the membrane is transferred to the PET confirms that the strain-sharing status remains identical throughout the entire dry transfer and TFT fabrication process. The amount of tensile strain in Si layers on any plastic host can now be engineered in the same manner as in the strain-sharing technique: one can tune the SiGe/Si thickness ratio and the Ge fraction in the alloy layer. The strain in Si thin films due to strain sharing is calculated from the $+0.11°$ angular shift, which gives the out-of-plane lattice constant of the strained Si layers. The in-plane strain is calculated by using $\epsilon_{||} = \epsilon_{\perp}(1 - v)/2v$, where v is the Poisson ratio for Si and is chosen as 0.277. The calculated strain in the Si layers is thus 0.362%, which agrees very well with the force-balance model in Mooney et al. [7].

Unstrained and strained Si TFTs on PET are electrically characterized in the same way: by using the coated ITO layer as the gate electrode, the SU-8 layer as the gate dielectric, and the metal pads on the surface as source and drain electrodes. The measurements of TFT device characteristics are performed at room temperature with an HP4155B semiconductor parameter analyzer. The gate length (L_G) is determined by the gap between the source and drain metal pads and is varied from 3 µm to 50 µm. The gate width (W_G) is chosen as 60 µm. To compare the intrinsic properties between the unstrained and the strained Si TFTs, fresh devices of both types were electrically characterized. Figure 3.4(a, b) shows representative output current–voltage $(I–V)$ curves of a 3 µm gate length TFT made on unstrained and strained Si active layers, respectively. N-type field-effect transistor (FET) characteristics are exhibited by both types of membrane. Since the active layers in both devices are undoped, the metal (Ti) contacts made for source and drain are Schottky contacts. For the strained Si active layer, a reduced Schottky barrier height may be present owing to the lowered conduction band energy (strain splits the conduction band and creates a lower energy $\Delta 2$ valley) in comparison to the unstrained Si. The lack of 'full' saturation in the strained Si TFT shown in Figure 3.4(b) is speculated to be due to the lower Schottky barrier height (equivalent to smaller parasitic source and drain resistance) of the strained Si TFTs. Figure 3.4(c, d) shows the comparison of drain current and transconductance between the unstrained and the strained Si TFTs with different gate lengths $(L_G = 3, 10, 20$ and 50 µm) at a low source–drain bias (V_{DS}) of 50 mV. The peak transconductance values for the unstrained and the strained Si TFTs were reached at around 12 V and 14 V, respectively (Figure 3.4d). Note that even though the strained Si TFTs have lower threshold voltage (2.5 V) than the unstrained counterparts (5 V), both the drain current and the transconductance of the TFTs made on strained Si are much higher than those made on unstrained Si. The I_{ON}/I_{OFF} ratio of the strained Si TFTs is more than 10^3 at $V_{DS} = 1.5$ V.

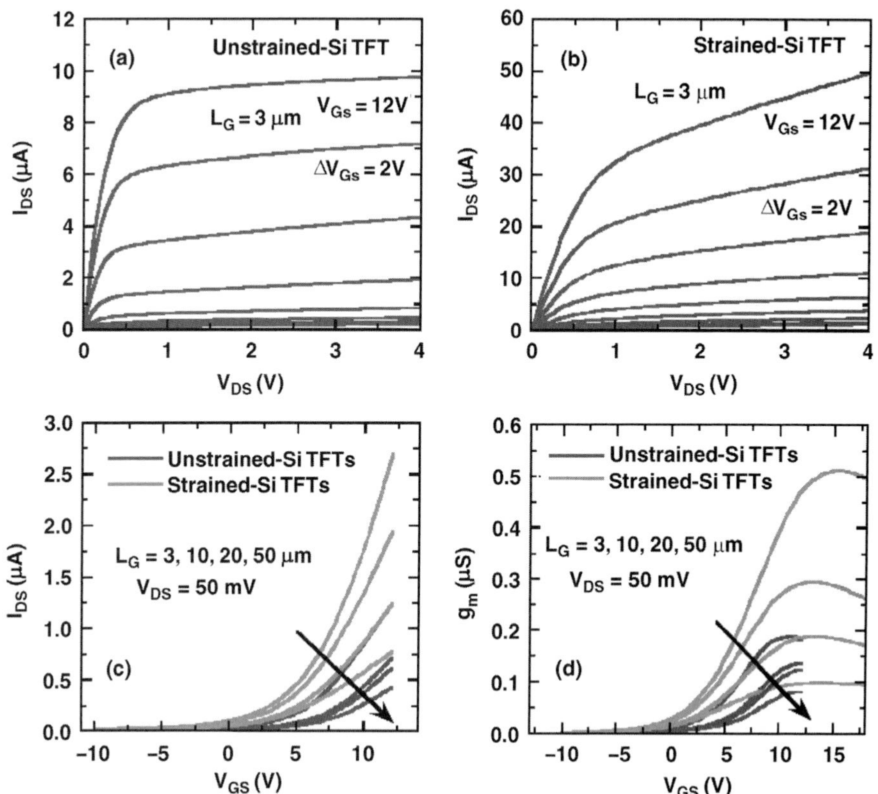

FIGURE 3.4 *DC characteristics of the thin-film transistors (TFTs). (a, b) Representative I–V curves of a 3 µm gate length TFT fabricated on unstrained and strained Si, respectively. The gate voltage (V_{GS}) varies from −4 to 12 V in 2 V intervals. (c) Drain current (I_{DS}), and (d) transconductance (g_m) comparison between unstrained and strained Si TFTs. The gate length from top to bottom is 3, 10, 20 and 50 µm for both samples. V_{DS} is 50 mV.*

The field-effective mobility (μ_{FE}) in the linear region is extracted using the formula $\mu_{FE} = L_G g_m/(W_G C_G V_{DS})$ [11], where L_G and W_G are the physical dimensions of the gate length and width, $V_{DS} = 50$ mV, and g_m is picked as the highest transconductance measured from the devices. The dielectric capacitance (C_G) is evaluated by measuring dummy patterns in the form of parallel capacitors and is 2.16 and 2.53 nF/cm^2 for unstrained Si and strained Si samples, respectively. The slight difference in the capacitance is attributed to the different final SU-8 thicknesses between these two samples and agrees well with the thickness measurement (1.8 µm and 1.4 µm for the unstrained and the strained samples). The extracted carrier mobility in TFTs made on

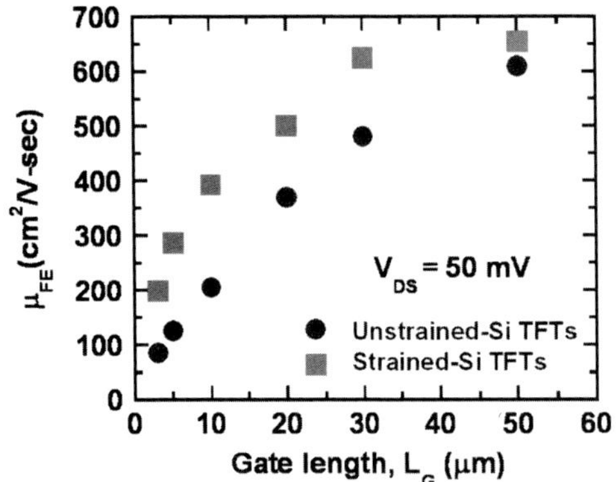

FIGURE 3.5 *Extracted field-effect mobility (μ_{FE}) of unstrained and strained Si thin-film transistors (TFTs) on flexible polymer hosts.*

both the unstrained and the strained Si with various gate lengths is shown in Figure 3.5. The extracted decreasing mobility with shorter gate length is the evidence of an unoptimized source/drain parasitic resistance caused by the high Schottky barrier of metal contacts. The parasitic resistance has more influence on the intrinsic channel resistance when the gate length is reduced, which results in lower source-to-drain current (I_D) and thus lower extracted mobility [12].

By comparing the strained and unstrained Si TFTs, it can be observed that the extracted mobility of strained Si TFTs shows remarkable enhancement relative to those made with unstrained Si. The mobility enhancement is the result of two combined effects both caused by strain. The biaxial tensile strain splits the Si conduction band into two valleys that not only increase the electron mobility but also decrease the Schottky barrier height of the metal contacts made on the strained Si. With the reduced Schottky barrier height, both thermionic emission and field emission are greatly enhanced, which results in higher source-to-drain current [13].

Summary

A simple, flip-transfer technique to transfer single-crystal Si membranes to flexible polymer hosts was described. The high transfer rate and the fact that the patterning alignment remains unchanged before and after transfer suggest that this process can be readily introduced into a device fabrication process flow. Of more importance, strained Si channel devices are realized through the elastic strain sharing between thin Si layers and a thicker SiGe alloy layer when multilayer membranes are released from the oxide. The crystalline quality does not deteriorate throughout the entire TFT fabrication and the strain-sharing status is also fully replicated onto the polymer host. The strained Si/SiGe/Si TFTs show much higher drain current and peak transconductance values than the identical unstrained Si TFTs. Both mobility and drain–current in the strained Si TFTs are enhanced owing to the effects of strain. Further exploration of these characteristics will lead to low power-consumption, high-speed and high-performance electronics on both rigid and flexible substrates.

DOPING OF SILICON NANOMEMBRANES

Several types of semiconductor material have been considered for flexible electronics, among which single-crystal semiconductor nanomembranes have much higher charge carrier mobility values than amorphous, polycrystalline and organic semiconductors. The high mobility values and the recently demonstrated transferable characteristics of these nanomembranes made them an enabling material for high-performance flexible electronics applications [2, 14–16]. To achieve low parasitics, such as low contact resistance and low sheet resistance values, and thus to enable the effective transformation of high charge carrier mobility values into high device speed, requires effective doping control in these transferable thin-film materials. For flexible TFTs made on low-temperature plastic substrates employing silicon nanomembranes (SiNMs) [17–19], the temperature constraint of the substrate poses a significant challenge to achieve effective doping. This challenge can be circumvented by predoping the Si template layer of SOI before releasing and transferring Si template layer (as a nanomembrane) to a plastic substrate. For the flip-transfer method of SiNMs described above, the back side of the SOI template layer is exposed on the plastic substrate and a high doping concentration is thus required on the back side with the predoping method to achieve low parasitic resistance values. Doping on the back side of the SiNMs is more difficult to control than on the front side. Compared to n-type doping, considering the fast boron diffusion into oxide and the fast transient enhanced diffusion (TED) [20], obtaining a high doping concentration of p-type at the interface of the Si/SiO_2 is particularly challenging. Furthermore, good crystal characteristics of the thin-film material and a good interface between metal and semiconductor are believed to be essential for the low sheet resistances and contact resistances [21]. Therefore, the doping and annealing condition should be well chosen to minimize the damage due to the ion implantation. In this section, detailed p-type boron-doping (n-type doping will be described in the next section) control experiments are conducted to overcome the back-side doping challenge for flexible single-crystal SiNMs.

Sample preparation

High-energy ion implantation is necessary to achieve a high doping concentration at the Si/SiO_2 interface. Low-energy doping is also performed to suppress the dopants from diffusing into the BOX layer. Therefore, the doping profiles are first simulated using the commercial software Athena (Silvaco) until the averaged (over the Si template layer thickness) doping concentrations of both low- and high-energy implantations after annealing are nearly the same.

As the highest hole mobility is available along Si (110)/<110>, SOI substrate with a Si (110) template layer of 190 nm, based on hybrid orientation technology (HOT), is used to investigate for the p-type doping conditions. The SOI samples are ion implanted under two conditions: (1) boron ([B]), 5 keV, 5×10^{15}/cm^2; and (2) [B], 70 keV, 5×10^{15}/cm^2, followed by furnace annealing at 950°C for 30 min. The samples are then released and flip-transferred [22] onto a 175 μm thick flexible PET plastic substrate with SU-8 spun on top serving as the adhesive layer. The transmission line model (TLM) patterns are defined and metal pads are deposited at room temperature with Ti/Au of 30/120 nm. The spacing between contacts varies from 10 to 50 μm.

Doping profiles

Secondary ion mass spectrum (SIMS) analyses are used to analyze the implanted samples. The doping profiles of both before and after annealing are plotted in Figure 3.6. The boron (B) and oxygen (O) were quantified in atoms/cm^3, and the Si in secondary ion counts. The sharp drop in ion intensity in the BOX is due to sample charging effects. The top surface is slightly oxidized after annealing because the N_2 ambient contains 5% O_2. It can be seen that the doping concentrations inside the Si layer, particularly near the interface of Si/SiO$_2$, are almost the same, 10^{20}/cm^3, for both conditions. Notice that before annealing, the low-energy implantation leaves a wider low-concentration region inside the top Si template layer than the high-energy one, indicating less damage in the deeper region. This property is important to the crystal characteristic, which will be discussed in the next section. It can also be predicted that the sheet resistance of low-energy doping should be higher, because the total amount of dopants in the Si layer is larger than in the high-energy one.

Doping results

TLM patterns are formed on the predoped flexible SiNMs after transfer onto plastic, and the resistances are measured between various contact pads. The resistances versus distances under two conditions are plotted in Figure 3.7. The sheet resistance and contact resistivity can be extracted and calculated from the slopes and intercepts along the y-axis, as listed in Table 3.1. As expected, the low-energy doped SiNMs show lower sheet resistance of $R_s = 100.95$ Ω/□ owing to the larger amount of dopants.

Moreover, it is worth noticing that despite the same doping concentrations at the Si/SiO$_2$ interface, the low-energy doped SiNMs also show lower contact

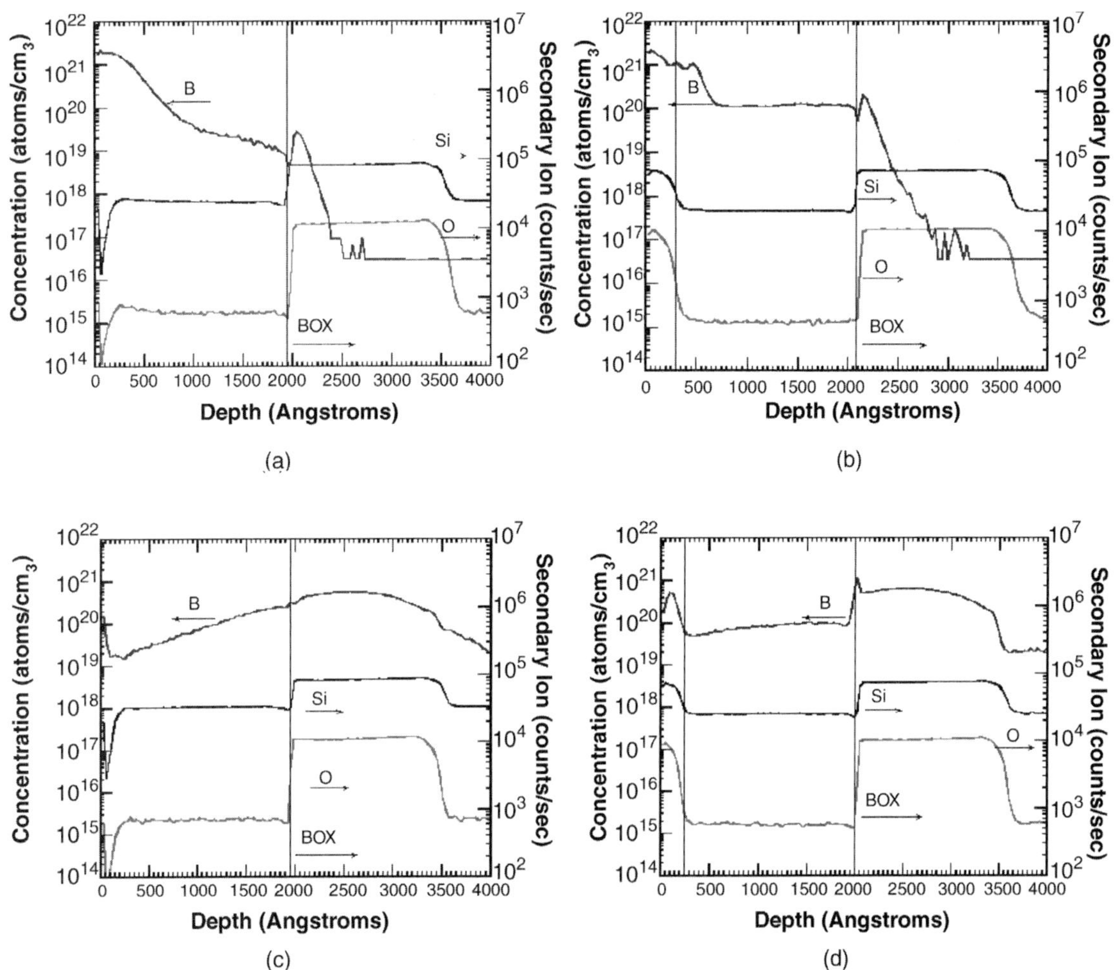

FIGURE 3.6 *Secondary ion mass spectrum (SIMS) results of low-energy doped (5 keV, 5 × 10^{15}/cm^2) Si nanomembranes (SiNMs); (a) before annealing; (b) after annealing; and high-energy doped SiNMs (70 keV, 5 × 10^{15}/cm^2); (c) before and (d) after annealing.*

resistivity ρ_c of 1.87×10^{-4} Ω/cm^2. It is believed that the reduced damage, i.e. the better crystalline quality of the low-energy case, is responsible for the lower resistance values [21]. Figure 3.8(a, b) shows the transmission electron microscope (TEM) results of the low- and high-energy doped SiNMs before annealing, respectively. The broad fringes across the bottom of the image in Figure 3.8(a) are due to a thickness gradient in this wedge-shaped sample. As can be seen, in the low-energy case, only a very thin top layer of the SOI template layer was damaged, leaving the rest of the material still single crystalline almost without defects that served as a seed layer and facilitated the

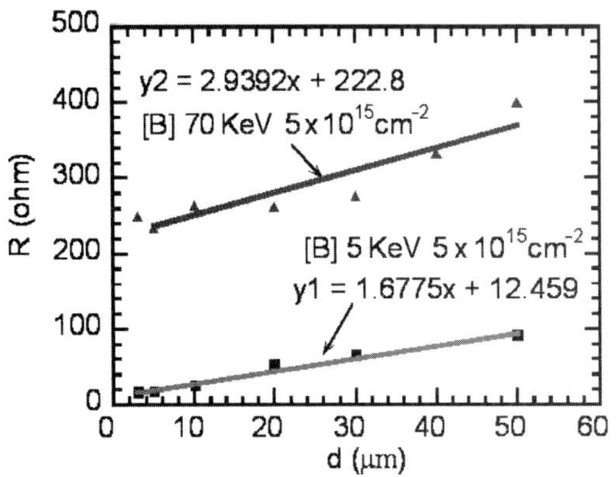

FIGURE 3.7 *Resistances (R) versus metal pad distances (d) of both low- and high-energy doped flexible single-crystal Si nanomembranes.*

subsequent recrystallization process during annealing. However, for the high-energy case, a large number of defects were clearly identified throughout the entire layer, which hardly provides the seed layer. Furthermore, the interface between top Si and the BOX layers suffers from more damage during the implantation than the low-energy doping, which may result in better contact with metal.

Summary

In this section, effective boron-doping control in flexible single-crystal SiNMs was illustrated for the purpose of reducing parasitics in devices. The p-type doping was used as an example to reduce both contact resistivity and sheet resistances. Two ion implantation conditions with a low and high energy were chosen and compared. The doping profile was verified with SIMS analysis.

Table 3.1 Calculated sheet resistance R_s and contact resistivity ρ_c under different doping conditions (after annealing)

Doping condition	R_s ($\Omega/\square$)	ρ_c (Ω/cm^2)
Low-energy	100.95	1.87×10^{-4}
High-energy	176.35	3.34×10^{-4}

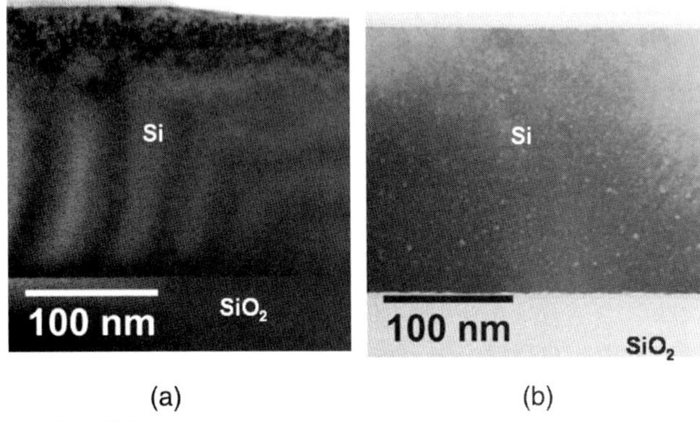

(a)   (b)

FIGURE 3.8 *Transmission electron microscope images of (a) low- and (b) high-energy doped Si nanomembranes (SiNMs) before annealing. More defects are found in the high-energy doped SiNMs.*

Contact resistivity and sheet resistances were extracted from TLM measurements. Low-energy doping shows lower sheet resistance values owing to the larger amount of dopants in the SiNMs than in the high-energy case. Moreover, since less damage is induced during the implantation, low-energy doping benefits from better crystal characteristics as well as better Si/BOX interface properties after the annealing process. Therefore, low-energy implantation is favored in the flip-transfer technique, which can be applied in flexible high-speed TFTs.

MICROWAVE THIN-FILM TRANSISTORS

Flexible microelectronics and macroelectronics that are lightweight, robust and capable of being folded or rolled up for easy carriage, storage and attaching to uneven surfaces have been an important pursuit of research in the past decade. Some of the potential applications such as flexible displays and RF identification tags require modest electrical performance of the active TFTs, while low cost and the capability of large-area fabrication are the primary concerns [23–25]. For these applications, flexible electronics using organic semiconductors, amorphous silicon (α-Si) and polycrystal silicon (poly-Si) have shown tremendous promise [26–28]. At the other end of the scale, however, lie applications that require reasonable cost but high-performance RF circuitry capable of running regimes of up to several hundred megahertz to gigahertz. One example is the large military active antennae, used for surveillance systems, unmanned aerial vehicles, etc. [29]. Flexible electronics are ideal for these applications since light weight, small stowage volume and the ability to be formed into complex shapes are crucial. Active circuitry on these antennae ought to be operated in the ultrahigh-frequency (UHF; $\sim$500 MHz) and higher frequency range. Such a demand on higher operation frequencies has posed a high bar for existing TFT technology since the organic semiconductors and α-Si exhibit very limited carrier mobility and poly-Si often requires a non-trivial crystallization process using, for instance, an excimer laser. Furthermore, extra care is needed when using such crystallization processes on low-temperature and low-cost polymer substrates [30, 31].

Currently, the large-area antennae mentioned above are built by hardwiring the high-performance, stand-alone units which are fabricated using highly developed integrated circuit (IC) technology. It is only recently that high-quality, single-crystal semiconductors have been monolithically integrated onto flexible polymer substrate [1–3, 32]. As shown earlier in this chapter (Strained silicon nanomembranes), electron mobility as high as 270 cm^2/V·s has been demonstrated on TFTs using single-crystal Si channel with improved ohmic contacts for source and drain [33], and further increase in the device performance using a strained Si channel [34] is presented. This section describes the fast, flexible single-crystal Si RF TFTs that are monolithically integrated on polymer substrate and capable of operating at microwave frequencies.

Device fabrication

To integrate single-crystal Si onto polymer substrate, commercially available (from Soitec) SOI substrates were used. These SOI substrates (6 inch diameter) are created by Smart-Cut® and have a top Si (001) template layer of 200 nm on a BOX layer, also of 200 nm. After the BOX was selectively removed, the released thin Si template layer (SiNM) was transferred to a polymer host using a flip-transfer technique (without using stamps as the transfer tool) as shown in the section Strained silicon nanomembranes. With the capability of integrating single-crystal Si with the polymer host, the greatest challenge remaining to make high-performance TFTs has been the temperature constraint. The high-temperature processes needed for bulk Si technology, such as dopant activation and gate dielectric layer formation, are inaccessible on polymer substrates. Therefore, a process is developed for fabricating top-gated Si TFTs on a polymer substrate that retains low parasitic resistance for source/drain contacts and enables the feasibility of using a thin gate dielectric layer. With this process, the highest temperature applied to the polymer substrate can be kept well below 120°C.

In the process of making n-type TFTs, the low-resistivity source and drain regions were first formed on the SOI substrate via phosphorus ion implantation at a dose of 4×10^{15} cm^{-2} and an energy of 40 keV, followed by annealing in a horizontal furnace at 850°C for 45 min in ambient N_2 to activate the implanted dopants. The current approach deviates from the self-aligned source/drain formation method that is commonly used in fabricating bulk (on rigid Si substrate) Si metal oxide semiconductor field-effect transistors (MOSFETs). However, by forming the source and drain regions before the formation of the gate-dielectric stack, the process dilemma mentioned previously can be solved: high-temperature procedures are needed to achieve low contact resistance but are not compatible with low-temperature polymer substrate. Since ion implantation is a mature technique, p-type (n-type is described herein) and thus complementary geometry can also be achieved for more functionality and low-power application on flexible substrates using this approach. After the thermal annealing, similar fabrication steps to those described for strained silicon nanomembranes (see above) were conducted. The Si template layer was patterned into strips or in the form of mesh holes followed by BOX removal in aqueous 49% HF. The polymer host, which is PET in this study, was rinsed with acetone, IPA and DI water, followed by a 10 s O_2 plasma treatment (Unaxis 790, 50 W/100 mt/20 sccm O_2). A thin layer of SU-8-2 epoxy (~1.8 μm, Microchem), which serves solely as an adhesive layer to facilitate the upcoming transfer, was then spun on at 4000 rpm. The SOI, now with the Si template layer settled down and well registered on the Si handling substrate via weak van der Waals' forces [6], was brought face to face against the epoxy layer. The Si template layer was transferred to the polymer host by gently pressing and peeling off the Si handling substrate. As described for strained silicon nanomembranes (see above), the positions of the released Si membranes or strips remain unchanged during the release and the transfer

[34], which facilitates the mask alignment for subsequent metallization on a large-area scale.

Finally, the SU-8-2 was exposed under UV light and became fully cross-linked after baking at 115°C. The device isolation was realized by a step of dry plasma etching (SF_6/O_2) down to the SU-8-2 layer following optical photolithography. The gate stack, consisting of 200 nm amorphous silicon monoxide (SiO), 40 nm Ti and 300 nm Au, was formed with electron-beam evaporation at a pressure under 10^{-6} torr followed by lift-off. Evaporated SiO thin film is widely used in Josephson junction and metal–insulator–metal (MIM) capacitors in microwave circuits [35], and can be deposited at room temperature [36]. As a result, this approach keeps the polymer substrate at a low temperature throughout the gate stack formation. The lift-off process ensures minimum gate real estate. Only the π-shape, two-finger gates are left on the substrate, so the undesired bending and curling of the flexible substrate possibly created by the large stress gate stack can be minimized. Finally, source and drain metal contacts consisting of 40 nm Ti and 500 nm Au were formed on the low-resistive source and drain regions followed by lift-off. Figure 3.9 shows the optical microscopic image and a schematic illustration of the cross-section of a finished TFT. The source-to-gate and drain-to-gate distances are both 1 μm in the finished devices.

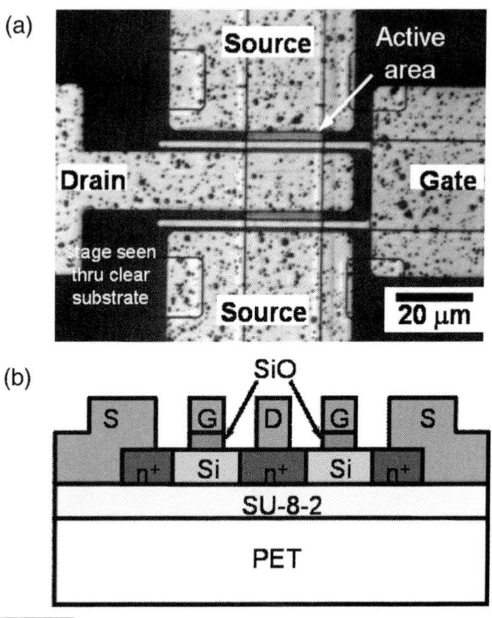

FIGURE 3.9 *(a) Optical microscopic image of a finished radiofrequency thin-film transistor (RF TFT) on polymer substrate.*

Device performance

The transfer length method is used to evaluate the contact resistance and the sheet resistance of the implanted regions. The extracted contact resistivity and sheet resistance were found to be 4.6×10^{-4} $\Omega{\cdot}cm^2$ and 94.5 $\Omega/\square$, respectively. Both values are lower than reported by Sun et al. [3]. Even lower resistance can be achieved with optimized dopant profile and annealing conditions. Figure 3.10(a) presents the normalized *I–V* characteristics of a two-gate-finger TFT that has a physical gate length of 2 μm and gate width of 2×30 μm. A much higher current level is achieved on these top-gated TFTs compared with

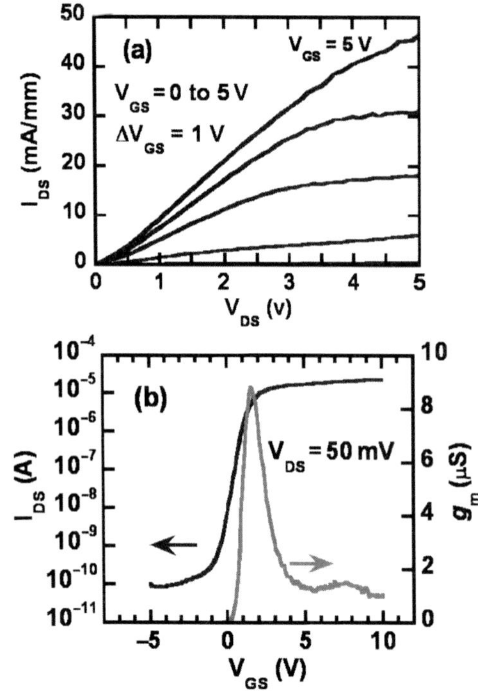

FIGURE 3.10 *Electrical characterizations of the top-gated radiofrequency thin-film transistor (RF TFT) on polymer substrate. (a) Normalized I–V characteristics of an RF TFT with physical $L_G = 2\,\mu m$ and $W_G = 60\,\mu m$. (b) Transfer characteristics of the TFT measured at $V_{DS} = 50\,mV$.*

the previously published results on bottom-gated single-crystal Si TFTs. Gate leakage current lower than 1 nA was achieved from the measurement. Figure 3.10(b) presents the transfer characteristics at $V_{DS} = 50$ mV. The I_{DS} is shown in the logarithmic scale and an I_{ON}/I_{OFF} ratio larger than 10^5 is demonstrated. Threshold voltage and subthreshold swing are extracted from the measurement results and are 0.5 V and 450 mV/decade, respectively. High subthreshold swing is an indication of high interface-trap density between SiO and Si. A dielectric constant for SiO of 5.8 is used in calculating the field-effect mobility using $\mu_{FE} = L_G g_m/(W_G C_{OX} V_{DS})$ [37]. The calculated electron mobility value is 230 cm^2/V·s, which agrees very well with other reported results.

To investigate the RF characteristics of the TFTs, small-signal S-parameters were measured on polymer substrate in the frequency range of 45 MHz to 40 GHz with an Agilent E8364A network analyzer using 100 µm pitch ground–signal–ground (GSG) probes (Cascade Microtech). The reference plane was calibrated to the probe's tip by short-open-load-through (SOLT) calibration using impedance standard substrate (GGB Industries) to eliminate possible errors from the network analyzer, cables and probes. Figure 3.11(a) presents the current gain ($|h_{21}|^2$) and the power gain (G_{max}) calculated from the de-embedded S-parameters as a function of frequency at bias values of $V_{DS} = 4$ V and $V_{GS} = 5$ V. An f_T as high as 1.9 GHz and f_{max} of 3.1 GHz are obtained from the RF TFT on polymer substrate. De-embedding procedure was also conducted to exclude the parasitics associated with the device GSG pad by using the open and short dummy patterns that were fabricated on the polymer substrate along with the active devices. No significant changes were observed on the de-embedded f_T and f_{max} values owing to the large device size. Figure 3.11(b, c) shows the bias dependence characteristics of the f_T and f_{max}. Both f_T and f_{max} peak at $V_{DS} = 4$ V and $V_{GS} = 5$ V.

Performance boost by layout optimization

Device layout design plays an important role in device speed. An improved layout is described here and is used to improve the flexible microwave TFT performance. The redesigned TFT (denoted TFT-2) fabrication process and the PET substrate used in this section are identical to those described for microwave TFTs (see above) (the previous device is denoted TFT-1 [18]). The process parameters and device dimensions for these two TFTs are compared in Table 3.2. For both TFTs, the source/drain regions are formed by ion implanting the designated regions on SOI substrates followed by high-temperature annealing (high T process). The 'predoped' Si template layer is then removed from the SOI substrates and integrated with flexible plastic host substrates via one-step printing. The rest of the process procedures, including the formation of active regions, gate electrodes and source/drain electrodes, are conducted at temperatures below 120°C (low T process). Since source/drain

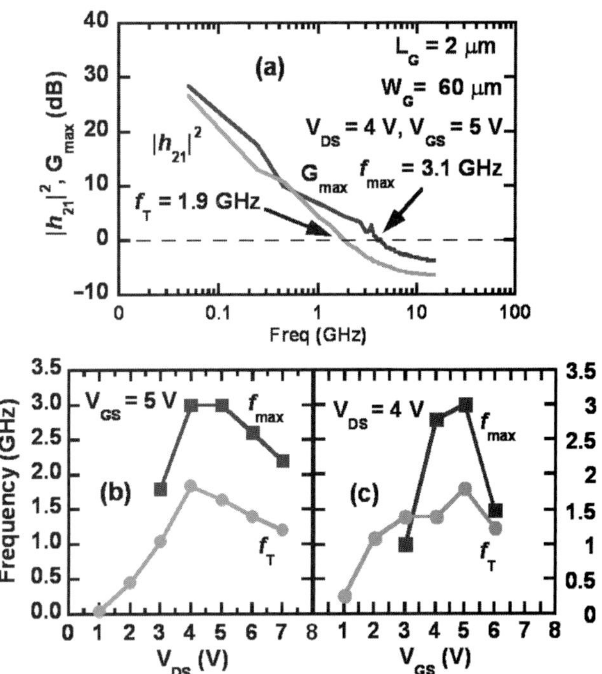

FIGURE 3.11 *AC small-signal radiofrequency (RF) characteristics of the top-gated RF thin-film transistor ($L_G = 2\,\mu m$ and $W_G = 60\,\mu m$) on polymer substrate. (a) Current gain ($|h_{21}|^2$) and power gain (G_{max}) as a function of frequency measured at $V_{DS} = 4$ V and $V_{GS} = 5$ V. (b) f_T and f_{max} as a function of V_{DS} at $V_{GS} = 5$ V. (c) f_T and f_{max} as a function of V_{GS} at $V_{DS} = 4$ V.*

Table 3.2 Comparison of process parameters and device dimensions between TFT-1 and TFT-2

Device name	TFT-1	TFT-2
Si orientation	(001)	(001)
Si thickness (nm)	200	200
SiO thickness (nm)	200	100
Gate width (µm)	2 × 30	2 × 20
L_{gate} (µm)	2	2.5
L_{sg} (µm)	1	−0.5
L_{dg} (µm)	1	−0.5

TFT: thin-film transistor.

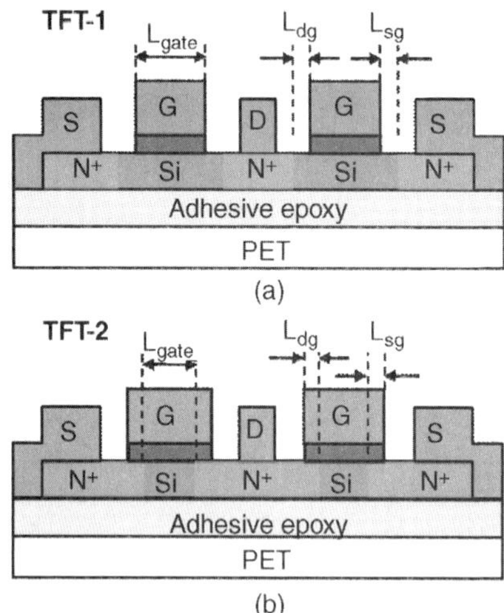

FIGURE 3.12 *Schematic cross-section of (a) TFT-1 and (b) TFT-2, illustrating the source/drain-to-gate separation on TFT-1 and source/drain-to-gate overlap on TFT-2. The gate dielectric (SiO) thickness of TFT-2 is about half of TFT-1. PET: polyethylene terephthalate.*

regions are formed before the gate stack, the self-alignment between the implanted source/drain regions and the gate that is commonly performed for regular MOS-FETs on bulk Si or SOI substrates is unavailable in this TFT fabrication process. Figure 3.12 schematically illustrates the cross-section of the two TFT layouts for comparison. TFT-1 (Figure 3.12a) has source-to-gate (L_{sg}) and drain-to-gate (L_{dg}) separations of 1 µm, which result in a large source access resistance (R_s). To reduce R_s, an overlap of 0.5 µm (indicated as negative values for L_{sg} and L_{dg} in Table 3.2) was designed between the gate and source/drain regions on TFT-2 (Figure 3.12b). The physical gate length (L_{gate}) for TFT-1 and TFT-2 is 2 µm and 2.5 µm, respectively.

Figure 3.13(a, b) shows a micrograph of a TFT-2 device with two-gate fingers in a π configuration (same as TFT-1 [8]) and an array of RF TFTs on a bent PET substrate, respectively. Figure 3.14 presents the normalized I–V and transfer characteristics of TFT-2. As shown in Figure 3.14, TFT-2 exhibits the typical n-channel FET characteristics, with a threshold voltage of -2 V, subthreshold swing of 480 mV/decade, I_{on}/I_{off} ratio higher than 10^5, and measured (extrinsic) maximum transconductance (g_m) of 70 µS (8.8 µS for TFT-1) at a gate-to-source voltage (V_{gs}) of -0.5 V and a drain-to-source voltage (V_{ds}) of 50 mV. Figure 3.15 (solid lines) shows the measured frequency response characteristics of TFT-2 with a physical gate length of 2.5 µm and gate width of 40 µm under $V_{gs} = 1$ V and $V_{ds} = 4$ V. The f_T and the maximum oscillation frequency f_{max} of TFT-2 are 2.04 and 7.8 GHz, respectively. Figure 3.16(a, b) shows the V_{ds} and V_{gs} dependence of f_T and the f_{max}, respectively. Note that even though TFT-2 has a thinner gate dielectric layer, it demonstrates similar f_T but much higher f_{max} compared with that of TFT-1. The influence of the TFT layout on the frequency response is elaborated below.

A small-signal equivalent circuit model (Figure 3.17) for these two TFTs was extracted from the measured scattering (S)-parameters at the bias conditions where the highest frequency responses of the two devices were measured. The extracted model parameters of TFT-1 and TFT-2 are summarized in Table 3.3.

The series parasitic resistance (R_s, R_d, R_g) and inductance (L_s, L_d, L_g) of each electrode were first determined by finding the linear regression between the real and the imaginary parts of the impedance parameters (Z-parameters) (converted from S-parameters), respectively [38]. The intrinsic TFT parameters were then extracted after subtracting the series parasitics [39]. The shunt admittances at the input and output, as well as the L_s, were omitted to simplify the extraction. Figure 3.15 presents the simulated current gain (H_{21}) and power gain (G_max) using the extracted small-signal equivalent circuit in comparison to the measured ones (solid lines) for TFT-2. Figure 3.18 shows the comparison of the simulated (dashed lines) and the measured (solid lines) results for TFT-1.

Fairly good agreements are exhibited with some minor discrepancies possibly due to the fact that some of the extracted parameters are slightly frequency dependent. Disregarding the minor discrepancies, the extracted parameters listed in Table 3.3 provide an important indication of how the TFT layout affects the frequency response. It shows that the intrinsic

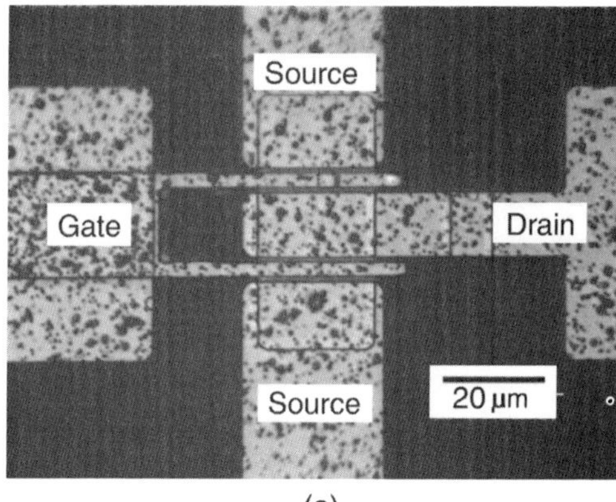

(a)

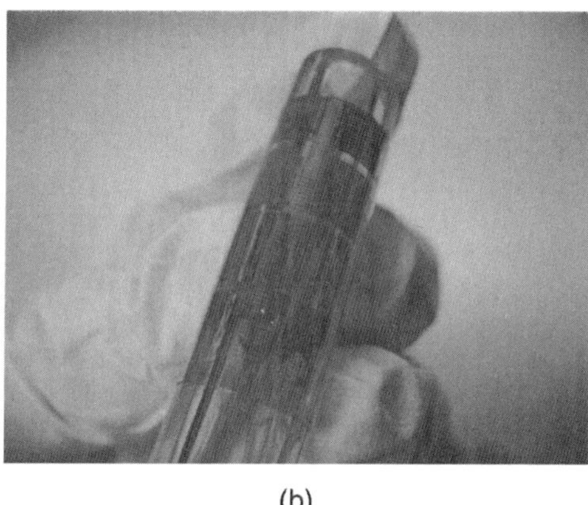

(b)

FIGURE 3.13 *(a) Photomicrograph of TFT-2. (b) Image of an array of TFT-2 on a bent flexible polyethylene terephthalate (PET) substrate.*

transconductance (g_mo) of TFT-2 does scale up with the thinner gate dielectric. However, the parasitic gate-to-source (C_gs) and gate-to-drain (C_gd) capacitances also increase because of the source/drain-to-gate overlap, which offsets the increase in g_mo. The intrinsic cut-off frequency (f_Ti) and f_max can be expressed as:

$$f_\text{Ti} = \frac{g_\text{mo}}{2\pi\left(C_\text{gs} + C_\text{gd}\right)} \tag{3.1}$$

$$f_\text{max} = \frac{f_\text{Ti}}{2\sqrt{\left(R_\text{g} + R_\text{s}\right)g_\text{ds} + 2\pi R_\text{g}C_\text{gd}f_\text{Ti}}} \tag{3.2}$$

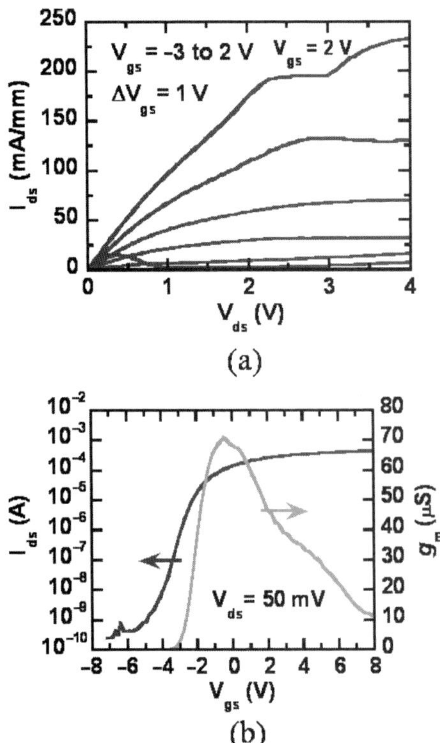

(a)

(b)

FIGURE 3.14 *(a) Current–voltage characteristics measured from TFT-2. (b) Transfer characteristics and transconductance measured from TFT-2.*

where g_{ds} is the output conductance and is the reciprocal of the output resistance (R_{ds}). In TFT-2, g_{mo} is about twice as high as in TFT-1. However, C_{gs} and C_{gd} of TFT-2 are also about twice as high as for TFT-1. For this reason, the thin gate dielectric of TFT-2 does not result in a much higher f_T. However, the overlapped source/drain-to-gate of TFT-2 greatly reduces the R_s and R_d, which are around five times lower than those of TFT-1 (Table 3.3). Furthermore, the overlapped drain and gate of TFT-2 also alleviates the channel length modulation compared with a separated drain and gate (in TFT-1), resulting in a higher R_{ds} (Table 3.3). The channel length modulation is also reflected on the measured I–V curves (Figure 3.14a), where the TFT-2 has flatter saturation regions than those of TFT-1. Through this analysis it is identified that the overlapped source/drain to gate results in higher C_{gs} and C_{gd} on TFT-2 that prevent the increase in f_T, but contribute to the reduction of R_s, R_d and g_{ds} that makes f_{max} much higher than that of TFT-1.

Summary

In summary, high-performance, microwave single-crystal Si RF TFTs on flexible polymer substrate have been described with an inexpensive fabrication process. In this process, gate stack was formed after high-temperature source and drain region formation, which allows one to achieve low-resistivity source and drain contacts and low sheet-resistance values via ion implant and subsequent high-temperature annealing procedures performed on the bulk SOI substrate. The metal/SiO gate

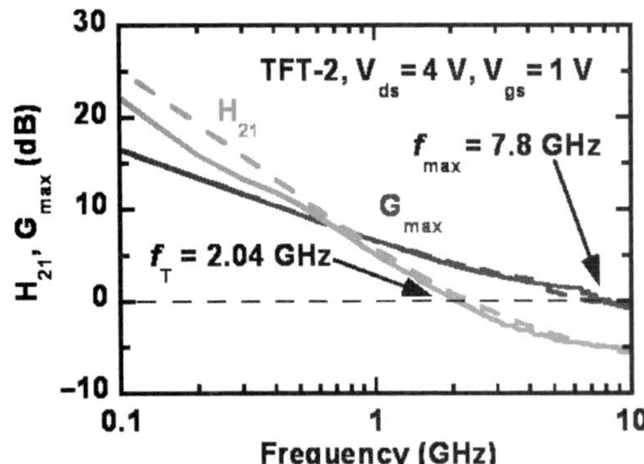

FIGURE 3.15 *Measured (solid lines) and simulated (dashed lines) frequency dependence of current gain (H_{21}) and power gain (G_{max}) of TFT-2 at bias conditions for the highest figure of merit.*

stack formed by evaporation and lift-off enables the temperature to be kept below 120°C in all the processing procedures and the residue stress to be kept to a minimum level after the Si layer has been transferred onto the polymer substrate. With this process, top-gated Si RF TFTs were fabricated on a flexible polymer substrate and demonstrated high electron mobility, high current drive capability and high frequency-response characteristics with record-high f_T of 1.9 GHz and f_{max} of 3.1 GHz. Further investigation into TFT layout design demonstrates that the source-to-gate and drain-to-gate distances play a crucial role in the high-frequency response of the TFTs. By properly overlapping the gate with the source and drain, a significant reduction in source/drain access resistance and output conductance has been achieved, which results in significantly improved f_{max}. Flexible TFTs with f_T of 2.04 GHz and f_{max} 7.8 GHz were realized with an optimized layout. The results suggest that the requirements of certain designated flexible RF applications may be met by suitably designing the TFT layout. These characteristics show the potential of high-frequency RF/microwave applications to bridge flexible electronics.

HYBRID COMPLEMENTARY THIN-FILM TRANSISTORS

So far, the fastest TFTs made on plastic have reached RF level cut-off/maximum oscillation frequency (f_T/f_{max}) values of 2.0/7.8 GHz [17], as shown in the

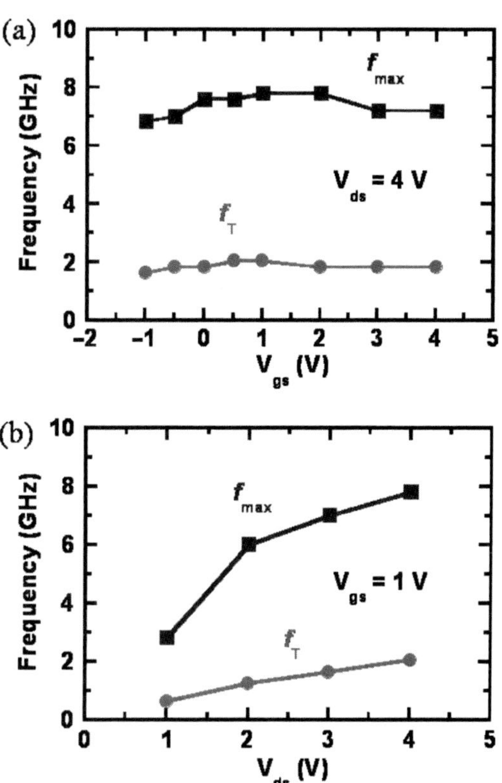

FIGURE 3.16 Measured bias dependence of f_T and f_{max} from TFT-2. (a) f_T and f_{max} dependence on V_{ds}. (b) f_T and f_{max} dependence on V_{gs}.

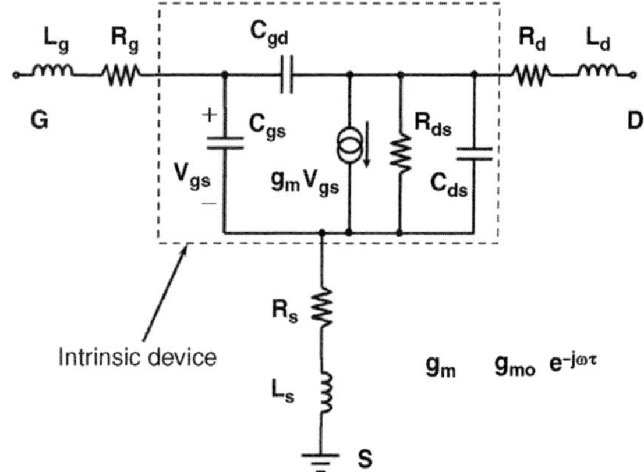

FIGURE 3.17 Small-signal equivalent circuit model used for thin-film transistor parameter extraction.

Table 3.3 Comparison of extracted device model parameters, FOM values and bias conditions between TFT-1 and TFT-2

	TFT-1	TFT-2
R_s (W)	208.6	40
R_d (W)	195.1	35
R_g (W)	8.2	0.1
L_s (nH)	0	0
L_d (nH)	0.47	0.35
L_g (nH)	0.08	0.06
C_{gd} (pF)	0.0137	0.04
C_{gs} (pF)	0.0285	0.0571
C_{ds} (pF)	0.0017	0.0033
R_{ds} (W)	2052	4500
g_{mo} (mS)	0.56	1.18
τ (ps)	4.3	4.8
Measured f_T (GHz)	1.9	2.04
Measured f_{max} (GHz)	3.1	7.8
Bias	$V_{ds} = 4$ V, $V_{gs} = 5$ V	$V_{ds} = 4$ V, $V_{gs} = 1$ V

FOM: figure of merit; TFT: thin-film transistor.

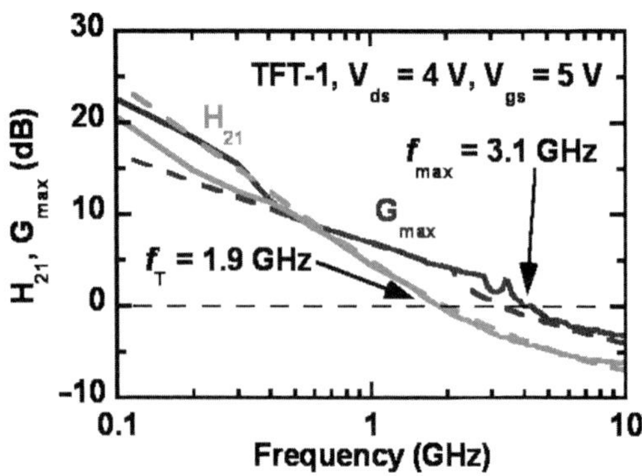

FIGURE 3.18 *Comparison of measured (solid lines) and simulated (dashed lines) frequency dependence of current gain (H$_{21}$) and power gain (G$_{max}$) of TFT-1 at bias conditions for the highest figure of merit.*

section Microwave thin-film transistors, above. Nevertheless, the device speed of the complementary TFTs is limited by the low hole mobility of silicon (001), which is the most commonly used orientation for flexible circuits. Therefore, HOT [40] has been used in flexible applications and complementary TFTs using Si (110) nanomembranes have been fabricated on plastic, showing high gains and large noise margins at low supply voltages [41]. To achieve the highest speed of the complementary logic circuit, it is necessary to transfer SiNMs with different orientations together, where the highest mobility of both electrons and holes can be integrated in one inverter. This section introduces the fabrication process with a two-time transfer technique of flexible hybrid orientation complementary inverters using both SiNMs, (001) and (110), on one plastic substrate. Initial DC performances of individual n- and p-TFTs were measured. The process described in this section shows potential to further enhance the logic speed by integrating SiNMs with different orientations.

Device fabrication

The fabrication of the TFTs in this study starts from two SOI wafers with lightly p-type doped 270 nm Si (001) and 190 nm Si (110) template layers on top of 200 nm BOX layers. According to the previous doping study in the section Doping of silicon nanomembranes, above, the source/drain regions were ion implanted with phosphorus with a dosage of 1×10^{16}/cm^2 at 12 keV and boron with a dosage of 5×10^{15}/cm^2 at 5 keV for n- and p-TFTs, respectively. A low-energy doping condition is chosen to suppress the ion damage, leaving more single-crystal material for a better recrystallization process, hence lower contact resistances and sheet resistances [21]. To form n-wells for p-TFTs, the p-type template layers were further lightly doped with 3×10^{11}/cm^2 phosphorus at 30 keV in designated regions to convert the channel into n-type. The samples were furnace annealed at 950°C for 45 min after the n-type implantation, and another 30 min after p-doping. Since the back side of the SiNMs will be exposed after the transfer, a shorter annealing time for p-type doping is preferred to prevent boron dopants from diffusing into the BOX, thus ensuring a high doping concentration at the Si/SiO$_2$ interface [20]. Both samples were then patterned into strips, and the BOX layers were selectively removed by HF.

The following transfer method was the critical step (Figure 3.19). First, released SiNMs on one of the samples

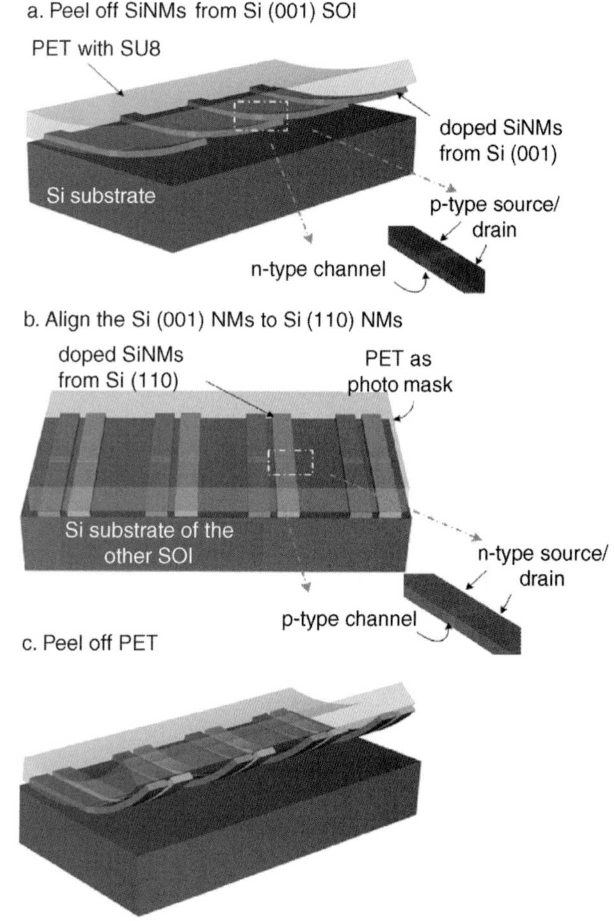

FIGURE 3.19 *Illustration of the transfer process: (a) transfer the p-type doped Si nanomembranes (SiNMs) from silicon-on-insulator (SOI) (001) wafer to polyethylene terephthalate (PET) substrate with SU-8 as glue layer; (b) using the PET as a photo mask, align SiNMs (001) to (110) on the n-type doped SOI (110) wafer; (c) peel off the PET from the substrate.*

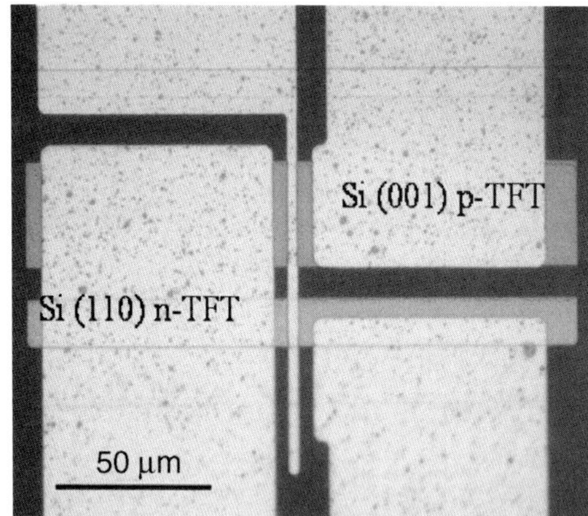

FIGURE 3.20 *Optical microscopic image of a finished hybrid complementary inverter on plastic substrate. The dark regions are the probing stage seen through the clear PET substrate. The top pink strip is the p-thin-film transistor (TFT) from the silicon-on-insulator (SOI) (001) source wafer, and the bottom green strip is n-TFT from SOI (110).*

[Si (001) in Figure 3.19a] were transferred onto the PET substrate with soft SU-8 on top as a glue layer. Next, the transparent plastic host with transferred Si strips was served as a photo mask, and the SiNMs of the other sample [Si (110) in Figure 3.19b] were aligned to it with two channel regions side by side. Then, the second sample was attached to the plastic 'mask' with firm bonding, and other arrays of SiNMs were transferred onto the same PET substrate. The plastic was then peeled off from the Si handling substrate again, with two types of SiNMs aligned next to each other, forming pairs of hybrid-oriented inverters (Figure 3.19c). After the two-time transfer, the SU-8 layer was cross-linked under UV exposure so that the SiNMs could stay on the plastic unsolvable. The gate stacks were electron-beam evaporated at room temperature followed by a standard lift-off procedure, consisting of 120 nm SiO, 30 nm Ti and 120 nm Au. Finally, source/drain metal pads were deposited with Ti/Au of 30 nm/150 nm. Figure 3.20 shows the optical microscope image of the finished complementary inverter. The top pink strip is from p-doped Si (001) while the bottom green one is from n-doped Si (110). Both types of device have gate lengths of 5 µm, while the gate widths were designed to be asymmetric, with 15 µm for n-TFT and 30 µm for p-TFT, to balance the switching speed.

Results and discussion

The DC characteristics of the individual TFTs and hybrid complementary inverters were measured using an Agilent 4155B semiconductor parameter analyzer. Figure 3.21 shows the transfer characteristics and I–V curves of individual n- and p-TFTs with gate width to gate length W/L of 30 µm /5 µm. The channel of n-TFT is along the Si (110)/<110> direction, while the p-TFT is along (001)/<110>. The threshold voltages V_T of n- and p-TFTs are 0.4 V and −0.776 V, respectively. The maximum transconductances g_m reach 16.26 µS for n-TFT and 0.875 nS for p-TFT, at a drain voltage $|V_D|$ = 100 mV. Note that the p-TFTs have less current drive force than those of n-TFTs; this is believed to be due to the unoptimized boron doping and poor interface between the semiconductor and the metal. These initial results indicate the potential to integrate in a hybrid manner the complementary TFTs with different orientations.

Summary

Hybrid-oriented complementary TFTs and inverters made on flexible plastic substrate using both single-crystal Si (001) and (110) together released from two SOI source wafers are described. The fabrication process requires two-time transfer to align the SiNMs from two different source SOI wafers side by side. The hybrid inverter is formed of n-TFT with channel direction along Si (110)/<110>, and p-TFT along Si (001)/<110>. Initial DC performances of individual n- and p-TFTs were investigated, and typical FET characteristics were achieved. The switching speed of the inverters may be further improved by combining n-Si (001)/<110> and p-Si (110)/<110> with the HOT and transfer approaches described in this chapter.

MICROWAVE FLEXIBLE PIN DIODES AND SWITCHES

High-speed flexible TFTs on a low-cost PET substrate, which can serve as the active amplification components needed for microwave frequency applications (e.g. in steerable antenna systems), have already been demonstrated [17, 18, 42], as described in previous sections. For constructing the high-performance integrated circuits, passive components that can be monolithically integrated with these high-frequency active TFTs on a flexible substrate and are also capable of microwave frequency operation are required. This section describes the fabrication and characterization of microwave switches placed on a flexible plastic substrate. Such switches are critical components for many wireless portable devices (such as cell phones) and for radar systems.

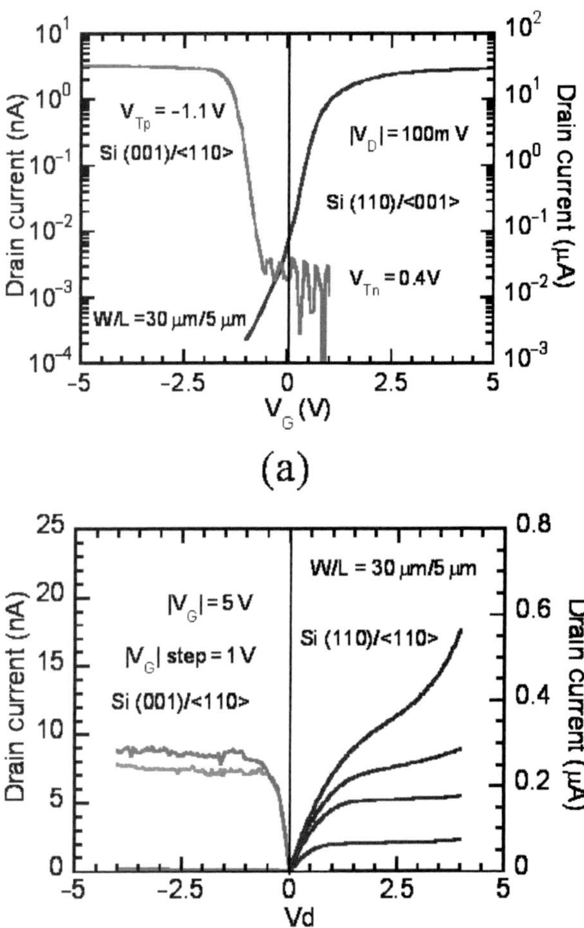

FIGURE 3.21 (a) Transfer and (b) I–V characteristics of individual flexible n- and p-thin-film transistors (TFTs) along Si (110)/<110> and Si (001)/<110>, respectively. The gate lengths and widths are 5 and 30 µm. The threshold voltage is 0.4 V for n-TFT and −0.776 V for p-TFT.

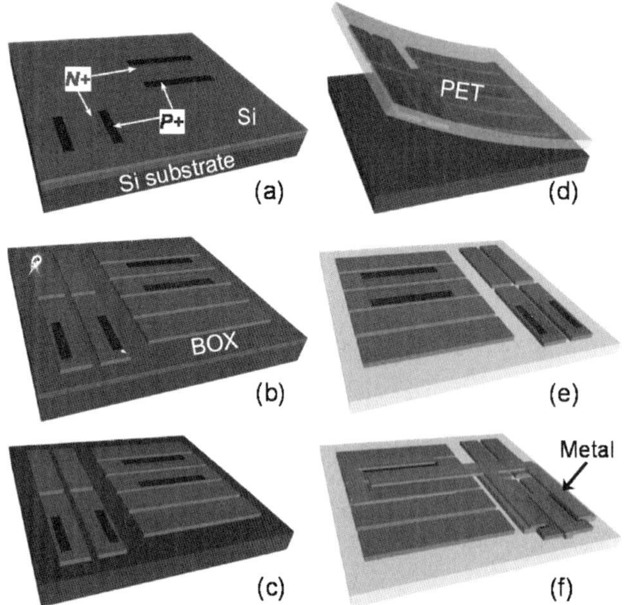

FIGURE 3.22 *Fabrication process illustration for series-shunt P-intrinsic (I, unintentionally doped)-N (PIN) diode radiofrequency (RF) switches. (a) Selective ion implantation for P+ and N+ regions was performed in two photolithography steps on a silicon-on-insulator (SOI) substrate, followed by thermal annealing. (b) The ion implanted top Si layer was then patterned into strips in another photolithography step followed by plasma dry etching. The buried oxide layer was exposed. (c) The SOI was immersed in hydrofluoric acid to release the top Si layer as Si nanomembrane (SiNM). SiNM settled down and became weakly bonded with the handling substrate ('in-place bonding'). (d) A polyethylene terephthalate (PET) substrate with SU-8 spun on was brought face to face with the Si substrate to pick up the SiNM strips due to stronger bonding force of SU-8 with SiNM, completing the flip transfer of SiNM. (e) The PET substrate was flipped with the original bottom side of SiNM being upside. (f) Metallization with electron-beam metal evaporation to complete the fabrication of the PIN diodes and RF switches.*

By using a combination of high- and low-temperature processes on transferable SiNMs to realize low parasitic resistance, high-performance flexible RF switches configured with P-intrinsic (I, unintentionally doped)-N (PIN) diodes with excellent bendability are shown to be achievable.

Device fabrication

The process of transferring SiNM and fabricating the PIN-diode switches is completely compatible with the process used to fabricate microwave TFTs [11, 12]. As shown in Figure 3.22, lightly doped p-type Si (001) UniBond SOI substrate with 200 nm silicon template and 200 nm BOX layers was used as the starting material. Heavily doped N^+ and P^+ regions were formed on designated regions defined by optical photolithography followed by ion implantation of 4×10^{15} cm^{-2} phosphorus at 40 keV, and of 4×10^{15} cm^{-2} boron at 25 keV, respectively (Figure 3.22a). The ion-implanted sample was then annealed in a horizontal furnace at 850°C for 45 min in N_2 ambient to form low contact- and sheet-resistance regions [17, 18]. The low parasitic resistances are critical in realizing high-performance devices (e.g. the insertion loss of the RF switches in this study). The subsequent SiNM release and flip-transfer processes are identical to those described in previous sections. In brief, 30 μm wide strips were formed by patterned dry-etching down to the BOX layer (Figure 3.22b). The BOX layer was then selectively removed by HF (Figure 3.22c). A PET substrate coated with unexposed SU-8 epoxy was then brought into contact with the released SiNM. After the PET substrate was lifted up, the SiNM flip-transfer step was finished (Figure 3.22d). A UV curing step was applied to the

PET to cure the SU-8 [11]. Using this direct flip-transfer method to transfer SiNMs from their original SOI substrate to a foreign host substrate, high-fidelity registration of SiNMs can be achieved.

After flipping the PET substrate (Figure 3.22e), now bearing the selectively ion-implanted strips, the fabrication of the PIN diodes was completed by forming a 40/500 nm Cr/Au metal stack on the predoped N^+ and P^+ regions (Figure 3.22f). As shown in Figure 3.22, unlike the conventional vertical PIN diode structures, the P, I and N regions in this work are arranged laterally on the SiNM. The thinness of the SiNM enables the planar-type structure to have high mechanical flexibility and allows severe bending during implementation. The width of the I region is determined by the separation between photolithographically defined P^+ and N^+ regions and was chosen to be 2 μm in this study to achieve a high-frequency response while maintaining proper breakdown voltages for power handling. Consequently, the active diode area is determined by multiplying the width of the P^+ (or N^+) region with the thickness of the SiNM (200 nm). Figure 3.23(a)

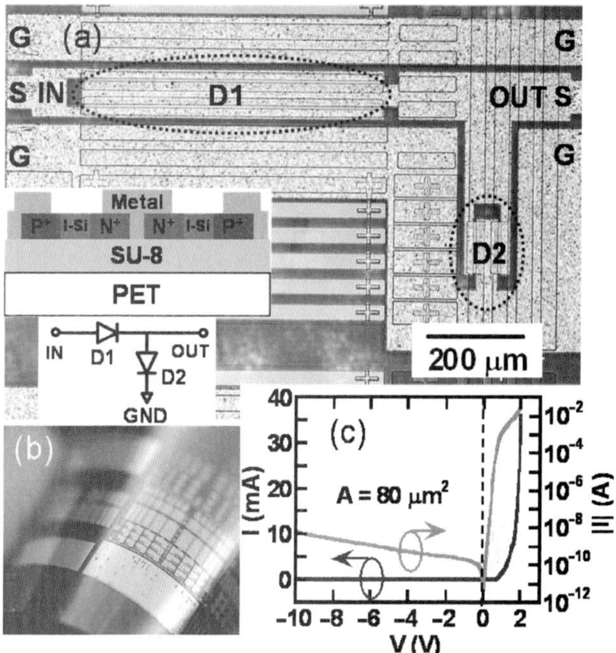

FIGURE 3.23 *(a) Optical microscope image of a finished shunt-series P-intrinsic (I, unintentionally doped)-N (PIN) diode single-pole single-throw switch. The diode area of D1 and D2 is 240 and 40 μm², respectively. The I-region width of all the PIN diodes is 2 μm. The top inset shows the cross-section of a lateral Si nanomembrane (SiNM) PIN diode on flexible polyethylene terephthalate (PET) substrate and the bottom inset shows the circuit schematic of the radiofrequency switch. (b) Optical image of finished PIN diode and switch arrays on a bent PET substrate. (c) Measured typical current–voltage (I–V) curves of an 80 μm² lateral SiNM PIN diode on flexible PET substrate.*

shows an optical microscope image of a finished shunt-series single-pole single-throw (SPST) switch employing two PIN diodes with the schematic cross-section of a lateral SiNM PIN diode; the corresponding circuit diagram is shown in the insets. Figure 3.23(b) shows an image demonstrating the flexibility of the finished PIN diodes and switches on a bent PET substrate.

Radiofrequency switch performance

The DC characteristics of the flexible SiNM PIN diodes were measured with an HP4155B semiconductor parameter analyzer and the RF characteristics of the SPST switches with an Agilent E8364A network analyzer. The *I–V* curves of an $80\,\mu m^2$ SiNM PIN diode on PET at room temperature are shown in Figure 3.23(c). Typical diode rectifying characteristics are demonstrated without apparent breakdown up to 10 V reverse bias. The ideal factor n is extracted to be 1.53. This relatively high value is probably due to high recombination velocity caused by the large periphery-to-area ratio inherent to the lateral PIN diode structure.

The SPST switch (Figure 3.23a) is turned to the ON state when D1 is forward biased (D2 can be zero or reverse biased). Under this state, an RF signal is transmitted from the IN port to the OUT port. The power ratio of the signal between the OUT port and the IN port through the switch is defined as insertion loss (S_{21}, in dB). High-performance RF switches typically require less than 1 dB insertion loss for operation frequencies of interest. When D2 is forward biased (D1 can be zero or reverse biased), the switch is turned to the OFF state. In this case, S_{21} (in dB) is defined as isolation. Isolation larger than 20 dB is generally required for most RF switches.

The ON-state small-signal scattering parameters of the RF switch were characterized with a forward bias current (I_f) of 10, 20 and 30 mA applied on D1. The OFF-state small-signal responses of the RF switch were measured at zero bias on D1 and a forward-bias current of 10 mA (I_2) applied on the shunt PIN diode D2. Figure 3.24(a) presents the measured insertion loss (ON state) and the measured isolation (OFF state) of the RF switch as a function of frequency under different bias conditions. As shown in Figure 3.24(a), lower insertion loss was measured as the bias on D1 increased, owing to the decreased equivalent on-resistance of D1. At 5 GHz, an insertion loss of 0.62 dB was measured at an I_f of 30 mA. At this bias level, as low as 0.93 dB insertion loss was measured up to 20 GHz, indicating the high performance of the RF switch. The high performance of the RF switch at the ON state is ascribed to the single crystal quality of the transferred SiNM and also to the very low parasitic resistance resulting from the effective doping process applied to the SiNM.

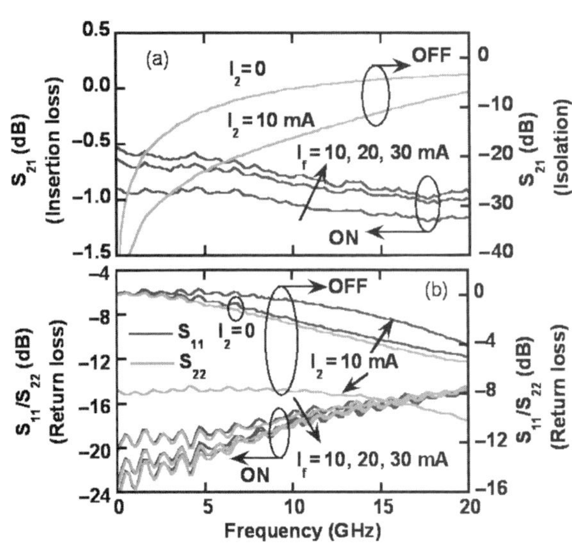

FIGURE 3.24 *(a) Measured insertion loss (ON state) and isolation (OFF state), and (b) return loss of a shunt-series Si nanomembrane P-intrinsic (I, unintentionally doped)-N (PIN) single-pole single-throw switch. The series and shunt PIN diodes have areas of 240 and 40 μm^2, respectively. The ON state is biased at $I_f = 10$, 20 and 30 mA. The OFF state is biased at zero bias on D1 and $I_2 = 10$ mA. (a) and (b) have the same x-axis scale.*

High isolation was also measured from the RF switch under the OFF state. Under zero bias ($I_2 = 0$ mA), 10.8 dB isolation was measured at 5 GHz and the value degraded to 3.4 dB at 20 GHz. Evidently, leaving D2 at zero bias does not provide sufficient isolation for the RF switch. The isolation characteristics were significantly improved when D2 was forward biased ($I_2 = 10$ mA). This is because, under forward bias, D2 has effectively created a short path for the signal from the OUT port to GND that otherwise would be transmitted back to the IN port. With the forward bias of D2, the resulting isolation is substantially improved from 10.8 to 22.9 dB at 5 GHz (Figure 3.24a). Although the isolation degrades quickly with the increase in frequency, at 20 GHz isolation of 6.9 dB was still achieved. Higher isolation may be achieved if the diode area of D2 is increased. However, the high impedance path created by D2 under its zero-bias (or reverse-bias) state may be sacrificed, which would increase the insertion loss under the ON state of the RF switch. As a result, there is a trade-off between the isolation under the OFF state and the insertion loss under the ON state, which can be optimized by properly selecting the relative diode area of D1 and D2.

The measured return loss (ratio between reflected power and input power at IN and OUT ports: S_{11} and S_{22}) of the RF switch is shown in Figure 3.24(b). Under the ON state, both S_{11} and S_{22} are very low and they degrade with the increase in frequency. As the bias applied to D1 is increased, improved return loss is obtained. In addition, S_{11} and S_{22} are largely the same under all the ON-state bias conditions used for D1, which is due to the low insertion loss of the RF switch. Under the OFF state, S_{11} and S_{22} are also about the same when D2 is zero biased ($I_2 = 0$ mA). However, when D2 is forward biased at $I_2 = 10$ mA, enormous discrepancy occurs between S_{11} and S_{22}, with S_{22} significantly improved and S_{11} slightly degraded. These changes correctly reflected the large and important influence of the shunt diode D2.

Summary

High-frequency microwave switches employing flexible single-crystal SiNM based PIN diodes are described in this section. The presented SiNM fabrication process is effective in achieving high-performance passive devices and the process compatibility will permit its further monolithic integration together with high-performance active components on low-cost and low-temperature plastic substrates. A SPST PIN shunt-series diode RF switch achieves insertion loss better than 0.62 dB and isolation higher than 22.9 dB up to 5 GHz, and 0.93 and 6.9 dB, respectively at 20 GHz. The RF switch demonstration has further illustrated the great potential of properly processed transferable single-crystal SiNM for high-frequency flexible electronics applications.

HIGH-FREQUENCY FLEXIBLE PASSIVES

With the technical barriers of using flexible electronics for high-speed applications broken by the introduction of high-mobility transferable single-crystal semiconductor nanomembranes, to implement these microwave flexible TFTs for high-speed applications, high-frequency microwave passive components, such as inductors, capacitors and transmission lines are needed to form functional circuits and systems. Of more critical importance, the passives also need to be mechanically flexible and robust and can be integrated together with the high-speed TFTs on low-temperature substrates. Some discrete inductors and capacitors were made on polyimide substrate and can work in a relatively low frequency range [43–45]. In this section, fabrication and characterization are presented for flexible high-frequency inductors and capacitors monolithically integrated on a PET substrate.

Device fabrication

The process began with optical photolithography on a PET substrate (5 mil). A 30/400 nm Ti/Au metal was deposited on the PET substrate by electron-beam evaporation followed by a lift-off (Figure 3.25a). This metal layer (M1) serves as the bottom electrode of MIM capacitors and the center lead metal of spiral inductors.

In conventional rigid chip-based integrated circuit technology, on-chip capacitors are conveniently implemented with multiple interconnect metal layers and plasma-enhanced chemical vapor-deposited (PECVD) dielectrics in between. The typical PECVD deposition temperature is about 300–350°C. Such a temperature is not suitable for PET substrates, since the Vicat softening temperature of PET is only 170°C. To form the MIM capacitors, a silicon monoxide (SiO) ($\varepsilon_r = 5.8$–10, depending on stoichiometry) layer with 200 nm thickness was electron-beam evaporated in vacuum on the top of the bottom electrode of the capacitors at room temperature. Another metal layer (M2) consisting of Ti/Au metal (30/400 nm) was then

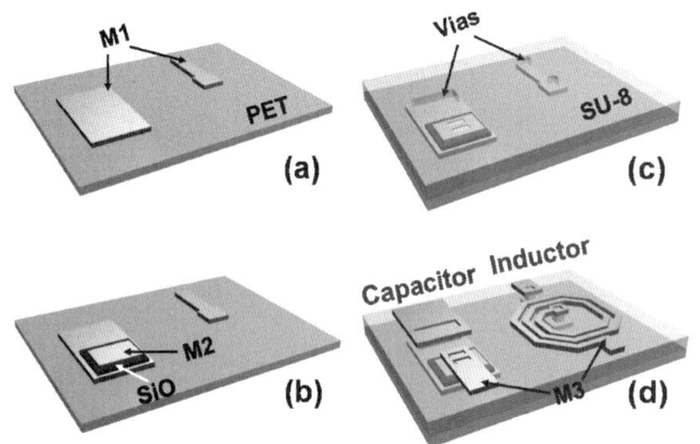

FIGURE 3.25 *Illustration of fabrication process for integrated flexible spiral inductors and metal–insulator–metal (MIM) capacitors. (a) M1 was evaporated on a polyethylene terephthalate (PET) substrate to form the bottom electrode of MIM capacitors and the center lead metal of inductors. (b) A 200 nm SiO layer was evaporated on the top of the bottom electrode as the capacitor high-k dielectric. M2 was evaporated on top of SiO to form the top electrode for capacitors. (c) A layer of 1.8 µm SU-8 was spun on to act as intermetal low-k isolation layer. Via holes were opened with lithography and SU-8 was cured to cross-link. (d) M3 was evaporated to form the spiral metals of inductors and interconnects.*

evaporated on the top of the patterned SiO dielectric layer as the top electrode to form the MIM structure of the capacitors (Figure 3.25b). The size of M2 was designed to be slightly smaller than that of the SiO layer to avoid short-circuiting with the bottom electrode. As a result, the MIM capacitance values are largely decided by the area of the top electrode, with the accuracy of the capacitance also being decided by the SiO thickness.

For the MIM capacitors, a higher dielectric constant (high k) and a smaller thickness are desired for high-density capacitors and thus reduced capacitor size. However, this requirement is contradictory to that for spiral inductors. A smaller intermetal dielectric constant (low k) will result in a smaller parasitic capacitance and thus higher resonance/operation frequencies for inductors. To solve this dilemma, relatively thick ($\sim$1.0 μm) SU-8 (MicroChem) (dielectric constant $\varepsilon_r = 3$) was used as the intermetal dielectric (also very flexible) for the spiral inductors. The SU-8 was spun on the sample surface followed by a photolithography patterning step (exposure for 2.8 s and developing) to open up via holes to access the bottom and the top electrodes of the capacitors and the center lead metal of the inductors. SU-8 was then cured to cross-link for 45 s under UV exposure and 115°C hard baking (Figure 3.25c). Finally, a 30 nm/1 μm Ti/Au top interconnect metal (M3) was evaporated to form the spiral metal of the inductors and to make interconnects (e.g. GSG) for both the inductors and the capacitors (Figure 3.25d). Thick metal was used for M3 to reduce the parasitic resistance (particularly for multiple-turn inductors) of the devices in order to realize high-frequency operations. The octagonal shape of the inductors was intended to reduce the sharp-bending microwave loss at high frequencies.

Overall, three metal layers and two dielectric layers are used to form the integrated passive components on a single PET substrate. The thin and planar structures of the devices make them highly robust to mechanical bending and are also suitable for achieving high-frequency operations. The thermal budget of the entire process was under 120°C (photoresist baking temperature), which can easily satisfy the tolerance of most low-temperature plastic substrates. Of more significance, the fabrication process is completely compatible with that used to fabricate microwave TFTs [17, 18]. Figure 3.26(a) shows optical images of a 4.5-turn inductor and Figure 3.26(b) a 6400 μm² MIM capacitor on a PET substrate. The spiral metal lines of the inductors have a width of 15 μm and a spacing of 4 μm. Figure 3.26(c) shows inductor and capacitor arrays on a bent PET substrate.

Results and discussion

The RF characteristics of the fabricated inductors and capacitors were measured with an Agilent E8364A performance network analyzer (45 MHz to

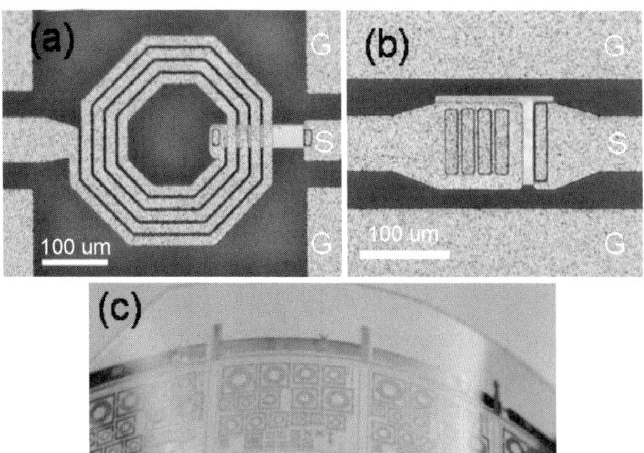

FIGURE 3.26 *Optical microscope images of (a) a 4.5-turn spiral inductor and (b) a 6400 μm² metal–insulator–metal (MIM) capacitor on a polyethylene terephthalate (PET) substrate. (c) An optical image of finished inductor and capacitor arrays on a bent PET substrate.*

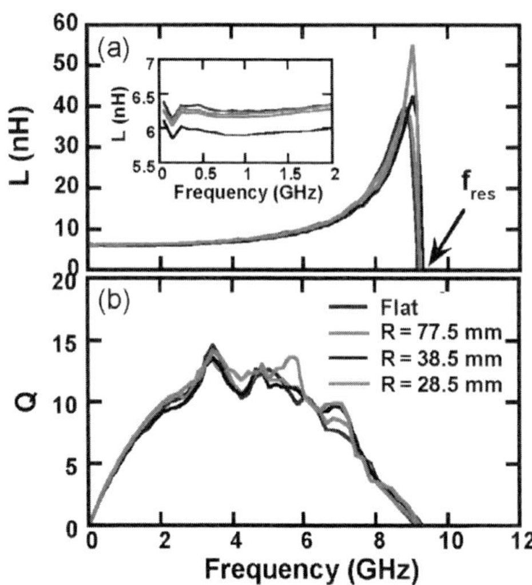

FIGURE 3.27 *Measured (a) L values and (b) Q values of a 4.5-turn spiral inductor as a function of frequency under flat and bending conditions. The spiral metal line width is 15 μm and the metal line spacing is 4 μm. The insert of (a) shows the zoom in of L values in the low frequency range. The f_{res} is indicated by the zero L values. (a) and (b) have the same x-axis scale.*

40 GHz) and the critical passive component parameters, such as inductance/capacitance (L/C) values, quality factor (Q) values and resonant frequency values, were extracted from measured scattering (S) parameters. The capacitance values of the MIM capacitors were also extracted with an HP 4284A RCL meter under low (1 kHz to 1 MHz) frequencies to eliminate the influences of parasitic inductances at high frequencies (e.g. introduced by interconnect metal lines).

The measured L and Q values of a 4.5-turn inductor as a function of frequency are plotted in Figure 3.27(a, b), respectively. A constant L value of ~6 nH was measured from 45 MHz up to ~5 GHz, with an f_{res} of 9.1 GHz. The peak Q of 10 was measured at 4.25 GHz (Figure 3.27b). Such a high Q and a high f_{res} make the spiral inductors implementable for RF circuits up to ~8 GHz, sufficiently meeting the active TFT operation frequency requirement [17].

Figure 3.27 also shows the results of bending tests (along the input-output direction) for the spiral inductors. In general, as the convex bending radius was reduced, the L values at low frequency slightly decrease, while a slight increase of both peak Q and f_{res} was observed. These results indicate that the spiral inductors fabricated on the plastic substrate exhibit very good mechanical robustness. Figure 3.28 shows the L values of the spiral inductors of different number of turns (same geometry as the 4.5-turn inductor) taken at 45 MHz (for a consistent comparison). As can be seen, the spiral inductors scale well with the size.

MIM capacitors with different periphery-to-area ratios (P/A) were also measured. The measured C values (taken from S-parameters) versus frequency for a $40 \times 40\,\mu m$ capacitor under both flat and bending states are shown in Figure 3.29. A value of 0.45 pF at 4 GHz was measured with a f_{res} of 13.5 GHz, even though the interconnect parasitics were not de-embedded. The Q value is measured as 6.4 at 8 GHz. In general, both the values of C and f_{res} were slightly increased as the convex bending radius is decreased. The unit-area capacitance values extracted at low frequencies using the RCL meter exhibit a good linear relation with respect to the capacitors' P/A. The linear relation will facilitate the implementation of the MIM capacitors (of any values) in flexible circuits.

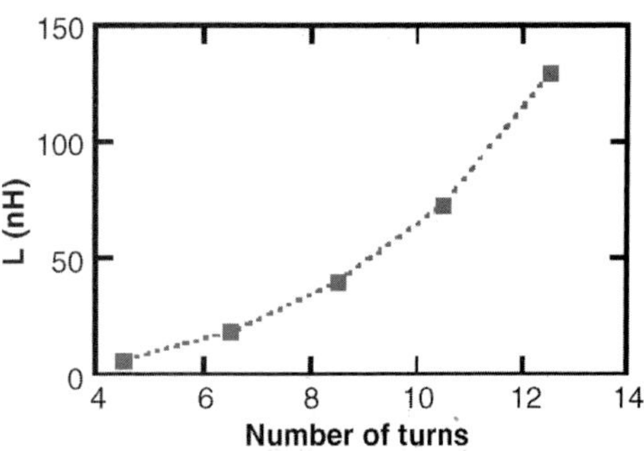

FIGURE 3.28 *Measured L values vs inductor size. The spiral metal line width is 15 µm and the metal line spacing is 4 µm for all inductors. The dashed line is used to guide the view of the data points.*

Summary

In summary, employing a process compatible with microwave TFTs, i.e. low-temperature evaporated SiO as high-k and SU-8 as low-k intermetal dielectric, high-frequency microwave flexible inductors and capacitors are demonstrated on low-temperature plastic substrates. Very high Q and high f_{res} were achieved from the passive devices. A 4.5-turn inductor shows a Q value of 10 and 9.1 GHz f_{res}, and a $1102\,\mu m^2$ capacitor shows 32.5 GHz f_{res} and a Q value of 6.8 at 8 GHz. Robust mechanical characteristics were also exhibited by these planar passives. The devices' characteristic values scale well with their size. Collectively, these passives can fulfill the frequency requirements of flexible RF circuits up to 8 GHz. Further integration of these passive components with high-speed flexible TFTs will eventually lead to flexible RF systems.

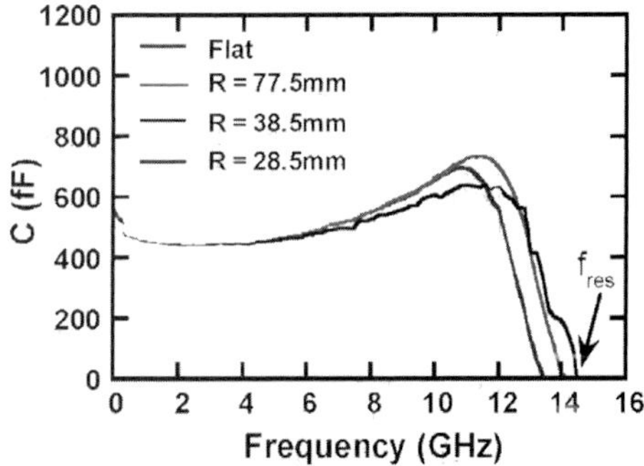

FIGURE 3.29 *Measured capacitance values of a $40 \times 40\,\mu m^2$ metal–insulator–metal (MIM) capacitor as a function of frequency under flat and bending states. The f_{res} is indicated by the zero capacitance values.*

REFERENCES

[1] Menard E, Lee KJ, Khang DY, Nuzzo RG, Rogers JA. A printable form of silicon for high performance thin film transistors on plastic substrates. Appl Phys Lett 2004;84:5398–400.

[2] Menard E, Nuzzo RG, Rogers JA. Bendable single crystal silicon thin film transistors formed by printing on plastic substrates. Appl Phys Lett 2005;86:093507.

[3] Sun Y, Kim S, Adesida I, Rogers JA. Bendable GaAs metal-semiconductor field-effect transistors formed with printed GaAs wire arrays on plastic substrates. Appl Phys Lett 2005;87:083501.

[4] Mack S, Meitl MA, Baca AJ, Zhu ZT, Rogers JA. Mechanically flexible thin-film transistors that use ultrathin ribbons of silicon derived from bulk wafers. Appl Phys Lett 2006;88:213101.

[5] Hoyt JL, Nayfeh HM, Eguchi S, Aberg I, Xia G, Drake T, et al. Strained silicon MOSFET technology. Tech Digest Int Electron Device Meeting 2002;23–6.

[6] Cohen GM, Mooney PM, Paruchuri VK, Hovel HJ. Dislocation-free strained silicon-on-silicon by in-plane bonding. Appl Phys Lett 2005;86:251902.

[7] Mooney PM, Cohen GM, Chu JO, Murray CE. Elastic strain relaxation in free-standing SiGe/Si structures. Appl Phys Lett 2004;84:1093–5.

[8] Cohen GM, Mooney PM, Chu JO. Free standing silicon as a compliant substrate for SiGe. Mater Res Soc Symp Proc 2003;765:141.

[9] Ismail K, LeGoues FK, Saenger KL, Arafa M, Chu JO, Mooney PM, Meyerson BS. Identification of a mobility-limiting scattering mechanism in modulation-doped Si/SiGe heterostructure. Phys Rev Lett 1994;73:3447–50.

[10] Roberts MM, Klein LJ, Savage DE, Slinker KA, Friesen M, Celler G, et al. Elastically relaxed free-standing strained-silicon nanomembranes. Nature Mater 2006;5:388–93.

[11] Schroder DKSemiconductor material and device characterization. 2nd ed. New York: Wiley; 1998.

[12] Luan S, Neudeck GW. An experimental study of the source/drain parasitic resistance effects in amorphous silicon thin film transistors. J Appl Phys 1992;72:766–72.

[13] Yagishita A, King T-J, Bokor J. Schottky barrier height reduction and drive current improvement in metal source/drain MOSFET with strained-Si channel. Jpn J Appl Phys 2004;43:1713–6.

[14] Ahn JH, Kim H-S, Lee KJ, Jeon S, Kang SJ, Sun Y, et al. Heterogeneous three-dimensional electronics by use of printed semiconductor nanomaterials. Science 2006;314:1754–7.

[15] Ahn J-H, Kim H-S, Menard E, Lee KJ, Zhu Z, Kim D-H, et al. Bendable integrated circuits on plastic substrates by use of printed ribbons of single-crystalline silicon. Appl Phys Lett 2007;90:213501.

[16] Kim D-H, Ahn J-H, Kim H-S, Lee KJ, Kim T-H, Yu C-J, et al. Complementary logic gates and ring oscillators on plastic substrates by use of printed ribbons of single-crystalline silicon. IEEE Electron Device Lett 2008;29:73–6.

[17] Yuan H-C, Celler GK, Ma Z. 7.8-GHz flexible thin-film transistors on a low-temperature plastic substrate. J Appl Phys 2007;102:034501.

[18] Yuan H-C, Ma Z. Microwave thin-film transistors using Si nanomembranes on flexible polymer substrate. Appl Phys Lett 2006;89:212105.

[19] Pang H, Yuan H-C, Lagally MG, Celler GK, Ma Z. Flexible microwave single-crystal Si TFTs with fmax of 5.5 GHz. IEEE 65th Device Research Conference Digest 2007. p. 15–6.

[20] Stolk PA, Gossmann H-J, Eaglesham DJ, Jacobson DC, Rafferty CS, Gilmer GH, et al. Physical mechanism of transient enhanced dopant diffusion in ion-implanted silicon. J Appl Phys 1997;81:6031–50.

[21] de Oliviera RM, Dalponte M, Boudinov H. Electrical activation of arsensic implanted in silicon on insulator (SOI). J Phys D Appl Phys 2007;40:5227–31.

[22] Bock K. Polymer electronics systems-polytronics. IEEE Proc 2005;93:1400–5.

[23] Garnier F, Hajlaoui R, Yassar A, Srivastava P. All-polymer field-effect transistor realized by printing technique. Science 1994;265:1682–6.

[24] Drury CJ, Mutsaers CMJ, Hart CM, Matters M, de Leeuw DM. Low-cost all-polymer integrated circuits. Appl Phys Lett 1998;73:108–10.

[25] Voss D. Cheap and cheerful circuits. Nature 2000;407:422–4.

[26] Baude PF, Ender DA, Haase MA, Kelley TW, Muyres DV, Theiss SD. Pentacene-based radio-frequency identification circuitry. Appl Phys Lett 2003;82:3964–6.

[27] Gelinck GH, Huitema HEA, van Veenendaal E, Cantatore E, Schrijnemakers L, Van der Putten JBPH, et al. Flexible active-matrix displays and shift registers based on solution-processed organic transistors. Nat Mater 2004;3:106–10.

[28] Chen Y, Au J, Kazlas P, Ritenour A, Gates H, McCreary M. Flexible active-matrix electronic ink display. Nature 2003;423:136.

[29] Reuss RH, Chalamala BR, Moussessian A, Kane MG, Kumar A, Zhang DC, et al. Macroelectronics: perspectives on technology and applications. Proc IEEE 2005;93:1239–56.

[30] Lee M-C, Han S-M, Kang S-H, Shin M-Y, Han M-K. Tech Digest. Int Electron Devices Meetings 2003;215.

[31] Inoue S, Utsunomiya S, Saeki T, Shimoda T. Surface-free technology by laser annealing (SUFTLA) and its application to poly-Si TFT-LCDs on plastic film with integrated drivers. IEEE. Trans Electron Devices 2002;49:1353–60.

[32] Tilke A, Rotter M, Blick RH, Lorenz H, Kotthaus JP. Single-crystalline silicon lift-off films for metal-oxide-semiconductor devices on arbitrary substrates. Appl Phys Lett 2000;77:558–60.

[33] Zhu Z-T, Menard E, Hurley K, Nuzzo RG, Rogers JA. Spin on dopants for high-performance single-crystal silicon transistors on flexible plastic substrates. Appl Phys Lett 2005;86:133507.

[34] Yuan H-C, Ma Z, Roberts MM, Savage DE, Lagally MG. High-speed strained-single-crystal-silicon thin-film transistors on flexible polymers. J Appl Phys 2006;100:013708.

[35] Ma Z, Mohammadi S, Lu L-H, Bhattacharya P, Katehi LPB, Alterovitz SA, Ponchak GE. An X-band high-power amplifier using SiGe/Si HBT and lumped passive components. IEEE Microwave Wireless Components Lett 2001;11:287–9.

[36] Greiner JH, Kircher CJ, Klepner SP, Lahiri SK, Warnecke AJ, Basavaiah S, et al. Fabrication process for Josephson integrated-circuits. IBM J Res Develop 1980;24:195–205.

[37] Hartman TE, Blair JC, Bauer R. Electrical conduction through SiO films. J Appl Phys 1966;37:2468–74.

[38] Raskin JP, Dambrine G, Gillon R. Direct extraction of the series equivalent circuit parameters for the small-signal model of SOI MOSFETs. IEEE Microwave Guided Lett 1997;7:408–10.

[39] Dambrine G, Cappy A, Heliodore F, Playez E. A new method for determining the FET small-signal equivalent circuit. IEEE Trans Microwave Theory Tech 1988; 36:1151–9.

[40] Yang M, Chan VWC, Chan KK, Shi L, Fried DM, Stathis JH, et al. Hybrid-orientation technology (HOT): opportunities and challenges. IEEE Trans Electron Devices 2006;53:965.

[41] Yuan H-C, Ma Z, Ritz CS, Savage DE, Lagally MG, Celler GK. Complementary single-crystal silicon TFTs on plastic. ECS Trans 2007;6:139–44.

[42] Sun Y, Menard E, Rogers JA, Kim HS, Kim S, Chen G, et al. Gigahertz operation in flexible transistors on plastic substrates. Appl Phys Lett 2006;88:183509.

[43] Lenihan T, Schaper L, Shi Y, Morcan G, Parkerson J. Embedded thin film resistors, capacitors and inductors in flexible polyimide. Electronic Components and Technology Conference; 1996. p. 119.

[44] Waffenschmidt E, Ackermann B, Wille M. Integrated ultra thin flexible inductors for low power converters. Power Electronics Specialists Conference; 2005. p. 1528–34.

[45] Lim HC, Zunino J, Federici JF. Flexible membrane LRC strain sensor fabricated using MEMS method. Sensors Transducers J 2008;91:39–46.

Thin Films of Single-Walled Carbon Nanotubes for Flexible Electronic Device Applications

Congjun Wang[1] and Qing Cao[2]

[1]National Energy Technology Laboratory, US Department of Energy, Pittsburgh, Pennsylvania, USA, and Parsons Project Services Inc., South Park, Pennsylvania, USA

[2]IBM TJ Watson Research Center, Yorktown Heights, New York, USA

INTRODUCTION

Single-walled carbon nanotubes (SWNTs) have attracted substantial research interest during the past decade, owing to their unique quasi-one-dimensional (1D) structure, nanometer size, as well as exceptional electrical, mechanical, optical, thermal and chemical properties [1–4]. Among those attributes, their mechanical and electrical properties are most noticeable. Nanotubes are among the strongest fibers in the world, with a Young's modulus in the range of 1~2 TPa and fracture stress of ~50 GPa, as determined from SWNT bundles as well as arrays of isolated SWNTs [5–8]. Depending on their chirality and diameter, SWNTs can behave as either a metallic material, i.e. a material with zero band gap (E_g), or a semiconductor material. SWNTs demonstrate very high mobility, ~10 000 cm^2/V·s, and ballistic transport at room temperature [9, 10]. Also, as nanometer size wires, they can withstand very high current density [11]. Therefore, semiconducting SWNTs (s-SWNTs) are envisioned as a very promising material candidate to replace or complement single-crystalline silicon (Si) in the next generation high-performance nanoelectronic devices, while metallic SWNTs (m-SWNTs) are proposed to replace copper wire as advanced interconnects in microprocessor chips. Great progress has been made in these directions, as exemplified by a molecular circuit based on an individual SWNT developed at IBM in 2006 and an integrated circuit (IC) chip with nanotubes as interconnects developed at Stanford

CONTENTS

Semiconductor Nanomaterials for Flexible Technologies

105

University and Hitachi in 2008 [12, 13]. However, since we still lack an effective method to control the structure and position of each nanotube in synthesis/high-yield large-scale assembly, it remains a daunting challenge to scale up to any realistic system.

Utilizing SWNT thin films, which are composed of a large number of nanotubes in the form of either random networks or aligned arrays, instead of individual SWNTs to construct electronic systems has the potential to avoid many scaling-up challenges. Since many nanotubes are involved in electron transport in such nanotube films, they can offer favorable statistics to minimize device to device variations even with a collection of electrically heterogeneous SWNTs, as the property of each device is relatively insensitive to the position/property of an individual nanotube in the film. Moreover, the device output can be adjusted through varying the geometry of nanotube films, offering necessary flexibility in circuit/system design. In the meantime, in an optimized layout, such as well-aligned nanotube arrays, these SWNT films can demonstrate properties approaching those of individual SWNTs [14]. Therefore, many believe that nanotubes in the thin-film format are the most realistic path to practical integrated electronic systems. Furthermore, their attractive mechanical properties may enable new applications with requirements on both electrical performances and mechanical robustness, like flexible/stretchable electronics. This chapter summarizes the progress made in this field. The first section reviews the controlled preparation of SWNT films. Then, experimental and theoretical work on understanding electron transport in such nanotube films is described, followed by examples of their various applications in electronic devices, e.g. flexible/transparent thin-film transistors (TFTs) and digital/analog circuits. Finally, the future perspectives of this field are discussed.

PREPARATION OF SINGLE-WALLED CARBON NANOTUBE THIN FILMS

Preparation of SWNT thin films with tuneable density (D, number of SWNTs within a unit area in the case of random network films or number of SWNTs within a unit length for aligned arrays) and coverage from submonolayer to multilayers on various substrates is essential to understanding the basic physics and tuning their properties for practical applications. In addition, the overall spatial layouts of the SWNTs, their lengths as well as their orientations need to be controlled, because the physical and electronic properties of SWNT films critically depend on these parameters. Furthermore, exquisite control of the diameters of the SWNTs and even the distribution of semiconducting and metallic SWNTs within a given sample can lead to drastic improvement in achieving desirable electronic properties. Finally, SWNT thin-film fabrication techniques should be compatible with large-scale processing to be cost-competitive and economically viable.

Thin films of SWNTs can be formed by either the deposition of SWNTs dispersed in solution onto solid substrates [15–17] or the direct growth of SWNT films using chemical vapor deposition (CVD) methods [18, 19]. SWNT films formed by CVD can also be easily transferred to a variety of substrates by a dry transfer process. Despite its potential of low-cost processing and compatibility with a wide range of substrates, solution deposition of SWNTs suffers from the fact that these procedures typically require harsh steps, such as high-power ultrasonication and strong acid treatment, which can severely degrade their electronic properties and their average lengths. In contrast, the CVD method provides SWNTs with high structural integrity and longer lengths which can result in enhanced electronic properties of the SWNT thin films. More importantly, CVD synthesis enables unprecedented control over a wide range of topological parameters which is unlikely to be achieved by solution deposition. The controllable growth of SWNTs by CVD is discussed in this section.

Chemical vapor deposition growth of SWNT random networks

With CVD, the control of D is achieved by the selection of catalysts, feed gas, catalyst concentration, growth time, and so on. For instance, high-density random network SWNT films with D in the range of several hundred SWNTs/μm^2 are obtained by using ethanol as a carbon source and Fe/Co/Mo catalyst loaded on silica nanoparticle supports (Figure 4.1a) [20, 21]. Because of the high D, m-SWNTs, which comprise one-third of all the SWNTs, are fully interconnected as a percolating network. These films therefore exhibit metallic properties and serve as conducting layers, which are referred to as metallic carbon nanotube networks (m-CNNs). Despite the high tube density, such m-CNNs can be used as transparent conductors because of their high transmittance in the visible spectral region and low sheet resistance (R_s)[22]. Alternatively, the use of Fe catalysts and methane as the feed gas can lead to the formation of SWNT films with intermediate to low D (Figure 4.1b). Using Fe nanoparticles deposited from different concentrations of ferritin solution as catalysts and methane as feed gas, D of SWNT films can be precisely tuned in a certain range (Figure 4.1c, d) [18]. These films display semiconducting properties because D is only high enough to allow percolation transport through pathways that involve at least one semiconducting SWNT (s-SWNT). Such semiconducting carbon nanotube networks (s-CNNs) can be incorporated into functional electronic devices as the active semiconductors, best

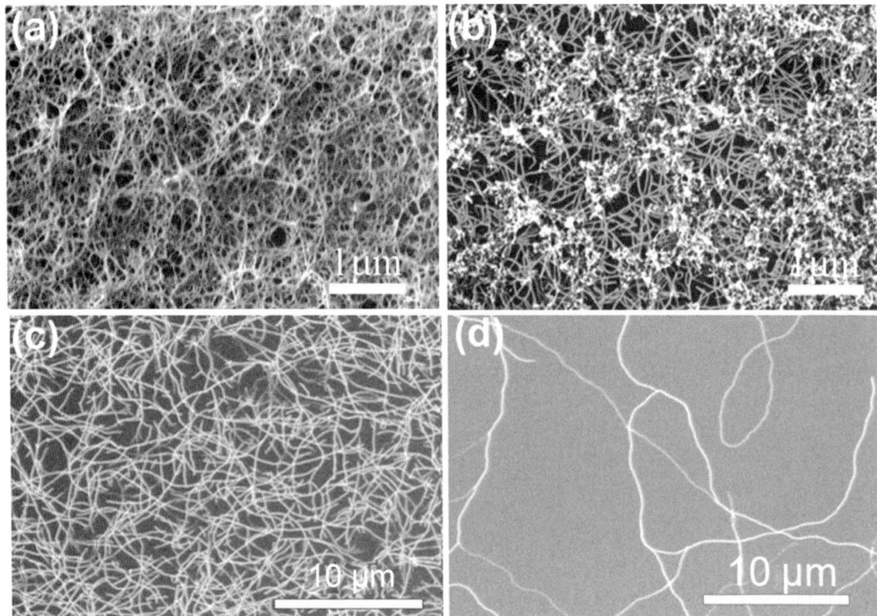

FIGURE 4.1 *Scanning electron microscopy images of single-walled carbon nanotube (SWNT) random network synthesized by chemical vapor deposition with (a) silica nanoparticles loaded with Fe/Co/Mo as catalyst and ethanol as feeding gas, (b) silica nanoparticles loaded with Fe/Co/Mo as catalyst and methane as feeding gas, (c) Fe nanoparticles deposited from concentrated ferritin solution and methane as feeding gas, (d) Fe nanoparticles deposited from diluted ferritin solution and methane as feeding gas.*

positioned to compete with amorphous Si for display applications and organic semiconductors for flexible electronics especially designed for applications with demanding requirements on mechanical robustness and compatibility with harsh environmental conditions [23, 24].

Growth of SWNT aligned arrays

Besides D, control over the orientation or alignment of SWNTs is necessary because the electronic transport in SWNT thin films is sensitive to tube–tube junctions. For example, if a s-SWNT and a m-SWNT cross, the m-SWNT can at least partially shield the s-SWNT from the gate field. The contact barriers at SWNT–SWNT junctions also inhibit transport, leading to degraded and inconsistent performances [25, 26]. Aligned SWNT arrays can minimize or eliminate the possibility of the formation of SWNT–SWNT junctions and take full advantage of the extraordinary intrinsic properties of SWNTs. Careful selection of certain crystalline substrates has enabled use of the orientationally anisotropic interactions between SWNTs and the substrates to achieve large-scale growth of aligned SWNTs [27–32]. The degree of the alignment is dependent on surface quality, and can be adjusted by various parameters including the annealing time and catalyst density [27]. Nearly perfect alignment on a wafer scale can be accomplished by confining catalysts in small areas such that the growth of SWNTs occurs in regions of the substrate where there is no interference from catalyst particles (Figure 4.2) [14]. The linearity of these perfectly aligned tubes is within ~5 nm over many micrometers and the deviation from parallelism is less than 0.1 degree (Figure 4.2d). The density of such aligned arrays can be varied from $2 \sim 5$ tubes/μm^2 to $25 \sim 50$ tubes/μm^2 [28, 33]. These aligned SWNT arrays are best suited for high-performance

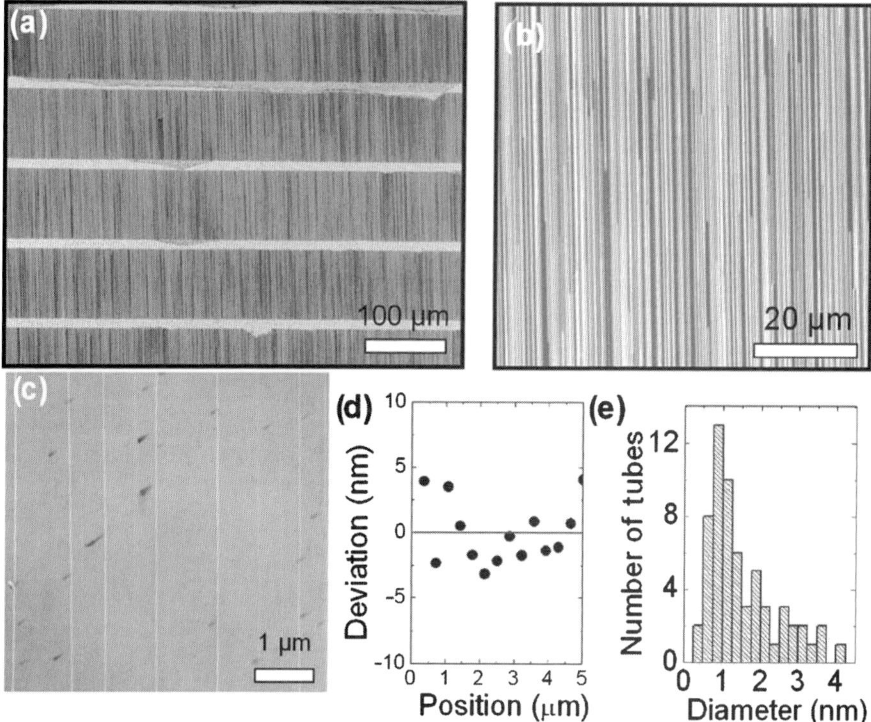

FIGURE 4.2 *Characterizations of aligned single-walled carbon nanotube (SWNT) arrays grown on single-crystalline quartz substrates. (a) SEM image of aligned SWNT arrays grown over a large area. The horizontal bright lines correspond to prepatterned catalyst regions. Magnified SEM image (b) and AFM image (c) of high-density SWNT arrays. (d) Deviation of the center of a SWNT along its axis from a perfect line. (e) Histogram showing the diameter distribution of aligned SWNTs synthesized by substrate-guided chemical vapor deposition growth.*

electronics applications, such as analog radiofrequency (RF) devices, where high D, linear configurations and a complete absence of SWNT–SWNT junctions are indispensable.

Forming SWNT films on flexible substrates

A major disadvantage of CVD methods is that they are typically carried out at high temperatures, usually above 600°C. This requirement severely limits the choice of compatible substrates. Several methods have been developed to transfer SWNTs from the growth substrate to a wide range of surfaces such as plastic sheets for flexible, low-cost electronics [34–38]. Transfer printing techniques therefore can avoid the limitations associated with the high-temperature CVD growth of SWNTs by separating the synthesis from materials that would not survive under the growth conditions. One such method uses polydimethylsiloxane (PDMS) stamps to remove SWNT films from a growth substrate such as SiO_2/Si after hydrofluoric acid (HF) etching of the oxide [34]. This approach is simple and maintains the excellent electronic properties of SWNTs during and after the transfer process. The high efficiency of the technique is evidenced by the fact that the values of D on the receiving surface are almost identical to those on the original growth substrate. Similar methods that use carrier films, which adhere strongly to the SWNTs and may serve as plastic substrates for subsequent device/circuit fabrication, can transfer tubes directly without undercut etching (Figure 4.3a) [35]. Such techniques can be used, for example, to transfer aligned SWNTs grown on quartz (Figure 4.3b). In both cases, a metal film, which is subsequently removed by wet-etching after transfer, can be applied on top of SWNT films to bind them together during the printing processes. Multiple transfer steps can enable further control of D and the SWNT layouts (Figure 4.3c) [35].

MODELING PROPERTIES OF SINGLE-WALLED CARBON NANOTUBE FILMS

Stick-percolation modeling of SWNT films

Understanding the fundamental physics of charge transport through SWNT films is critical to interpreting and optimizing the performance when they are used as the functional components of next generation electronics. The classical percolation theory only describes homogeneous infinite networks. For application in transistors, the electronic heterogeneity of the SWNTs, their anisotropic alignment and the finite extent of the thin films require the development of non-linear, finite-size percolation models, for predictive assessment of the properties [39–41]. The key geometric parameters for

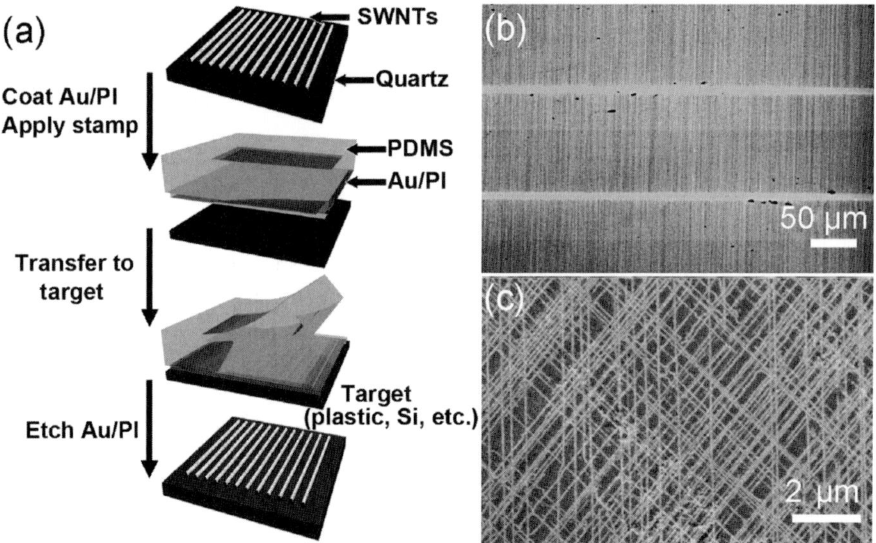

FIGURE 4.3 *Transfer printing of single-walled carbon nanotube (SWNT) films. (a) Schematic showing the printing process to transfer SWNT film grown by chemical vapor deposition onto other substrates, e.g. plastic substrates. (b) SEM image of aligned SWNT arrays transferred on plastic substrates. (c) SEM image of SWNT triangle structures formed through two consecutive printing processes.*
(Reproduced with permission from [35] © 2007 American Chemical Society.)

such modeling include D, average tube length (or stick length, L_S), channel length L_C, and transistor channel width (W) or the width of the strips defined in SWNT networks (W_S), where D and L_S can be extracted by digitizing scanning electron microscope (SEM) images of nanotubes. The degree of nanotube alignment can be then described with an anisotropic parameter, R, defined as $R = {L_\parallel}/{L_\perp} = \sum_{i=1}^{N} |L_{S,i} \cos\theta_i| / \sum_{i=1}^{N} |L_{S,i} \sin\theta_i|$, where θ_i is the angle between each nanotube and the overall alignment direction. The simulated nanotube film can be formed by populating a two-dimensional (2D) grid, whose dimension is defined by L_C, W and W_S, with sticks according to measured L_S, D and R. The electrical properties of the nanotube film can be calculated through self-consistently solving drift-diffusion equation, Poisson equation and current continuity equation [42, 43]. This simulation tool can successfully describe the charge transport in nanotube films with dramatically different topological parameters. Figure 4.4 shows that for nanotube films with L_S ranging from 5 to 40 µm, and R from 2.9 to 21.4, simulation results (shown as lines) clearly predict the unique scaling behavior of TFTs based on various nanotube thin films, and agree well with experimental data (shown as symbols). This result demonstrates that the

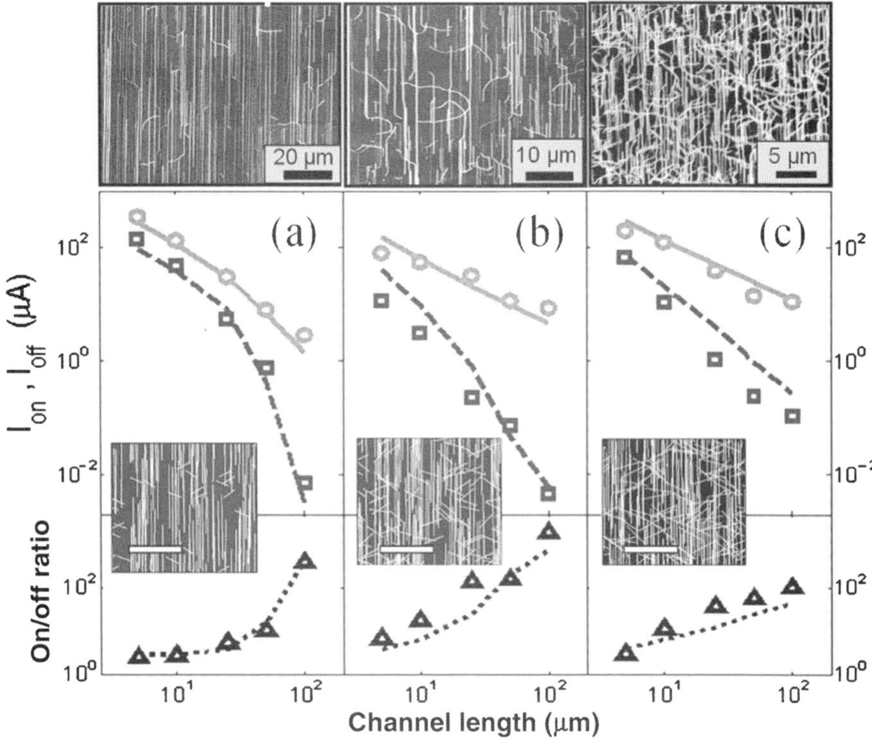

FIGURE 4.4 *Percolative modeling of measured (symbols) and simulated (lines) properties, i.e. on-state current (I_{on}), off-state current (I_{off}) and on/off ratio, of thin-film transistors (TFTs) built on aligned (a), partially aligned (b), and high tube density partially aligned (c) single-walled carbon nanotube (SWNT) films. Top parts: SEM images of respective SWNT films. Inset: Images of simulated SWNT films used in modeling. Scale bar: 10 μm.*
(Reproduced with permission from [42] © 2007 American Chemical Society.)

charge transport in nanotube film can be described quantitatively, enabling us to facilitate the optimization of layout design and properties of nanotube thin films with the help of numerical simulation capability, which will be further illustrated in the next section [44, 45].

Theory and practice of striping scheme to improve device on/off ratio

It has been shown that the topology of SWNT thin films can significantly affect their electrical properties, and their correlation can be successfully predicted by heterogeneous percolative modeling. Therefore, it is possible to

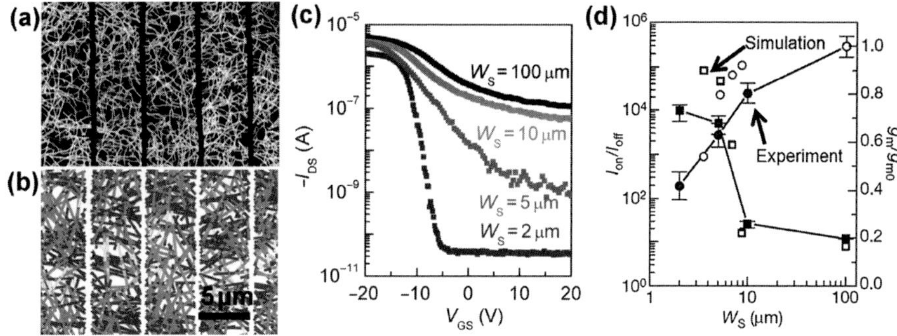

FIGURE 4.5 *'Striping' scheme to enhance device on/off ratios. (a) SEM image of a single-walled carbon nanotube (SWNT) random network after striping. (b) Image of simulated current density in a striped SWNT film (yellow: high; red: medium; blue: low). (c) I_{DS}–V_{GS} characteristics of a set of SWNT thin-film transistors (TFTs) with identical device geometries (W = L = 100 µm), but different strip width (W_S), which changes from 100 µm, 10 µm, 5 µm, 2 µm from top to bottom. (d) Measured (filled symbols) and simulated (hollow symbols) device parameters (left axis, black: device on/off ratio; right axis, blue: normalized transconductance, where g_{mo} represents the response without strip) as a function of W_S for a set of devices with L = 100 µm.*
(Reproduced with permission from [46] © 2008 Nature Publishing Group.)

better engineer the electrical properties of SWNT films by modifying their geometries. For example, the influence of m-SWNT contaminates in a s-CNN can be minimized by patterning an anisotropic SWNT film into an array of parallel strips using a top–down method (Figure 4.5a, b) [46]. Device results show that there is a dramatic increase in device on/off ratio but only a small decrease in device on current with decreasing W_S (Figure 4.5c). This behavior can be understood based on percolation theory [44]. When the nanotube films are cut into strips, the effective L_S is reduced. Therefore, the percolation threshold (N_C), which is defined by L_S according to $L_S \sqrt{\pi N_C} = 4.236$, is increased. N_C is the critical D of SWNT films, and if D is less than N_C the film is not continuous. Since m-SWNTs only account for 33% of tubes in as-synthesized SWNT films, this striping scheme allows the device on/off ratio to be increased without any purification process through eliminating the presence of purely m-SWNT pathways between device source/drain (S/D) electrodes by setting N_C higher than m-SWNT density but less than the overall SWNT density. Meanwhile, their influence on device current can be minimized if this striping scheme is implemented in optimized geometry. The full details of striped systems can be captured with heterogeneous percolative modeling (Figure 4.5d) [44]. The good

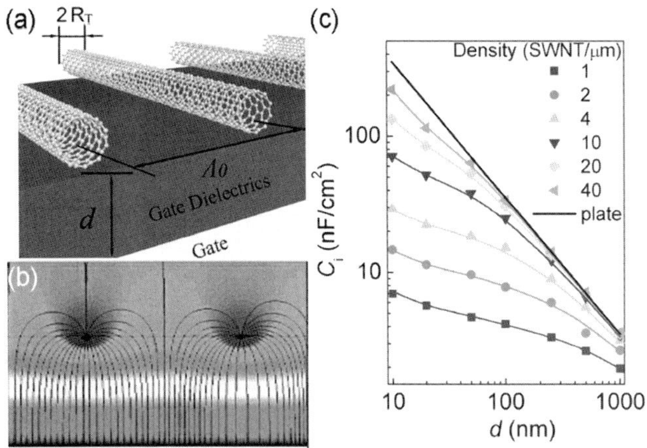

FIGURE 4.6 *Capacitance coupling between single-walled carbon nanotube (SWNT) aligned array and a planar electrode. (a) Schematic showing the model system used. R_T: nanotube radius; Λ_0: average tube–tube distance; d: dielectric thickness. (b) Electrical potential distribution simulated with finite-element method. The black lines depict field lines. (c) Capacitance per unit area as a function of dielectric thickness calculated using either analytical formula (lines) or finite element method simulations (symbols), with different nanotube densities.*

agreement between simulation results and experimental data suggests that one can rely on numerical simulation to design W_S of each TFT in a circuit, offering the possibility to uniformly obtain high performance and high on/off ratio devices without any additional process to enrich s-SWNTs or remove m-SWNTs.

Capacitance coupling of SWNT films

The capacitance coupling between SWNT film and the gate electrode in transistors can be quite different from that of conventional metal oxide semiconductor field-effect transistors (MOS-FET) owing to the 1D structure of each nanotube and the 2D mesh structure of the nanotube film [47, 48]. A simplified model system, a parallel array of evenly spaced SWNTs, can help in the semiquantitative analysis of capacitance coupling in SWNT TFTs (Figure 4.6a). The major geometric parameters are nanotube diameter (R_T), dielectric thickness (d) and average distance between neighboring nanotubes (Λ_0). Finite element simulation suggests that the electrical field distribution is different from that of a standard parallel plate capacitor owing to the presence of fringing field and electrical screening between each nanotube in the arrays (Figure 4.6b). An analytical expression is proposed to estimate the gate capacitance (C_i):

$$C_i = \left(\frac{\frac{2}{\varepsilon} \log \frac{\Lambda_0}{R_T} \sin^{\pi} 2d/\Lambda_0}{\pi} + C_Q^{-1} \right)^{-1} \Lambda_0^{-1}$$

where C_Q is quantum capacitance. For a wide range of Λ_0 and d, the formula yields results very similar to finite element simulations. The validity is further confirmed with direct experimental measurements performed on SWNT TFTs [47]. Clarifying gate capacitance coupling in SWNT TFTs is critical for comparing mobility values obtained from devices with different SWNT film topology and dielectrics. It is also critical for estimating the transient behavior of devices and circuits based on SWNT TFTs.

SINGLE-WALLED CARBON NANOTUBE FILM AS A FLEXIBLE TRANSPARENT CONDUCTIVE COATING

The use of SWNT films as conducting materials appears to be a simple idea. However, the overall electrical properties are intricately dependent on a variety of parameters such as the average tube length and diameter, deposition method, concentration of m-SWNTs and adventitious doping from the ambient [49, 50]. For conductive films, long and relatively large-diameter SWNTs, which can minimize the role of SWNT–SWNT junctions in transport and the bandgap of s-SWNTs, are desired [51, 52]. Ideally, the deposition method should enable the assembly of uniform films at high throughput on any substrate, with precise control of D. Several solution deposition approaches and transfer printing techniques have demonstrated attractive capabilities of forming large-area transparent conductive SWNT films on various substrates [15, 17, 21, 49]. A 50 nm thick film covering a 4 inch diameter wafer with sheet resistance $(R_S) < 100 \ \Omega/sq$ and transmittance $>70\%$ in the visible spectral range is shown in Figure 4.7(a). R_S of this m-CNN is comparable to the properties of indium–tin oxide films with similar thickness [49]. The conductance can be further improved by doping s-SWNTs with strong acid/oxygen or by hybridizing with Au nanoparticles [53–55]. Films prepared with m-SWNTs enriched by ultracentrifugation can exhibit sheet resistances as much as 10 times smaller than those of films made with identical procedures using unsorted SWNTs [56]. In addition, films comprised of m-SWNTs with narrow diameter distributions display a colored appearance, demonstrating the possibility for applications in conductive optical filters [56].

Combining the conductive SWNT films with semiconducting SWNT films has enabled the realization of flexible transparent transistors using SWNTs for all of the current-carrying layers [21, 57]. An optical image of an array of this type of all-tube transparent TFTs is shown in Figure 4.7(c). These devices are fabricated by sequential transfer printing of CVD nanotube networks with different D onto a plastic sheet. High D m-CNNs with high conductance form the S/D and gate electrodes, while moderate D s-CNNs function as the semiconductor. The optical transmittance, even in the most opaque S/D electrodes areas, is above 75%, comparable to some of

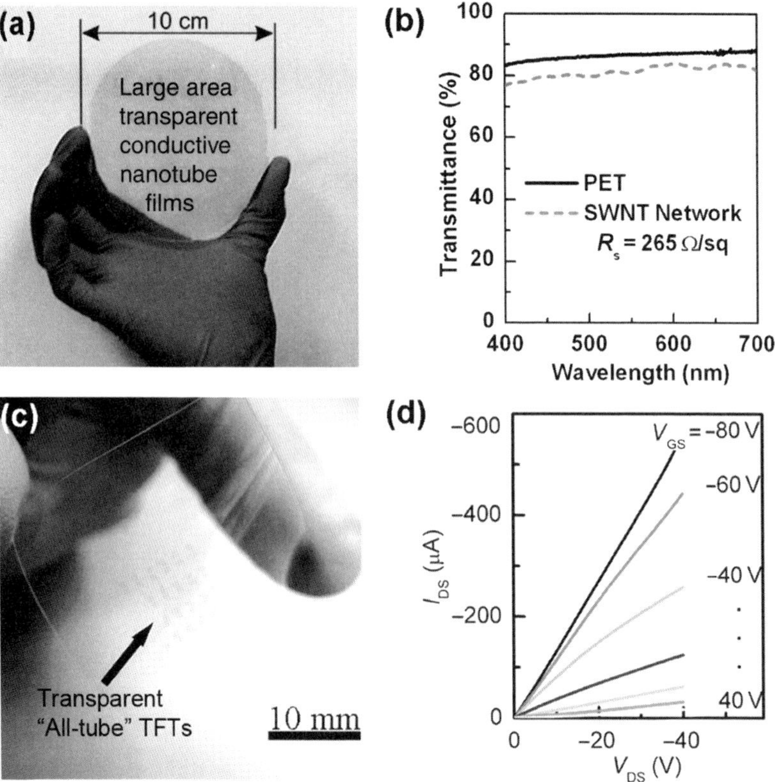

FIGURE 4.7 *Optical image, transmittance and electrical performances of optically transparent devices based on single-walled carbon nanotube (SWNT) films. (a) Optical image of a transparent conductive SWNT film uniformly coated on a quartz substrate. (Reproduced with permission from [49] © 2004 AAAS.) (b) Optical transmittance of an SWNT film, whose sheet resistance is ~265 Ω/sq, coated on a Mylar substrate. PET: polyethylene terephthalate. (c) Optical image of an array of transparent thin-film transistors (TFTs) using nanotube films as only electrical active materials. (d) I_{DS}–V_{DS} characteristics of transparent all-tube TFTs.*
(Reproduced with permission from [21] © 2006 Wiley-VCH.)

the best transparent transistors based on oxides. These all-tube transparent devices exhibit very attractive electrical properties, with effective mobilities of ~30 cm^2/V·s, similar to or somewhat larger than those of typical amorphous semiconducting oxides (Figure 4.7d) [58]. These performance attributes suggest potential use in applications that are more advanced than switching transistors in active-matrix liquid crystal displays. When combined with mechanically robust elastomeric dielectrics, the devices can withstand tensile strains up to 3.5% without degradation. Beyond this limit, the dielectrics fail, but the SWNT films remain conductive/semiconducting.

FLEXIBLE SINGLE-WALLED CARBON NANOTUBE THIN-FILM TYPE ELECTRONIC DEVICES

Gate dielectrics for flexible SWNT thin-film transistors

Conventional microfabrication techniques or printing approaches can be applied to SWNT films on plastic to form flexible devices and circuits (Figure 4.8a). An important component of SWNT TFTs is the gate dielectrics. Ideally, the dielectrics should offer high capacitances for low-voltage and hysteresis-free operation, together with low leakage current densities for power efficiency. Deposition methods that are compatible with flexible plastic substrates can also be critical, depending on the application. Certain classes of 3D cross-linked organic multilayers ($\sim$16 nm) formed by room-temperature self-assembly processes are attractive, because of the large capacitances ($\sim$170 nF/cm^2), excellent insulating properties (leakage current densities less than 10^{-9} A/cm^2) and smooth surface morphologies [59]. A different approach uses inorganic oxides (2–5 nm) prepared by atomic layer deposition (ALD), with spin-cast cross-linked epoxies ($\sim$10 nm) on top to serve as adhesive layers for transfer printing, if necessary. Such bilayer nanodielectrics, similar to organic multilayers, provide high capacitance (up to $\sim$330 nF/cm^2) as well as low leakage current density, interface charge density, interface state density and dissipation factors [60]. These high-capacitance dielectrics also greatly reduce the subthreshold swing (S) of

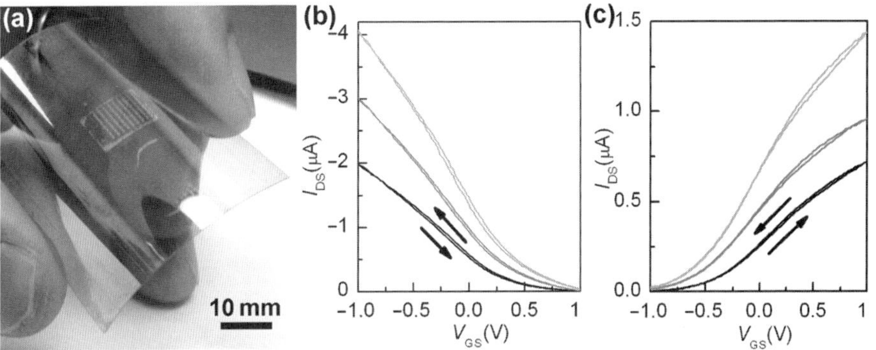

FIGURE 4.8 Optical image (a) and I_{DS}–V_{GS} characteristics of both p-channel (b) and n-channel (c) single-walled carbon nanotube thin-film transistors on plastic substrates.

SWNT TFTs, which enables operation at voltages even lower than those that would be inferred from the differences in capacitance. Therefore, hysteresis for both p-channel and n-channel devices built on bilayer nanodielectrics or organic multilayers is much smaller than that of devices on more widely explored thick oxide or polymer dielectrics, possibly owing to a reduction in the electrical fields near the SWNTs as a result of lower operating voltage and fewer traps in dielectrics (Figure 4.8b, c) [59, 60].

Integrated circuits based on flexible SWNT thin-film transistors

Despite numerous achievements in optimizing various aspects of isolated SWNT TFTs, these devices are only of practical value when integrated into circuits. Since the polarity of SWNTs can be controlled by charge-transfer doping methods, a complementary metal-oxide semiconductor (CMOS) type inverter, which represents an important element in digital circuits, can be constructed by connecting a p-channel and an n-channel bottom-gate devices, doped by oxygen and polymers, respectively (Figure 4.9a) [59–61]. With high-capacitance gate dielectrics to enhance the transconductance and S, voltage gains approaching 10 can be achieved, which is comparable to single-tube inverters based on local bottom-gated devices (Figure 4.9b) [60, 62]. Similar circuits, such as CMOS NAND gates, have also been recently demonstrated [63, 64].

For more complex structures/functions, top-gate configuration is adopted for these devices to facilitate multilayer interconnects. A p-channel metal oxide semiconductor (PMOS) inverter, the building block for PMOS-type logic circuits, is fabricated on polyimide substrates with two separately addressable SWNT TFTs [46]. The static transfer characteristics can be successfully predicted by simulations. The voltage gain is much larger than unity, and can thus be used to drive subsequent logic gates without losing logic integrity. The ability to achieve functional logic gates, such as inverters, NOR gate, NAND gate, which

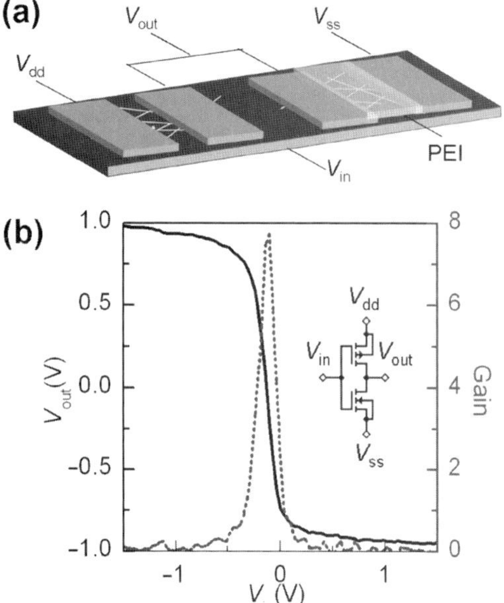

FIGURE 4.9 *Schematic (a), circuit diagram (b, inset), and static transfer characteristics (b) of a complementary metal-oxide semiconductor (CMOS) inverter composed of a p-channel and an n-channel single-walled carbon nanotube thin-film transistor. PEI: polyethylenimine.*
(Reproduced with permission from [60] © 2006 Wiley-VCH.)

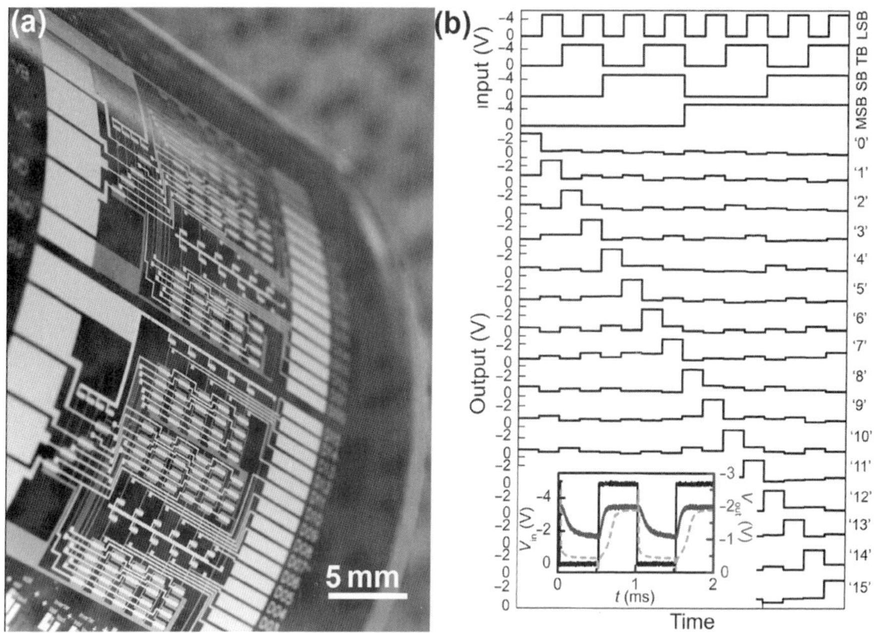

FIGURE 4.10 *(a) Optical image and (b) input–output characteristics of a four-bit row decoder circuit composed of 88 single-walled carbon nanotube thin-film transistors on plastic substrates. In (b), the first four traces, labeled most significant bit (MSB), second bit (SB), third bit (TB) and least significant bit (LSB), are input lines; the remaining traces, labeled from 0 to 15, show output voltages of 16 individual output lines. Inset: Measured (blue solid line) and simulated (red dashed line) dynamic response (right axis) of one output line with 1 kHz square wave input (black solid line, left axis).*
(Reproduced with permission from [46] © 2008 Nature Publishing Group.)

are building blocks for digital ICs, opens the way to realize circuits with more complex functions. Figure 4.10(a) shows an optical image of a chip featuring a four-bit decoder composed of 88 SWNT TFTs, which represents the largest SWNT circuit achieved to date [46]. This circuit can successfully decode a binary-encoded input of four data bits into 16 individual data output lines, where one output is enabled, depending on whether the encoded value corresponds to the data line number (Figure 4.10b). Because of the high mobility of the SWNT thin films, these decoder circuits can successfully operate in the kilohertz regime, even with critical dimensions ($\sim$100 μm) that are sufficiently coarse to be patterned by techniques such as screen printing (Figure 4.10b, inset). This attribute is important for their potential applications in low-cost, printed electronics.

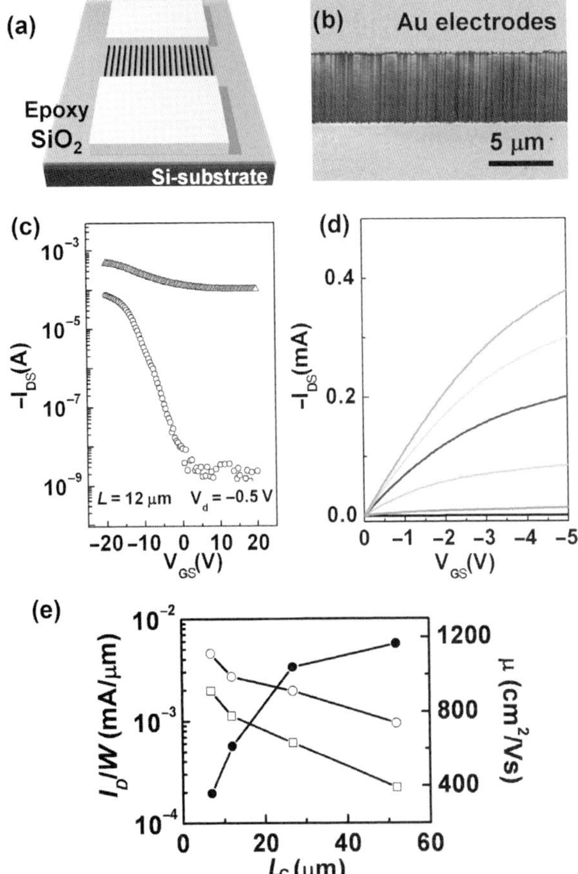

FIGURE 4.11 *Image and characteristics of thin-film transistors (TFTs) based on well-aligned single-walled carbon nanotube (SWNT) arrays. (a) Schematic showing the structure of a back gate TFT with transfer-printed SWNT aligned array as channel material. (b) SEM image of the device channel region. (c) I_{DS}–V_{GS} characteristics of a TFT before (top line) and after (bottom line) selective electrical breakdown, and its corresponding (d) I_{DS}–V_{DS} characteristics. (e) Channel width normalized on-state current (hollow symbols, left axis) and effective mobility (filled symbols, right axis) as a function of L_C for TFTs based on aligned SWNT arrays. (Reproduced with permission from [14] © 2007 Nature Publishing Group.)*

HIGH-PERFORMANCE DEVICES BASED ON ALIGNED SINGLE-WALLED CARBON NANOTUBE ARRAYS

SWNT thin-film transistors based on aligned arrays

Compared with s-CNNs, aligned arrays provide dramatic improvements in device mobility, owing to the long, straight geometries of the tubes and the absence of SWNT–SWNT junctions (Figure 4.11a, b) [14]. In all cases with aligned array devices, L_C is much smaller than the average tube length such that each tube bridges the S/D electrodes directly. As with the s-CNN, p-channel operation is typically observed in aligned array TFTs with high work function contact metals, with the possibility for ambipolar or n-channel behavior via the use of doping techniques [14]. The effective device mobility, extracted at long L_C, is ~1000 cm²/V·s when computed in the linear regime with a simple parallel-plate approximation for the capacitance coupling to the gate; this value is a 10-fold improvement over the mobility of similar devices that use s-CNNs. The per-tube mobilities, defined as values that consider only the capacitance of the s-SWNTs, computed in a rigorous manner, can exceed 2000 cm²/V·s. This value is only somewhat lower than expectation based on measurements on devices that incorporate single SWNTs (~3000 cm²/V·s), for this range of diameters [28]. Compared to devices based on s-CNNs, especially with optimized strip structures, those based on aligned arrays have small on/off ratios (typically in a range between 2 and 5) even for long L_C, and their on/off ratios cannot be increased with striping schemes because each SWNT directly bridges the

S/D electrodes and one-third of them are m-SWNTs. Selective electrical break-down can be used to increase the on/off ratio up to 10^5 (Figure 4.11c, d) [14]. In this process a large current is passed through the m-SWNTs while the p-channel s-SWNTs are switched off through applying a positive gate voltage. This procedure, which irreversibly eliminates transport pathways through metallic channels in the device, can also be applied to devices built on s-CNNs, although the process tends to be more reproducible with the arrays, owing perhaps to their relatively simple layouts [65]. Unlike s-CNN devices, the effective mobility decreases with L_C for aligned arrays (Figure 4.11e), suggesting that the contact resistance can be a significant fraction of the overall resistance even for devices with L_C in the micrometer range, mainly because of the lowered channel resistances in this case compared to that of the random network devices. Scaling analysis, similar to that described for the network devices, shows that the gate voltage modulates mainly the channel and has comparably small influence on the contact resistance for devices with L_C values in the micrometer range [14]. Chemical doping techniques or the use of certain metal carbides between the SWNT and S/D electrodes, as demonstrated in single tube devices, may help to reduce this contact resistance [66, 67].

Frequency response of SWNT thin-film transistors

SWNTs exhibit room-temperature carrier mobilities, particularly the hole mobility, higher than those typical of single-crystalline Si. SWNTs also have low intrinsic capacitance and, unlike many other materials envisioned for next generation electronics and devices, they are compatible with standard MOSFET structures [68, 69]. Furthermore, recent work suggests the possibility of highly linear operation as high-frequency amplifiers [70]. These properties indicate that SWNT arrays may eventually be competitive with Si or III–V semiconductors in next generation high-frequency devices, with the possibility for operation in the terahertz regime [71]. Transistors that use aligned arrays have much lower device output impedance than single tube devices, thereby allowing evaluation of scattering parameters (S-parameters) directly with a network analyzer to determine their frequency response [72]. Figure 4.12(a) shows current gain (H_{21}) and maximum available power gain (G_{max}), derived from the S-parameters for an SWNT array device with $L_C = 4\ \mu m$ and a 50 nm thick high-k gate dielectric. The S/D electrode layouts match conventional signal–ground–signal microwave probes (Figure 4.12a, inset). These results demonstrate unity current gain frequency f_t and unity power gain frequency f_{max} of 2.5 GHz and 1.1 GHz, respectively. L_C scaling studies indicate that cut-off frequencies can be further increased following

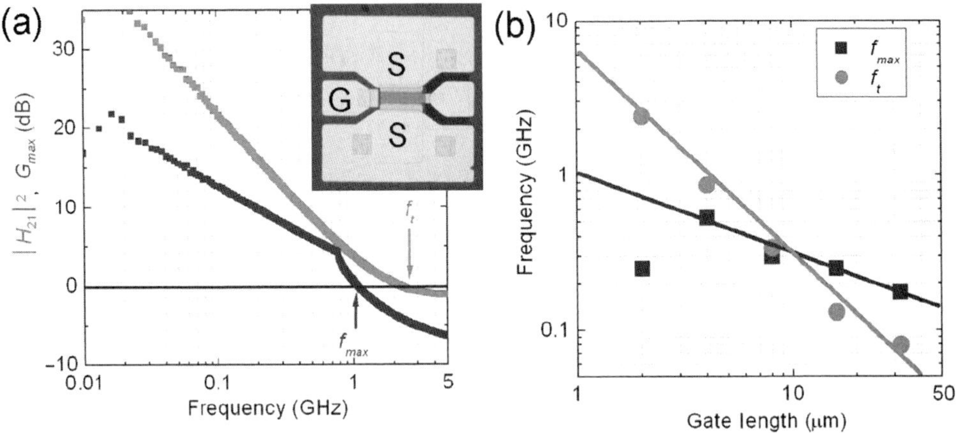

FIGURE 4.12 *Frequency responses of thin-film transistors (TFTs) based on aligned single-walled carbon nanotube (SWNT) arrays. (a) Current gain ($|H_{21}|^2$) and power gain (G_{max}) as a function of frequency for a SWNT TFT with L_C of 4 µm. (b) $|H_{21}|^2$ and G_{max} as a function of L_C for SWNT TFTs based on aligned nanotube arrays.*
(Reproduced with permission from [72] © 2008 National Academy of Science, USA.)

direct geometric scaling routes (Figure 4.12b). With further reduction of L_C to submicrometer regime, ~700 nm, f_{max} can be as high as 10 GHz [73]. Such results, as obtained with devices that have relatively low D, ~5 tubes/µm, and long L_C, above ~700 nm, suggest promise for achieving much higher speeds in optimized devices. In addition, the frequency behavior can be successfully reproduced by use of conventional small-signal models, with remarkably good agreement between measured and modeled results over two decades of frequency up to 50 GHz [73]. These outcomes illustrate the suitability of established modeling tools to guide future device designs.

SWNT transistor radio: demonstration of a functional system

Analog circuits represent key components of communications and other systems in widespread, growing commercial use. High-speed transistors are essential to the operation of such circuits. The modest complexity of the analog circuits (e.g. hundreds of devices), and the relaxed requirements on on/off ratio and other aspects, make SWNT array transistors well suited to this application area, especially novel flexible analog devices. As described above, RF transistors based on SWNT arrays have been demonstrated at frequencies up to 10 GHz, with relatively coarse device dimensions. They can therefore be further configured as RF power amplifiers and oscillators, two of the most important building blocks of analog electronic systems [72]. As an example, these devices can then be connected and impedance matched to generate functional radio receiving systems, where the SWNT TFTs provide all of the

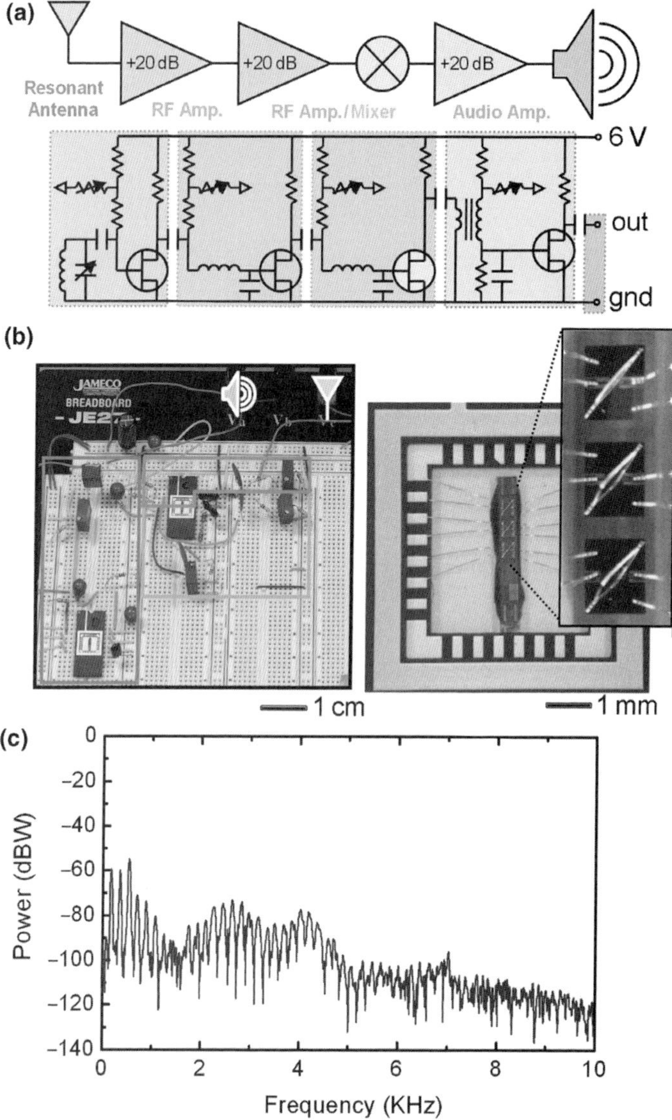

FIGURE 4.13 *(a) Block and circuit diagram, (b) optical image, and (c) output power spectrum of a radio system using single-walled carbon nanotube thin-film transistors as all of the active components.*
(Reproduced with permission from [72] © 2008 National Academy of Science, USA.)

active components, even including the audio output amplifier stage (Figure 4.13a). An optical image of a completed SWNT radio is shown in Figure 4.13(b) and the SWNT chips are connected to an external antenna and a speaker through wire bonding. This relatively simple system, and others like it, is fully functional and capable of receiving, demodulating and amplifying signals broadcast by commercial radio stations. Figure 4.13(c) presents the power spectrum recorded during a weather/traffic report that is directly audible through headphones connected to the system.

CONCLUSION

Individual SWNTs offer tremendous opportunities to understand electronic properties in the 1D regime. For realistic applications that can take advantage of the extraordinary electronic, mechanical and optical properties of SWNTs, SWNT thin films offer the best potential, especially in emerging areas such as next generation flexible and multifunctional electronics. Compared with existing technology and other potential materials envisioned for similar applications, SWNT films exhibit exceptional and unique properties, and at the same time are compatible with conventional semiconductor device materials and fabrication techniques. Transparent flexible and/or large area devices have been fabricated using SWNT films. The potential of using aligned SWNT arrays for high-frequency applications is also demonstrated. Future effort will be focused on improving the electronic homogeneity of SWNTs by either postsynthesis treatment or the development of electronically selective growth strategies. Controlled doping of s-SWNTs and enhanced deposition methods will also be investigated. How to overcome these challenges quickly and effectively may ultimately determine the success of SWNTs in the competition with other materials. Nonetheless, the many possibilities arising from the intrinsic extraordinary properties of SWNTs for both fundamental research and breakthrough technologies will undoubtedly keep SWNTs as fascinating as they have been for years to come.

ACKNOWLEDGMENT

Reference in this work to any specific commercial product is to facilitate understanding and does not necessarily imply endorsement by the U.S. Department of Energy.

REFERENCES

[1] Avouris P. Molecular electronics with carbon nanotubes. Acc Chem Res 2002;35:1026–34.

[2] Avouris P, Chen J. Nanotube electronics and optoelectronics. Mater Today 2006;9(10):46–54.

[3] Ouyang M, Huang JL, Lieber CM. Fundamental electronic properties and applications of single-walled carbon nanotubes. Acc Chem Res 2002;35:1018–25.

[4] Charlier JC, Blase X, Roche S. Electronic and transport properties of nanotubes. Rev Mod Phys 2007;79:677–732.

[5] Yu MF, Files BS, Arepalli S, Ruoff RS. Tensile loading of ropes of single wall carbon nanotubes and their mechanical properties. Phys Rev Lett 2000;84:5552–5.

[6] Treacy MMJ, Ebbesen TW, Gibson JM. Exceptionally high Young's modulus observed for individual carbon nanotubes. Nature 1996;381:678–80.

[7] Khang DY, Xiao JL, Kocabas C, MacLaren S, Banks T, Jiang HQ, et al. Molecular scale buckling mechanics on individual aligned single-wall carbon nanotubes on elastomeric substrates. Nano Lett 2008;8:124–30.

[8] Walters DA, Ericson LM, Casavant MJ, Liu J, Colbert DT, Smith KA, Smalley RE. Elastic strain of freely suspended single-wall carbon nanotube ropes. Appl Phys Lett 1999;74:3803–5.

[9] Zhou XJ, Park JY, Huang SM, Liu J, McEuen PL. Band structure, phonon scattering, and the performance limit of single-walled carbon nanotube transistors. Phys Rev Lett 2005;95:146805.

[10] Javey A, Guo J, Wang Q, Lundstrom M, Dai H. Ballistic carbon nanotube field-effect transistors. Nature 2003;424:654–7.

[11] Yao Z, Kane CL, Dekker C. High-field electrical transport in single-wall carbon nanotubes. Phys Rev Lett 2000;84:2941–4.

[12] Chen ZH, Appenzeller J, Lin YM, Sippel-Oakley J, Rinzler AG, Tang JY, et al. An integrated logic circuit assembled on a single carbon nanotube. Science 2006; 311:1735.

[13] Close GF, Yasuda S, Paul B, Fujita S, Wong HSP. A 1 GHz integrated circuit with carbon nanotube interconnects and silicon transistors. Nano Lett 2008;8:706–9.

[14] Kang SJ, Kocabas C, Ozel T, Shim M, Pimparkar N, Alam MA, et al. High-performance electronics using dense, perfectly aligned arrays of single-walled carbon nanotubes. Nat Nanotechnol 2007;2:230–6.

[15] Meitl MA, Zhou YX, Gaur A, Jeon S, Usrey ML, Strano MS, Rogers JA. Solution casting and transfer printing single-walled carbon nanotube films. Nano Lett 2004;4:1643–7.

[16] Park JU, Meitl MA, Hur SH, Usrey ML, Strano MS, Kenis PJA, Rogers JA. In situ deposition and patterning of single-walled carbon nanotubes by laminar flow and controlled flocculation in microfluidic channels. Angew Chem Int Ed 2006;45:581–5.

[17] Hu L, Hecht DS, Grüner G. Percolation in transparent and conducting carbon nanotube networks. Nano Lett 2004;4:2513–7.

[18] Li YM, Kim W, Zhang YG, Rolandi M, Wang DW, Dai HJ. Growth of single-walled carbon nanotubes from discrete catalytic nanoparticles of various sizes. J Phys Chem B 2001;105:11424–31.

[19] Franklin NR, Dai HJ. An enhanced CVD approach to extensive nanotube networks with directionality. Adv Mater 2000;12:890–4.

[20] Zhang GY, Mann D, Zhang L, Javey A, Li YM, Yenilmez E, et al. Ultra-high-yield growth of vertical single-walled carbon nanotubes: hidden roles of hydrogen and oxygen.. Proc Natl Acad Sci USA 2005;102:16141–5.

[21] Cao Q, Hur S-H, Zhu Z-T, Sun Y, Wang C, Meitl MA, et al. Highly bendable, transparent thin film transistors that use carbon nanotube based conductors and semiconductors with elastomeric dielectrics. Adv Mater 2006;18:304–9.

[22] Gruner G. Carbon nanotube films for transparent and plastic electronics. J Mater Chem 2006;16:3533–9.

[23] Snow ES, Novak JP, Lay MD, Houser EH, Perkins FK, Campbell PM. Carbon nanotube networks: nanomaterial for macroelectronic applications. J Vac Sci Technol B 2004;22:1990–4.

[24] Cao Q, Rogers JA. Ultrathin films of single-walled carbon nanotubes for electronics and sensors: a review of fundamental and applied aspects. Adv Mater 2009;21:29–53.

[25] Fuhrer MS, Nygard J, Shih L, Forero M, Yoon YG, Mazzoni MSC, et al. Crossed nanotube junctions. Science 2000;288:494–7.

[26] Odintsov AA. Schottky barriers in carbon nanotube heterojunctions. Phys Rev Lett 2000;85:150–3.

[27] Kocabas C, Hur SH, Gaur A, Meitl MA, Shim M, Rogers JA. Guided growth of large-scale, horizontally aligned arrays of single-walled carbon nanotubes and their use in thin-film transistors. Small 2005;1:1110–6.

[28] Kocabas C, Kang SJ, Ozel T, Shim M, Rogers JA. Improved synthesis of aligned arrays of single-walled carbon nanotubes and their implementation in thin film type transistors. J Phys Chem C 2007;111:17879–86.

[29] Kocabas C, Shim M, Rogers JA. Spatially selective guided growth of high-coverage arrays and random networks of single-walled carbon nanotubes and their integration into electronic devices. J Am Chem Soc 2006;128:4540–1.

[30] Ismach A, Kantorovich D, Joselevich E. Carbon nanotube graphoepitaxy: highly oriented growth by faceted nanosteps. J Am Chem Soc 2005;127:11554–5.

[31] Ismach A, Segev L, Wachtel E, Joselevich E. Atomic-step-templated formation of single wall carbon nanotube patterns. Angew Chem Int Ed 2004;43:6140–3.

[32] Han S, Liu XL, Zhou CW. Template-free directional growth of single-walled carbon nanotubes on a- and r-plane sapphire. J Am Chem Soc 2005;127: 5294–5.

[33] Ding L, Yuan DN, Liu J. Growth of high-density parallel arrays of long single-walled carbon nanotubes on quartz substrates. J Am Chem Soc 2008;130: 5428.

[34] Hur S-H, Park OO, Rogers JA. Extreme bendability in thin film transistors that use carbon nanotubes transferred from high temperature growth substrates. Appl Phys Lett 2005;86:243502.

[35] Kang SJ, Kocabas C, Kim HS, Cao Q, Meitl MA, Khang DY, Rogers JA. Printed multilayer superstructures of aligned single-walled carbon nanotubes for electronic applications. Nano Lett 2007;7:3343–8.

[36] Bradley K, Gabriel J-CP, Grüner G. Flexible nanotube electronics. Nano Lett 2003;3:1353–5.

[37] Hines DR, Mezhenny S, Breban M, Williams ED, Ballarotto VW, Esen G, et al. Nanotransfer printing of organic and carbon nanotube thin-film transistors on plastic substrates. Appl Phys Lett 2005;86:163101.

[38] Liu XL, Han S, Zhou CW. Novel nanotube-on-insulator (NOI) approach toward single-walled carbon nanotube devices. Nano Lett 2006;6:34–9.

[39] Alam MA, Pimparkar N, Kumar S, Murthy J. Theory of nanocomposite network transistors for macroelectronics applications. MRS Bull 2006;31:466–70.

[40] Kumar S, Murthy JY, Alam MA. Percolating conduction in finite nanotube networks. Phys Rev Lett 2005;95:066802.

[41] Kumar S, Pimparkar N, Murthy JY, Alam MA. Theory of transfer characteristics of nanotube network transistors. Appl Phys Lett 2006;88:123505.

[42] Kocabas C, Pimparkar N, Yesilyurt O, Kang SJ, Alam MA, Rogers JA. Experimental and theoretical studies of transport through large scale, partially aligned arrays of single-walled carbon nanotubes in thin film type transistors. Nano Lett 2007;7:1195–202.

[43] Pimparkar N, Cao Q, Kumar S, Murthy JY, Rogers J, Alam MA. Current–voltage characteristics of long-channel nanobundle thin-film transistors: a 'bottom–up' perspective. IEEE Electron Device Lett 2007;28:157–60.

[44] Pimparkar N, Cao Q, Rogers JA, Alam MA. Theory and practice of 'striping' for improved on/off ratio in carbon nanonet thin film transistors. Nano Res 2009;2:167–75.

[45] Pimparkar N, Kocabas C, Kang SJ, Rogers J, Alam MA. Limits of performance gain of aligned CNT over randomized network: theoretical predictions and experimental validation. IEEE Electron Device Lett 2007;28:593–5.

[46] Cao Q, Kim HS, Pimparkar N, Kulkarni JP, Wang CJ, Shim M, et al. Medium-scale carbon nanotube thin-film integrated circuits on flexible plastic substrates. Nature 2008;454:495–500.

[47] Cao Q, Xia MG, Kocabas C, Shim M, Rogers JA, Rotkin SV. Gate capacitance coupling of singled-walled carbon nanotube thin-film transistors. Appl Phys Lett 2007;90:023516.

[48] Wunnicke O. Gate capacitance of back-gated nanowire field-effect transistors. Appl Phys Lett 2006;89:083102.

[49] Wu Z, Chen Z, Du X, Logan JM, Sippel J, Nikolou M, et al. Transparent, conductive carbon nanotube films. Science 2004;305:1273–6.

[50] Li ZR, Kandel HR, Dervishi E, Saini V, Xu Y, Biris AR, et al. Comparative study on different carbon nanotube materials in terms of transparent conductive coatings. Langmuir 2008;24:2655–62.

[51] Kaempgen M, Duesberg GS, Roth S. Transparent carbon nanotube coatings. Appl Surf Sci 2005;252:425–9.

[52] Zhang DH, Ryu K, Liu XL, Polikarpov E, Ly J, Tompson ME, Zhou CW. Transparent, conductive, and flexible carbon nanotube films and their application in organic light-emitting diodes. Nano Lett 2006;6:1880–6.

[53] Kong BS, Jung DH, Oh SK, Han CS, Jung HTSingle-walled carbon nanotube gold nanohybrids: application in highly effective transparent and conductive films. J Phys Chem C 2007;111:8377–82.

[54] Geng HZ, Kim KK, So KP, Lee YS, Chang Y, Lee YH. Effect of acid treatment on carbon nanotube-based flexible transparent conducting films. J Am Chem Soc 2007;129:7758–9.

[55] Geng HZ, Kim KK, Song C, Xuyen NT, Kim SM, Park KA, et al. Doping and dedoping of carbon nanotube transparent conducting films by dispersant and chemical treatment. J Mater Chem 2008;18:1261–6.

[56] Green AA, Hersam MC. Colored semitransparent conductive coatings consisting of monodisperse metallic single-walled carbon nanotubes. Nano Lett 2008;8:1417–22.

[57] Artukovic E, Kaempgen M, Hecht D, Roth S, Grüner G. Transparent and flexible carbon nanotube transistors. Nano Lett 2005;5:757–60.

[58] Fortunato E, Barquinha P, Pimentel A, Goncalves A, Marques A, Pereira L, Martins R. Fully transparent ZnO thin-film transistor produced at room temperature. Adv Mater 2005;17:590–4.

[59] Hur SH, Yoon MH, Gaur A, Shim M, Facchetti A, Marks TJ, Rogers JA. Organic nanodielectrics for low voltage carbon nanotube thin film transistors and complementary logic gates. J Am Chem Soc 2005;127:13808–9.

[60] Cao Q, Xia MG, Shim M, Rogers JA. Bilayer organic–inorganic gate dielectrics for high-performance, low-voltage, single-walled carbon nanotube thin-film transistors, complementary logic gates, and p–n diodes on plastic substrates. Adv Funct Mater 2006;16:2355–62.

[61] Hur S-H, Kocabas C, Gaur A, Shim M, Park OO, Rogers JA. Printed thin film transistors and complementary logic gates that use polymer coated single-walled carbon nanotube networks. J Appl Phys 2005;98:114302.

[62] Javey A, Kim H, Brink M, Wang Q, Ural A, Guo J, et al. High-kappa dielectrics for advanced carbon-nanotube transistors and logic gates. Nat Mater 2002;1:241–6.

[63] Nouchi R, Tomita H, Ogura A, Kataura H, Shiraishi M. Logic circuits using solution-processed single-walled carbon nanotube transistors. Appl Phys Lett 2008;92:253507.

[64] Yu WJ, Kim UJ, Kang BR, Lee IH, Lee EH, Lee YH. Adaptive logic circuits with doping-free ambipolar carbon nanotube transistors. Nano Lett 2009;9:1401–5.

[65] Zhou YX, Gaur A, Hur SH, Kocabas C, Meitl MA, Shim M, Rogers JA. p-channel, n-channel thin film transistors and p–n diodes based on single wall carbon nanotube networks. Nano Lett 2004;4:2031–5.

[66] Chen J, Klinke C, Afzali A, Avouris P. Self-aligned carbon nanotube transistors with charge transfer doping. Appl Phys Lett 2005;86:123108.

[67] Klinke C, Chen J, Afzali A, Avouris P. Charge transfer induced polarity switching in carbon nanotube transistors. Nano Lett 2005;5:555–8.

[68] Appenzeller J. Carbon nanotubes for high-performance electronics – progress and prospect.. Proc IEEE 2008;96:201–11.

[69] Avouris P, Chen ZH, Perebeinos V. Carbon-based electronics. Nat Nanotechnol 2007;2:605–15.

[70] Baumgardner JE, Pesetski AA, Murduck JM, Przybysz JX, Adam JD, Zhang H. Inherent linearity in carbon nanotube field-effect transistors. Appl Phys Lett 2007;91:052107.

[71] Zhong ZH, Gabor NM, Sharping JE, Gaeta AL, McEuen PL. Terahertz time-domain measurement of ballistic electron resonance in a single-walled carbon nanotube. Nat Nanotechnol 2008;3:201–5.

[72] Kocabas C, Kim HS, Banks T, Rogers JA, Pesetski AA, Baumgardner JE, et al. Radio frequency analog electronics based on carbon nanotube transistors. Proc Natl Acad Sci USA 2008;105:1405–9.

[73] Kocabas C, Dunham S, Cao Q, Cimino K, Ho XN, Kim HS, et al. High-frequency performance of submicrometer transistors that use aligned arrays of single-walled carbon nanotubes. Nano Lett 2009;9:1937–43.

CHAPTER 5

Flexible Field Emitters Based on Carbon Nanotubes and Other Materials

Soo-Hwan Jeong[1] and Kun-Hong Lee[2]

[1]*Department of Chemical Engineering, Kyungpook National University, Daegu, Korea*
[2]*Department of Chemical Engineering, Pohang University of Science and Technology, Pohang, Korea*

INTRODUCTION

Semiconductors, secondary batteries and flat panel displays (FPDs) have recently become the three star items in the electronics industry. FPDs are the latest entry in this list, and enormous research efforts have been devoted to developing bigger, lighter and brighter displays.

Among FPDs, field emission displays (FEDs) have many favorable features, in theory, over liquid crystal displays (LCDs) or plasma display panels. Therefore, work on field emitter array (FEA) devices has evolved into a fast-growing new area in microelectronics, thanks to progress in lithographic techniques in micrometer-sized device fabrication (for selected reviews, see [1–3]). FEAs are electron sources that have the form of an array of microfabricated sharp tips, typically with an integrated extraction electrode (gate). When the gate is biased to a large enough positive potential with respect to the tip, electron emission takes place from the tip into a vacuum via electron tunneling through the solid-vacuum potential barrier – a phenomenon known as field emission. FEAs have found a wide range of applications, including field emission electron microscopes, pressure gauges and sensors, microwave amplifiers and, most importantly, field emission FPDs. Having been aggressively pursued by a number of major companies worldwide, such as PixTech, Candescent, Motorola, Futaba and Samsung, this new FED technology is expected to be an alternative to LCD in the high-information-content FPD market. Its capacity to offer cathode ray tube (CRT)-like true color and gray scale, high brightness, wide viewing angle, fast response, low power consumption and

CONTENTS

Semiconductor Nanomaterials for Flexible Technologies

129

a broad range of operating temperature distinguishes the FED from other FPDs.

Earlier in the development of FEAs, a common way of obtaining a large local field of a few V/nm was to use a very sharp tip by chemical etching to a several hundred nanometers [4]. Most of these studies had the objective of higher local-field generation at the emitter tip. These efforts, in combination with progress in microfabrication techniques, led to the development of Spint tip emitters in the late 1960s [5]. Typically, high melting point metals, such as Mo and W, were used as the field emitter tip. Since the earliest reports by Baker et al. [6] and Lea [7], who noted that carbon materials showed a considerable reduction in pressure-dependent instability of emission current under a relatively moderate vacuum, there have been numerous reports regarding the use of various carbon-based materials, such as diamond-like carbon (DLC) [8, 9], graphite [10, 11] and diamond [12–14], as field emission sources. Although these materials are technologically important owing to their relatively good electron emission properties, potentially even more interesting are the carbon nanotubes (CNTs).

Since the first discovery of CNTs by Sumio Iijima of the NEC Corporation in 1991 [15], the interest in CNTs and the breadth of research activities across the world on their application to field emission devices have been extraordinary [16, 17]. CNTs possess the following properties favorable for field emitters: (1) high aspect ratio, (2) small radius of curvature at their tip, (3) high chemical stability, and (4) high mechanical strength. Motivated by these attractive features of CNTs, numerous studies on the measurement of field emission from CNTs have been undertaken by different groups. It has been shown that CNTs exhibit better emission characteristics, i.e. low turn-on and threshold field, than other materials. Most of the early efforts were devoted to growing aligned CNTs in the desired position, i.e. on the patterned substrate.

Nowadays, flexible displays are playing an increasingly important role in the global high-technology industry, serving as the crucial enabling technology for a new generation of portable devices that are designed to combine mobility with compelling user interfaces. Flexibility is necessary to electronic displays in many situations, such as outdoor use, use in public space and displaying over a large area. Therefore, there have been many reports regarding the development of CNT-based flexible FEAs. Flexible electronics such as a flexible field emission device have long been a dream and have been under extensive development in the past decade. Flexible FEDs in the FPD technologies offer many potential benefits, including reductions in weight and thickness, improved ruggedness and non-linear form factors. These features make flexible displays attractive for a variety of electronic products ranging from cell phones and personal digital assistants (PDAs) to computers, toys, electronic books and wearable displays.

The main goal of this chapter is to give a representative overview of the present status of flexible field emitters based on CNTs, with a special emphasis on recent development. The application of CNTs to other flexible devices, such as sensors [18, 19], solar cells [20], supercapacitors, and transistors [21–25], is beyond the scope of this chapter and discussed in chapters 4, 8 and 9.

The text in this chapter is organized as follows. Following this introduction, a brief description of the basic and general principle of field emission is given and field emission properties of CNTs are outlined. Then, the current state of the art concerning the three approaches to the flexible FEAs based on CNTs is reviewed. The ongoing research in this area can be classified into three major groups: the attachment method, composite/matrix approach, and direct growth method for the fabrication of flexible FEAs. In addition, flexible FEAs based on other materials are introduced. Finally, an attempt is made to sketch some perspectives concerning the development of this field in the near future.

FIELD EMISSION AND CARBON NANOTUBES AS FIELD EMITTERS

Principles of field emission

Besides the higher available current densities, field emission-based vacuum devices have several important advantages over their solid-state counterparts in particular applications [3]. First, because the electron transport in the vacuum device is ballistic, power dissipation does not occur in the transport medium as it does in the case of the collision-dominated transport in semiconductors. Second, operation of the vacuum device is temperature independent. Third, the vacuum device is radiation resistant. An excellent review covering the physical background of field emission phenomena and the current status of FEAs can be found in Temple [1]. The classic book by Gomer, 'Field emission and field ionization' [26], is also recommended for further understanding of the physics related to this topic. The following paragraphs summarize the fundamental principles of and factors affecting field emission.

Field emission is a quantum-mechanical phenomenon in which electrons tunnel through a potential barrier at the surface of a solid as a result of the application of a large electric field ($\sim 10^7$ V/cm). In contrast to thermionic emission or photoemission, in which electrons acquire sufficient energy via heating or energy exchange with photons to overcome the potential barrier, electrons need not obtain a large amount of energy to be emitted. In thermionic emission, the current density can be shown to be:

$$I = 120T^2 \, e^{\phi kT} \, \text{A/cm}^2 \tag{5.1}$$

where e is the elementary charge of an electron, ϕ is the work function of metal, and k is the Boltzmann constant, and temperatures upward of 1500 K are required for electrons to be released [26]. In photoemission, the metal surface should be irradiated with light of energy $hv >> \phi$ for appreciable current density. In field emission, however, the presence of an external electric field not only makes the width of the potential barrier finite but also lowers the height of the potential barrier. If the potential barrier is thin and low enough (for low energy electrons around 2 nm), penetration will occur with finite probability and emission current can be measured. The thickness and lowering of potential barrier are given by

$$\text{Thickness} = (\phi - \Delta\phi)/eE \tag{5.2}$$

$$\Delta\phi = \left(\frac{eE}{4\pi E_o}\right)^{1/2} \tag{5.3}$$

where E is the applied external field and ε_o is the vacuum permittivity.

Figure 5.1 depicts a schematic situation at a metal surface under field emission conditions. On the left the density of occupied electron states inside the metal is represented by the Fermi–Dirac distribution at 300 K. In the presence of a strong electric field the potential barrier becomes triangular and the height is lowered owing to the external electric field, E. The slope strongly depends on the amplitude of the local electric field at the surface. Therefore, the local field is considerably enhanced if a very sharp and protruding emitter is placed in the electric field. Electrons close to the Fermi energy can escape into vacuum by tunneling. Inside the metal, electrons occupy the energy band up to Fermi energy level. The potential energy outside

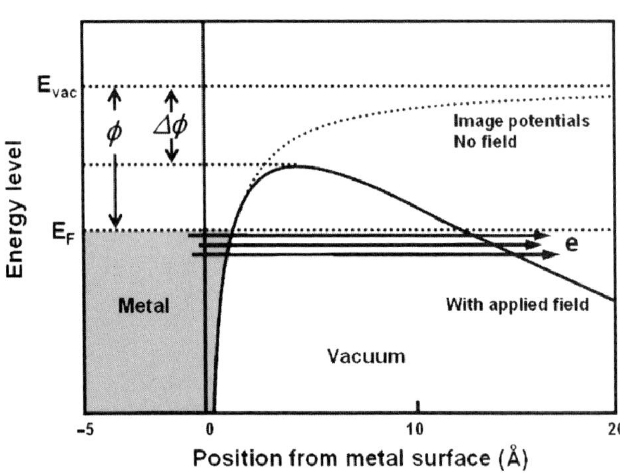

FIGURE 5.1 *Potential energy of electrons at the metal surface. Under the presence of a strong electric field the shape of the potential barrier becomes triangular shape and the height is lowered. The slope strongly depends on the amplitude of the local electric field at the metal surface.*

the metal is regarded as entirely due to the image forces, where x is the distance from the surface of the metal. When a high electric field is applied, the field contribution to the potential energy is $-eFx$, where F is the strength of the electric field. Then, the effective barrier can be described by the potential function $U(x) = -(e^2/4x) - eFx$. The resulting energy distribution of the emitted electrons is very narrow. For a metal, a sharp peak positioned at the Fermi energy of the emitter is observed.

Fowler–Nordheim equation

The field emission current density can be expressed by the Fowler–Nordheim (F–N) equation:

$$J = \frac{e^3 E^2}{8\pi h \phi t^2(y)} \exp\left[\frac{-8\pi(2m)^{1/2}\phi^{3/2}}{3heE}v(y)\right] \tag{5.4}$$

where $y = \phi - \Delta\phi$ with $\Delta\phi$ given by Eqn (5.3), h is Planck's constant, m is the electron mass, and $t(y)$ and $v(y)$ are the Nordheim elliptic functions. The formula could be described in different forms according to approximation of these elliptic functions. If $t(y)$ and $v(y)$ are taken to be unity corresponding to a triangular surface potential barrier, the F–N equation becomes [27]:

$$J = \frac{1.56 \times 10^{-6} E^2}{\phi} \exp\left(\frac{-6.83 \times 10^7 \phi^{3/2}}{E}\right) \tag{5.5}$$

while, when the first approximation $t^2(y) = 1.1$ and $(y) = 0.95 - y^2$ are used, a somewhat different form can be obtained [3]

$$J = 1.42 \times 10^{-6}\frac{E^2}{\phi} \exp\left(\frac{10.4}{\phi^{1/2}}\right) \exp\left(\frac{-6.44 \times 10^7 \phi^{3/2}}{E}\right) \tag{5.6}$$

where J is in units of A/cm^2, E is in units of V/cm, and ϕ is in units of eV. Equations (5.5) and (5.6) show that current densities of 10^2–10^3 A/cm^2 can be expected for fields of 3–6 $\times$ 10^7 V/cm. This is the range most frequently occurring in emission work. Although Eqns (5.5) and (5.6) apply strictly to temperature equal to 0 K, it can be shown that $J(300$ K$) = 1.03J(0$ K$)$ while $J(1000$ K$) = 1.5J(0$ K$)$ using low-temperature expansion of the F–N equation; therefore, the error involved in the use of the equation for moderate temperatures ($\sim$300 K) can be neglected [26]. Plotting $\ln(J/E^2)$ versus

$1/E$ results in a straight line and this is often called the Fowler–Nordheim plot.

In field emission, external electric fields in the order of 10^7 V/cm are required for appreciable electron current as stated above. To generate this extremely high electric field, one has to use the effect of field enhancement at tip-like structures. Near the surface with high curvature, the local electric field E is enhanced by the relation $E = \beta \times E_o$, where β is the field enhancement factor and E_o is the global electric field given by $E_o = V/d$, where V is the applied voltage and d is the vacuum gap in the diode type field emission configuration. It can be simply understood that the higher the aspect ratio of the emitter, the larger the field enhancement effect can be obtained.

In the F–N plot, the slope is equal to either $-6.83 \times 10^7 \phi^{3/2}$ for Eqn (5.5) or $-6.44 \times 10^7 \phi^{3/2}$ for Eqn (5.6) if the local electric field E is known. However, usually in a field emission measurement one can measure only the global electric field E_o. Therefore, substituting the local electric field E by the global electric field E_o by the relation shown above and rearranging the equation, one can obtain an F–N plot with a slope of $-6.83 \times 10^7 \phi^{3/2} \beta^{-1}$ or $-6.44 \times 10^7 \phi^{3/2} \beta^{-1}$. Consequently, the work function can be determined from the slope of the F–N plot if the field enhancement factor is known and vice versa. That is, the work function and field enhancement factor cannot be determined independently. In general, the work function for a planar metal surface is used and the field enhancement factor is calculated, but for a surface with a very small radius of curvature the work function value can be significantly different from the one in the planar case because of the dependence of the work function on the image charge potential. To avoid this problem, Küttel et al. [28] proposed a method for independently determining the work function and field enhancement factor by a combined measurement of the field emitted electron energy distribution and the current–voltage (I–V) characteristics of an emitter.

Field enhancement and work function effects

As can be seen from Eqns (5.5) and (5.6), the field emission current density is a strong function of field enhancement factor β and work function ϕ. A simulation for emission-current density dependence on β and ϕ clearly shows that increasing β or lowering ϕ causes a dramatic increase in the emission current density [1]. This implies that for an effective FEA, an emitter material with a low work function should be chosen and the device structure must be carefully designed to exhibit a high field enhancement factor, since the work function ϕ is an electronic property of the emitter surface while the value of the field enhancement factor β depends on geometric parameters of the device

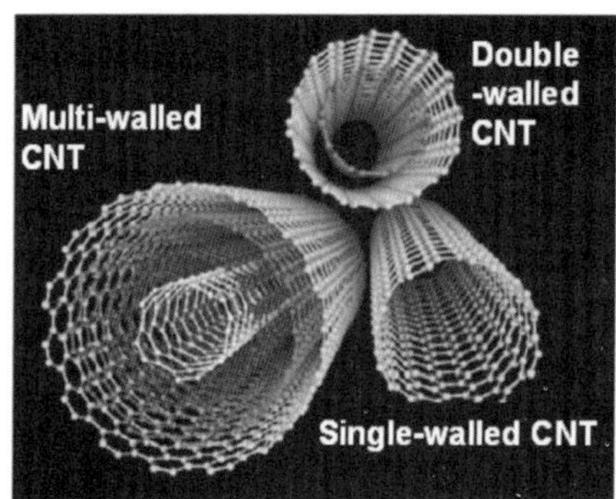

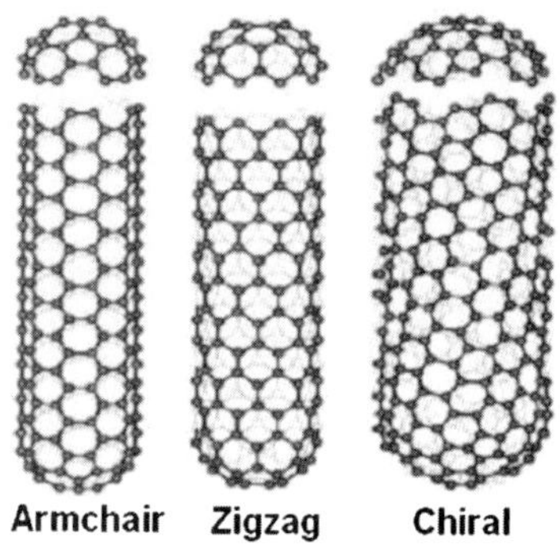

FIGURE 5.2 *Classification of carbon nanotubes (CNTs).*

structure. The affecting geometric parameters on the field enhancement factor β are summarized in Temple [1].

Field emission from carbon nanotubes

As shown in Figure 5.2, the CNT is a unique form of carbon, configurationally equivalent to a two-dimensional graphene sheet rolled into a tube. In the ideal case, a CNT consists of either one cylindrical graphene sheet [single-walled carbon nanotube (SWNT)] or several nested cylinders with an interlayer spacing of 0.34–0.36 nm that is similar to the typical interspacing of turbostratic graphite [multiwalled carbon nanotube (MWNT)]. SWNTs can be either semiconducting or metallic-like depending on their diameter and chirality, whereas MWNTs show semimetallic behavior. Several synthesis methods have been used to produce CNTs. The three most commonly used methods are based on the arc, laser and chemical vapor deposition (CVD) techniques. An excellent review covering the synthesis of CNTs can be found in Sinnott and Andrews [29].

Since the pioneering work in the 1990s [30–32], regarding the utilization of CNTs as a field emission source, CNTs have drawn much attention to the application of field emitters in FED. Owing to their geometric structure, diameters down to 10 Å and lengths up to several millimeters resulting in an extremely high aspect ratio, greater than 10 000, nanotubes have been recognized as very promising field emitting structures which can enhance the local electric field at the emitter surface to a great extent. In addition to these geometric characteristics of small radius of curvature and high aspect ratio, good electric conductance together with high mechanical strength and chemical inertness are required for a desirable field emitter [33]. In particular, the chemical inertness of carbon field emitters is one of the most important

Table 5.1 Turn-on (E_{to}) and threshold (E_{thr}) field (V/µm) for various film field emitters

Emitter material	E_{to}	E_{thr}	Ref.
Si nanowires	N/A	13	[37]
Diamond	24	40	[38]
Diamond B doped	16	30	[38]
Diamond N doped	1.5	>> 8	[39]
Diamond tips (gold-coated)	3	22	[40]
SWNT	1.5	3.9	[41]
MWNT	1.1	2.2	[41]

E_{to}: Turn-on field for which emission current density of 10 µA/cm^2 is obtained; E_{thr}: threshold field for which emission current density of 10 mA/cm^2 is obtained; SWNT: single-walled carbon nanotube; MWNT: multiwalled carbon nanotube; N/A: not applicable.

advantages over metal (typically Mo) or silicon (Si) microtips, which suffer emission degradation due to ion bombardment and chemical contamination, and therefore require a high-vacuum environment for operation [34]. It has been reported that CNT field emitters show stable emission currents over 8000 h under a moderate vacuum condition of 10^{-8} torr, while Mo microtip field emitters require ultrahigh-vacuum conditions of 10^{-10} torr [35, 36]. Although there are disputes over the emission mechanism of DLC or CVD diamond film, the field emission from CNTs is believed to rely on their geometric field-enhancing properties. In comparison with other field emitters, the applied voltages needed for field emission with CNTs are far lower for a comparable emitted current (Table 5.1) [37–41]. Therefore, numerous reports on the field emission characteristics of CNTs have been released. Most of the efforts in the early development period were devoted to growing aligned nanotubes in the desired position, i.e. on the patterned electrode. Catalytic CVD is the most commonly used method, to directly grow high-yield nanotubes and to avoid any cumbersome postprocessing step. However, the inherently high density of nanotube arrays directly grown by catalytic CVD exhibits undesirable emission characteristics rather than showing high current density due to the large number of emission sites. Nillson et al. [42] presented a simulation result on field penetration between parallel standing tip-like structures, which verified their experimental observations: when the nanotube density is too high, resulting in too close spacing between neighboring tubes, emission current density is lower than that from medium-density nanotube arrays. This is caused by field screening, in which the applied electric field cannot penetrate into the small intertip spacing and consequently high field enhancement does not occur. The limit of zero distance between the tubes would correspond to a flat metal surface without field penetration. The authors predicted that an intertube distance of about twice the height of the nanotubes would optimize the emission current density from the calculation result, which showed a rapid decrease in β value when the intertip spacing was smaller than twice the length of the tips, while for larger spacing the β values remained almost unaffected.

Several demonstrations of CNT field emission devices have been presented [4, 36]. For example, Saito and Uemura [36] fabricated a CRT lighting element using bundles of MWNT that operated at 10 kV anode voltage. Bonard et al. [43] realized a luminescent, mercury-free tube by fabricating a cylindrical CNT-based field emission diode (Figure 5.3). Samsung [4] has presented 38 inch full devices with CNTs.

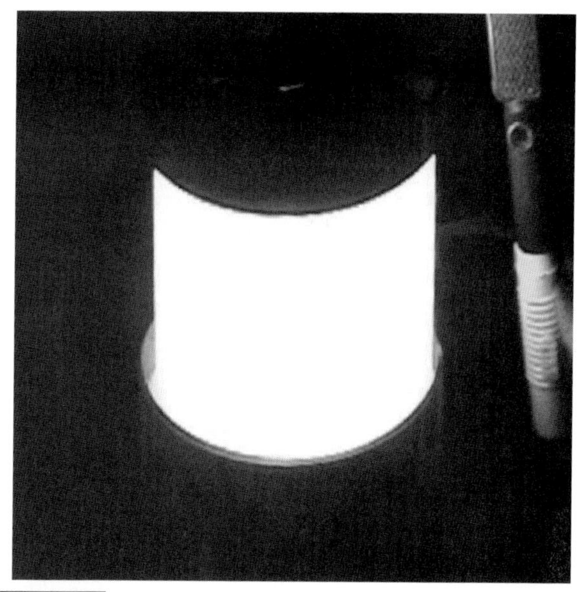

FIGURE 5.3 *Luminescent tube realized with a multiwalled carbon nanotube cathode. The applied voltage was 5.4 kV, and the emitted current density 0.5 mA/cm^2 on the cathode and 0.06 mA/cm^2 on the anode. The emitted light intensity amounts to 10 000 cd/m^2. (Reproduced with permission from [43] © 2001 American Institute of Physics.)*

FLEXIBLE FIELD EMITTERS BASED ON CARBON NANOTUBES

Attachment method

Although numerous methods have been proposed to grow CNTs, such as arc-discharge [15], laser ablation [44] and CVD [45], only a few limited methods allow controlled and patterned growth of CNTs on a substrate. It has been demonstrated that plasma-enhanced CVD is one of the best routes to selective and aligned growth of CNTs on Si and glass substrates [4, 45, 46]. In general, despite the high level of control in plasma-enhanced CVD, this method typically requires high CNT growth temperatures, which limit the choice of materials for flexible substrates.

Attaching readily available CNTs on the flexible substrate is an intuitive method and offers an alternative route to avoid the high processing temperature of CNT growth, which significantly limits the choice of substrate materials and fabrication processes. In recent years, several research groups have been involved in attaching CNTs for this purpose, using self-assembly monolayer (SAM) [47, 48], hydrogen bonding [49], electrophoretic [50] and microwave welding methods [51].

It is well known that SWNTs can be attached using a SAM of thiol-based chemicals on a gold surface [52, 53]. The utilization of the SAM technique for successful preparation of patterned flexible substrate was reported by Lee et al. [48]. Purified SWNT bundles were chopped in an acidic solution. During the chopping process, the edge sites of graphene layers of SWNT were functionalized with carboxyl groups. Then, the functionalized SWNTs were dispersed in a basic solution. A commercially available Ti-metallized polymer film was chosen as the flexible substrate. To make a patterned array for SWNT field emitters on a flexible polymer substrate, patterned Au dot arrays with a diameter of 230 µm were deposited on the metallized organic polymer substrate through a shadow mask by a magnetron sputter. Next, the prepared substrate was dipped into ethanol solution of 11-mercaptoundecanoic acid to form SAM on the patterned Au dot arrays. After treatment of the thiolized substrate in $ZnSO_4$ solution, the substrate was dipped into the prepared basic solution containing the SWNTs to link the carboxyl groups at the end of SWNT to the Zn ion attached to the 11-mercaptoundecanoic acid. Their optical microscopic image of the substrate and schematic drawing of the flexible CNT emitter array are shown in Figure 5.4. The field emission currents from SWNTs attached to the flexible polymer substrate were measured. The emission

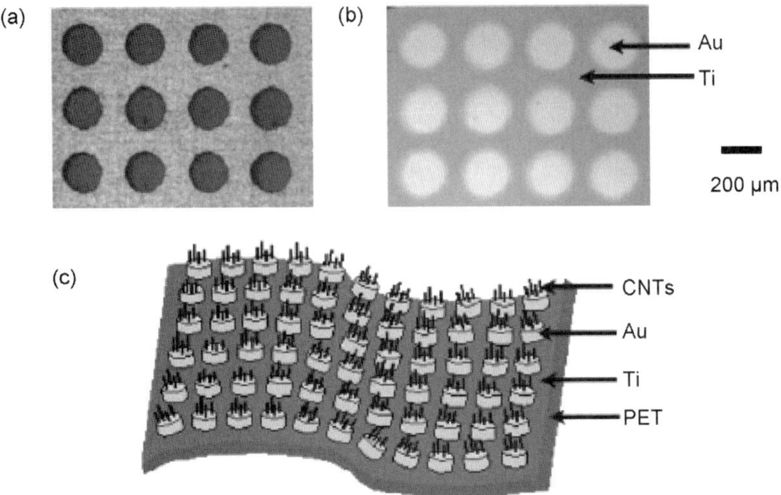

FIGURE 5.4 *Organic polymer substrate for flexible field emission display. (a) Optical microscopic image of shadow mask; (b) optical microscopic image of sputtered substrate; (c) schematic diagram of flexible substrate and aligned carbon nanotubes (CNTs). PET: polyethylene terephthalate.*
(Reproduced with permission from [48] © 2003 American Institute of Physics.)

current density reached 1.6 mA/cm^2 at 6.0 V/µm. The reported turn-on field at an emission current density of 10 µA/cm^2 and the field enhancement factor estimated from the F–N plot were 3.88 V/µm and 875, respectively. This simple process has no fundamental limitations in being scaled up to a flexible field emission device of a large area. However, it should be mentioned that the presence of an insulating SAM between SWNTs and Au electrode could have a contrary effect on the performance of the field emission device. Although an insulating layer between the cathode electrode and emitter can present a uniform emission [54], a drawback of this SAM strategy may be the limited capability of transport of electrons from the Au layer to SWNTs.

In another study [49], a dip-processing attachment method using hydrogen bonding between oxidized CNTs and a partially oxidized surface carbon fiber fabric was suggested. To obtain stable aqueous dispersions, high-purity MWNTs were functionalized by refluxing in a mixture of concentrated sulfuric and nitric acid. This resulted in the formation of oxygen-containing acid groups, such as a carboxylic acid group, to the MWNTs. The formed polar groups on the surface of CNTs readily interact with water molecules via hydrogen bonding. After the carbon fiber fabric was made hydrophilic by oxygen plasma treatment, it was dipped into the oxidized CNT dispersion. The hydrogen bonding resulted in strong adhesion between the carbon fabric and oxidized CNTs. Because the field emission is a surface phenomenon and strongly depends on the work function of materials [55–57], the CNT surface was modified with alkali metal cations of Li$^+$ to reduce the work function of the emitters. The reported threshold field at an emission current of 1 nA was 0.25 V/µm. This relatively low threshold value was attributed to the reduction in work function. It should be noted that surface modification of the field emitter with alkali metals, such as Cs, considerably increases the field emission current by lowering the work function of the emitter [58–60].

Another complementary strategy of CNT attachment on the substrate from prepared CNT powder is electrophoretic deposition of CNTs on a flexible polymeric substrate. Electrophoretic deposition is a high-throughput industrial process that has been widely used for coating of colloidal particles [61]. There have been reports regarding CNT films on conducting substrate using electrophoretic methods [62–64]. This process enables facile and efficient deposition of uniform CNT coatings on conducting substrates with high-level control of film thickness and morphology [63]. Moreover, the technique allows for a high level of flexibility in the shape and size of the substrates on which the thin CNT film is formed. Ma et al. recently used this technique to prepare the CNT-deposited flexible films and thus measure field emission properties [50]. Initially, the ground CNT powder was dispersed in Mg(NO$_3$)$_2$·6H$_2$O dissolved isopropyl alcohol solution. The role of Mg^{2+} ions in the CNT suspension during the electrophoretic process has been explained in detail elsewhere [65]. In brief, the CNTs in solution became positively charged as a result of the adsorption of Mg^{2+}, and hence under an electrical field moved towards the negative electrode and deposited there to form a thin film. A polyimide (PI) film coated with titanium was chosen as a cathode in their work. The prepared CNT deposited flexible film showed a turn-on field of about 2.96 V/μm and current density of 200 μA/cm^2 in an electric field of 5.9 V/μm.

Unlike the attachment of CNTs on a flexible substrate, with the aid of thiol-based chemicals and hydrogen bonding between CNTs and flexible substrates, direct bonding of CNTs on a polymer substrate by microwave irradiation and its application to flexible field emitter devices has been developed [51]. The microwave absorption capability of MWNTs is well known: this phenomenon led to polymer welding, in which microwave-irradiated MWNTs could readily release heat to melt the polymer and strongly bond to the nearby polymeric substrate in a extremely short period [66, 67]. Based on this technique, Wang et al. [51] developed the flexible MWNT field emitter on a polycarbonate plate, providing very strong bonding between CNTs and the polymeric substrate. To make the flexible CNT field emitter, first the paste is prepared by roller mixing CNTs with polymer solution and a dispersant. The prepared paste is screen printed and then microwave welding is performed. To remove surface dispersant and loosely bonded MWNTs, ultrasonic cleaning in water is carried out. The resulting flexible emitter arrays in Figure 5.5 exhibits good electron field emission properties, such as a very low turn-on field of 0.8 V/μm, a high emission current of around 6 mA/cm^2 in the applied field of 1.5 V/μm, and emission reproducibility for five cycles. These results are explained by the formation of direct electron-conducting paths through the CNTs to the tips of the CNT emitters. Under bending, compression and tension up to 45 degrees, an ohmic contact is kept because there is no interlayer binder or buffer between the CNT emitter and substrate, nor does an additional conducting layer serve as the electrode.

Composite/matrix method

CNTs have many outstanding properties, such as low density, high aspect ratio, high electrical conductivity, elastic moduli in the TPa range and high fracture strain [17, 68]. These properties make them attractive candidates for making advanced composite materials with multifunctional features. Therefore, CNT–polymer composites have been one of the most important topics for many potential applications requiring the combination of mechanical, electronic and/or optical properties of CNTs and polymer materials [69–73]. There have also been many attempts to incorporate CNTs in polymer matrices to tailor electrical properties suitable for different applications.

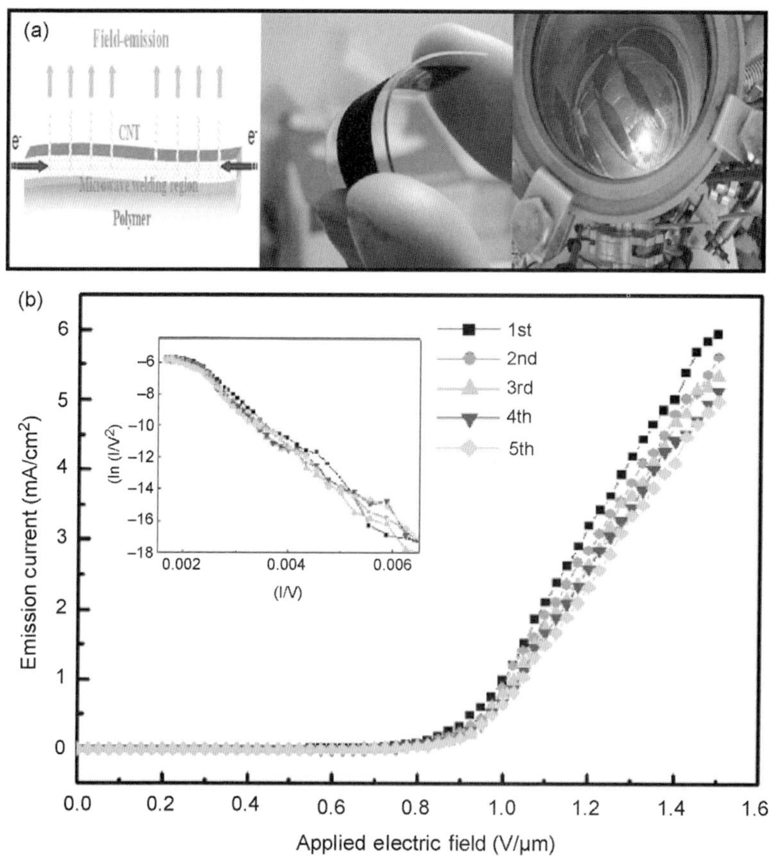

FIGURE 5.5 *Flexible field emission devices. (a) Flexible field-emission device made of multiwalled carbon nanotubes microwave-welded on a polycarbonate substrate. CNT: carbon nanotube. (b) Field-emission property of the device; the inset shows the corresponding Fowler–Nordheim plots.*
(Reproduced with permission from [51] © 2007 American Institute of Physics.)

For the purpose of application of CNT–polymer composite to flexible electronic devices, Jung et al. developed an effective method for fabricating CNT–polymer composites by incorporating vertically aligned MWNTs into a poly-(dimethylsiloxane) (PDMS) matrix [74]. The proposed technique appears to be worthy of notice because (1) CNT architecture of any shape and dimension can easily be transferred completely into a polymer matrix, keeping the CNT alignment, and (2) the resulting CNT–polymer composite film has remarkable flexibility and shows conductivity under severe strain. The process used in their research is shown in Figure 5.6. Their strategy begins with the synthesis of vertically aligned CNTs on prepatterned SiO_2 dot arrays using thermal

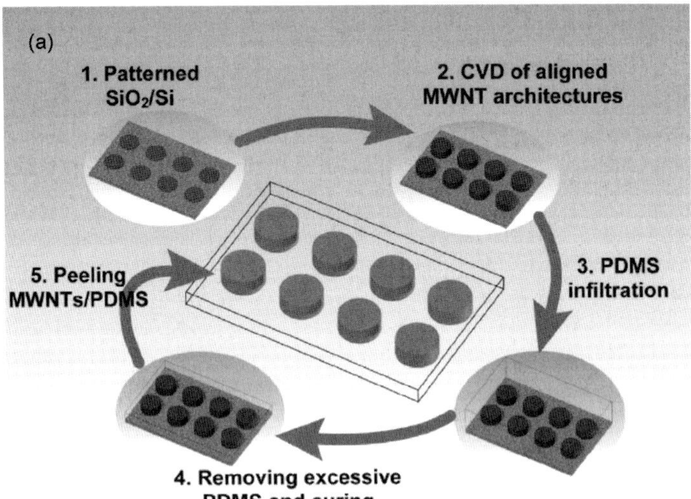
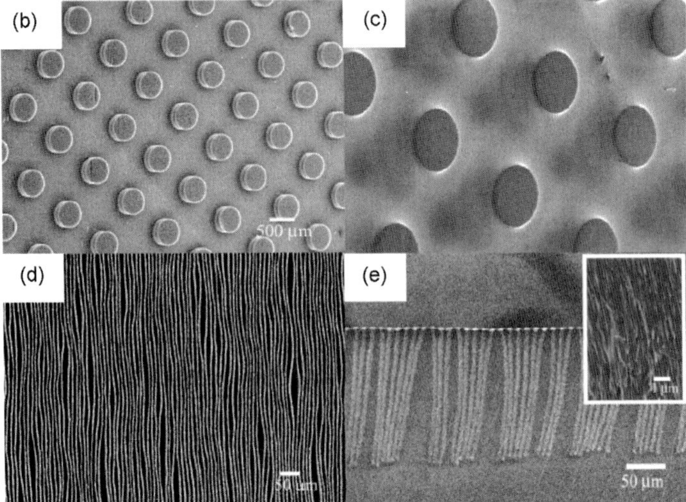

FIGURE 5.6 *(a) Schematics illustrating the fabrication steps of the aligned multiwalled carbon nanotube–poly(dimethylsiloxane) (MWNT–PDMS) array structures. CVD: chemical vapor deposition. SEM images showing the MWNT architectures before and after PDMS infiltration in two different scales and shapes: arrays of nanotube pillars (500 μm diameter and 100 μm height) (b) before and (c) after infiltration. (d) Top view of nanotube walls (7 μm line width, 7 μm distance between walls, 100 μm height) before PDMS polymerization. (e) Cross-sectional SEM image of the same nanotube walls after infiltration. Inset: high-magnification image showing that the nanotube pattern and alignment remain intact after infiltration.*

(Reproduced with permission from [74] © 2006 American Chemical Society.)

vapor deposition [75]. Next, the aligned CNT structures on the substrate are soaked in PDMS prepolymer solution. After removing any excessive PDMS solution to obtain the optimum thickness for CNT–PDMS composite film, a thermal curing process follows. Finally, composite films of CNT–PDMS are peeled off. The composite films detached from the substrate are very flexible and could be deformed easily into various shapes without disturbing the CNT alignment. Suitable control of the amount of PDMS used while preparing the CNT–PDMS composite could result in films with low-density exposed CNTs on the top surface to reduce the field screening effect [76]. After making an electrical contact at the bottom part of the composite, field emission measurement was carried out. The current density of 1 mA/cm^2 was obtained at the threshold field range of 0.76–2.16 V/µm, whereas the turn-on field was 0.5–0.87 V/µm at 1 nA. The estimated field enhancement factor was in the range of 8000–19 100. This good emission behavior could be attributed to the inherently good contact between CNT and the cathode electrode, low-density aligned CNTs and insulation of neighboring CNTs. Similar synthetic approaches based on the above-mentioned method to make flexible CNT field emitters were also reported by other groups [77, 78].

Based on screen printing, a widely used high-throughput process, by Samsung [79], another preparation approach to flexible CNT emitters in the composite processing method was reported [80]. Instead of the glass frit, which is typically used as an inorganic binder in the screen-printing process and requires a high annealing temperature (>450°C), spin-on-glass (annealing temperature of 150°C) was used for the flexible purpose. The printing and annealing process was conducted on indium–tin oxide (ITO) coated polyethylene terephthalate (PET) film. Screen printing has many advantages in the commercial realization of CNT emitter devices in terms of a simple manufacturing process, large area production and low cost.

Direct growth method

Although the extension of the above-mentioned CNT attachment technology for the fabrication of flexible field emitters does not seem to have any fundamental difficulty, there are several noticeable problems such as the chemical stability of the substrate against various solvents during the fabrication, the interfacial resistance between CNTs and substrate, and the long-term stability of the attached CNTs.

Direct growth of CNTs on flexible substrates is a straightforward way to obtain flexible FEAs without damaging the substrates. Starting from the very early publications regarding the efforts of lowering the growth temperature of CNTs on solid substrates [81–85], the growth of CNTs at reduced temperatures and/or on plastic substrates at reduced temperatures has been paid a lot of attention [86–88]. The former has focused on compatibility with current semiconductor and display processes, and the latter on developing the flexible devices.

For example, a two-stage heating technique was successfully adapted to grow CNTs at a relatively low temperature (550°C) on a large area of sodalime glass substrate by thermal CVD. The hydrocarbon was preheated at 550°C in the first heating zone and CNTs were synthesized at 550°C in the second heating zone; CNTs usually grow only at relatively high temperatures (600°C or above) with a CVD process [32, 45, 89].

Although a few alternative methods have been used to grow CNTs at low temperatures, including supercritical and sonochemical methods [90, 91], only a few methods allow controlled growth directly on a flexible substrate. Recently, flexible CNT field emitter devices on low-cost polymer substrates have been proposed. Hofmann et al. showed that plasma enhancement could reduce the CNT growth temperature to as low as 120°C, and extended this method to the preparation of aligned CNT FEAs on polymer film [83, 92]. In their typical synthesis, they deposited a 70 nm thick Cr layer and subsequently a 6 nm thick Ni catalyst. The roles of the Cr layer were conducting electrodes for emission measurement and providing an adhesion layer between the PI foil and Ni catalyst layer. The Ni layer was patterned by shadow masks for larger feature size (10 μm) or by electron-beam lithography for smaller feature size (100 nm). Then, CNTs were grown on the prepared substrate at 200°C using plasma-enhanced CVD. The scanning electron microscopy images of flexible FEAs of aligned CNTs grown from electron-beam patterned samples are shown in Figure 5.7. They also investigated the field emission properties of CNT FEAs on PI substrate. From the J–E curve shown in Figure 5.7(e), the turn-on field, i.e. the field for which J is 10^{-9} A/cm^2, and threshold field, i.e. the field for which J is 10^{-6} A/cm^2, were 3.2 and 4.2 V/μm, respectively. These emission properties compare well to those reported for CNTs grown on Si wafer [4, 93, 94]. Assuming a work function of CNT as 5 eV, they estimated that the field enhancement factor β from the unsaturated region of the F–N plot in the inset was 850.

However, despite a high level of control, such as degree of alignment and selective growth, CVD-based methods still limit the choice of flexible substrate materials and integration process. To overcome these problems, a different strategy of direct synthesis of CNTs on organic polymer substrates by microwave heating of catalysts was developed [95, 96]. The method is very simple and exploits the different dielectric properties of the catalyst particle and polymer substrate. Microwave heating, in which the microwave energy is delivered to the materials through molecular interactions with the

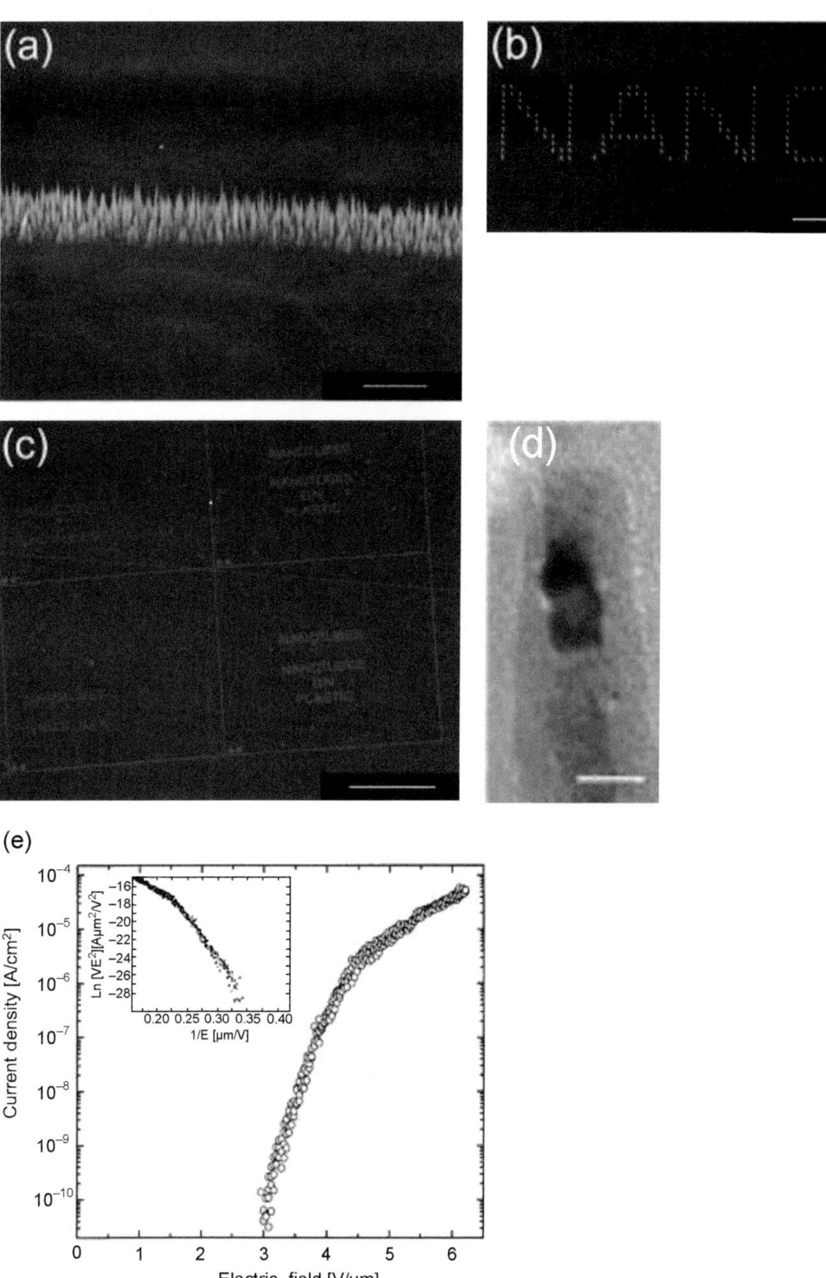

FIGURE 5.7 *(a)–(c) SEM photographs of vertically aligned carbon nanotubes (CNTs) grown from electron-beam patterned single 100 nm wide lines and 100 nm diameter dots of Ni onto Cr-covered plastic foil. (d) TEM image of as-grown CNT [scale bars: (a) 1, (b) 5, (c) 100 μm, (d) 15 nm]. (e) Emission current density as a function of the applied electric field for the CNT emitters on Cr-covered polyimide foil measured in parallel plate configuration. The anode area was 0.25 cm². Inset: corresponding Fowler–Nordheim plot. (Reproduced with permission from [92] © 2003 American Institute of Physics.)*

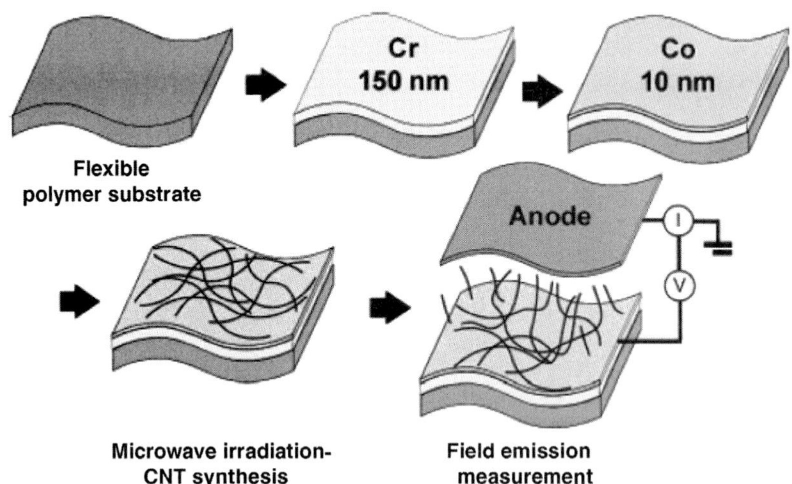

FIGURE 5.8 *Scheme of experimental procedure for the fabrication of a flexible carbon nanotube (CNT) field emitter by microwave irradiation.*
(Reproduced with permission from [96] © 2005 American Chemical Society.)

electromagnetic field, has potential advantages of uniform, rapid and volumetric heating [97, 98]. Furthermore, selective heating is possible by microwaves owing to the difference in dielectric properties of the materials [98]. Therefore, only the catalyst particles would be heated to the temperature of CNT synthesis, without increasing the temperature of the polymeric substrate on which the catalysts lie. As shown in Figure 5.8, a typical example of carrying out direct synthesis of CNT FEAs on polymer substrate is as follows. A thin layer of Cr is deposited on a 200 μm thick Teflon sheet by a magnetron sputter. Then, a 10 nm thick catalyst layer is prepared on Cr-deposited Teflon by the same method. The catalyzed substrate (Co/Cr/Teflon) is placed in a quartz reactor, and microwaves (300 W, single mode) are irradiated on it with a flowing reactant gas mixture of C_2H_2 and Ar for 5 s. This experiment results in highly uniform and well-graphitized MWNTs on Teflon substrate. The field emission measurement for CNTs on the Co/Cr/Teflon revealed that the turn-on field, i.e. the electric field to produce 10 μA/cm^2, was 3.6 V/μm. A CNT-based field emitter should have a current density of 80 μA/cm^2 to be used as a cold emission source [42]. The present sample generated this current density at 4.1 V/μm. The field enhancement factor, β, calculated from the slope of the F–N plot, was 1112–1546, depending on the range of the emission data.

Although the above-mentioned approaches have many advantages, such as lower cost of producing the emitter and the ability to process any shape and geometry, the flexibility test and field emission behavior under stress conditions were not carried out. Recently, Sim et al. demonstrated a carbon

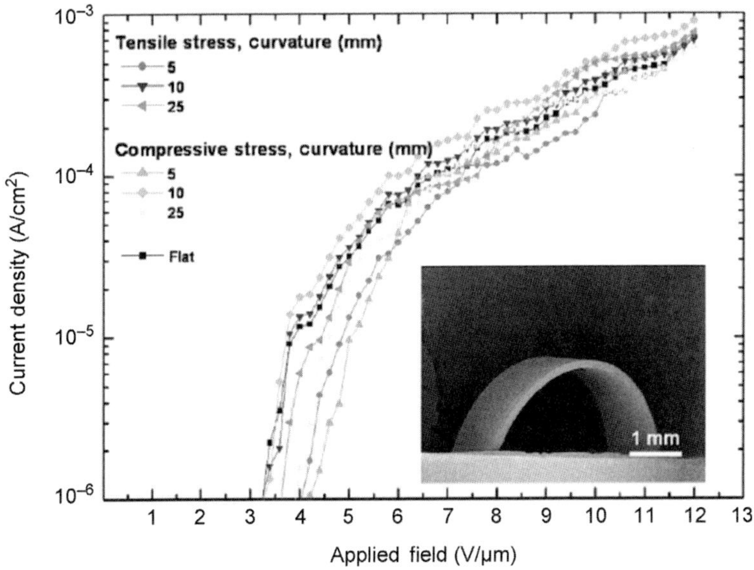

FIGURE 5.9 *Field emission measurement of the flexible carbon nanofiber emitters at different bending curvatures. Inset: bent flexible carbon nanofiber field emission device. (Reproduced with permission from [99] © 2007 American Institute of Physics.)*

nanofiber-based all-plastic FEA and performed reliability testing under substantial tensile and compressive stress conditions [99]. Their method is based on the ion-beam method. It is well known that oblique Ar^+ ion bombardment on a carbon surface results in the growth of carbon nanofibers even at room temperature [100–103]. Using the ion-induced growth technique, carbon nanofibers were grown on commercially available PI film. After coating a carbon layer on the film, the carbon-coated PI substrate was bombarded with incident Ar^+ ion beam at a relatively low vacuum pressure for 1 h. This resulted in flexible FEAs of carbon nanofiber-tipped cone-like nanostructures on the PI substrate. From the field emission measurements of samples under various stressing conditions (Figure 5.9), it was reported that the threshold field without strain was around 3.2 V/μm, whereas the threshold fields under stress conditions were in the range of 3.1–4.2 V/μm for an emission current density of 1 μA/cm². It should be noted that the changes in the value of the threshold field under stress conditions resulted from the change in gap size between emitter and anode during the bending or compression. The reliability measurement in flat and stressed conditions at an applied field of 10 V/μm for 16 h duration, and the degradation of field emission current under both conditions were relatively low and well maintained. This implies that the flexible FEAs are very stable and robust irrespective of planar or bent conditions.

The approaches mentioned in this section have been mainly focused on the development of growth of CNTs on polymeric substrate, such as local heating of a catalyst by microwave irradiation and low-temperature growth with the aid of plasma-enhanced CVD or ion-beam techniques, and their field emission properties. Unlike to these approaches, another strategy of direct synthesis CNTs on flexible substrate has been reported. This method is based on CNT growth on high melting-point substrates of flexible materials, such as carbon cloth and carbon fabrics under conventional high-temperature CNT growth conditions. Details of these methods are reported elsewhere [104, 105].

In this section, general approaches, such as attachment, composite/matrix and direct growth method, to CNT-based flexible field emitters were outlined. Table 5.2 lists the preparation methods and field emission performance of the emitters.

FLEXIBLE FIELD EMITTERS BASED ON OTHER MATERIALS

One-dimensional (1D) nanostructures with a high aspect ratio, such as nanotubes and nanowires, have been considered to be ideal field emitters that can emit electrons in a low electric field. Although CNTs have been intensively investigated for field emission source, other 1D nanomaterials have also been investigated as electron field emitters [106–114]. Among these, ZnO nanowires have been expected to achieve stable field emission because they have negative electron affinity, structural rigidity and chemical stability [115, 116]. To date, a few efforts have been made to characterize the emission properties of ZnO nanostructures prepared by low-temperature methods [117, 118]. The field-emission properties of ZnO nanowires are fairly comparable to those of CNTs [115, 116]. Recently, ZnO field emitters on plastic substrates were reported [117, 118]. For example, Yang et al. [117] prepared ZnO nanoneedles on PI foils by ion-beam irradiation and measured the field emission properties of the ZnO nanoneedles on PI films (Figure 5.10a). The adapted synthetic procedure for ZnO growth on polymeric substrate is very similar to the synthesis of CNTs by ion-beam irradiation [99, 100]. As shown in Figure 5.10b, a threshold field, i.e. the field for which current density is 1 mA/cm^2, is 4.1 V/μm and the emission current density reaches 1 mA/cm^2 at an applied field of 9.6 V/μm. Compared to the emission properties from ZnO nanowires on Si substrate [116], this emission behavior is better and comparable to that observed for CNTs grown on PI substrate. These emission properties might be attributed to the sharp tip of the ZnO nanoneedles. Electrochemical methods also enable ZnO nanostructures to be prepared on flexible substrates [118].

Table 5.2 Examples of reported data illustrating preparation methods and field emission properties of carbon nanotube-based flexible field emitters

Category	Preparation technique (flexible emitter)	Turn-on field	Threshold field or emission performance	Field enhancement factors	Ref.
Attachment method	Self-assembly monolayer (SWNT on PET film)	3.88 V/μm at 10 μA/cm^2	1.6 mA/cm^2 at 6.0 V/μm	875	[48]
	Hydrogen bonding (Li$^+$ MWNT on carbon fiber)	0.25 V/μm	N/A	N/A	[49]
	Microwave welding (MWNTs on PC film)	0.8 V/μm	6 mA/cm^2 at 1.5 V/μm	N/A	[51]
	Electrophoresis (MWNT PI film)	2.96 V/μm	200 μA/cm^2 at 5.9 V/μm	N/A	[50]
Composite /matrix method	Dipping (MWNT–PDMS composite)	0.5–0.87 V/μm at current change of ∼1 nA	0.76–2.16 V/μm at 10 mA/cm^2	8000–19 100	[74]
	Dipping (MWNT–PDMS composite)	2.38 V/μm	1.5 mA/cm^2 at 6.0 V/μm	N/A	[77]
	Screen printing using SOG (MWCN–SOG composite on PET film)	2.9 V/μm at 10 μA/cm^2	115 μA/cm^2 at 4.25 V/μm	N/A	[80]
	Dispersion (SWCT, MWNT–natural rubber composite)	N/A	1.0 V/μm	10 000	[78]
Direct growth method	Microwave irradiation (MWNT on Teflon)	3.6 V/μm at 10 μA/cm^2	80 μA/cm^2 at 4.1 V/μm	1112–1546	[96]
	Plasma-enhanced CVD (MWNT on PI film)	3.2 V/μm at 10^{-9} A/cm^2	4.2 V/μm at 10 μA/cm^2	850	[92]
	Ion beam method (carbon nanofiber on PI film)	N/A	∼3.65 V/μm at 1 μA/cm^2	2500	[99]
	Plasma-enhanced CVD (MWNT on carbon fiber)	0.25 V/μm	∼30 μA/cm^2 at 0.6 V/μm	∼17 000	[105]

SWNT: single-walled carbon nanotube; MWNT: multiwalled carbon nanotube; PC: polycarbonate; PDMS: poly(dimethylsiloxane); SOG: spin-on-glass; CVD: chemical vapor deposition; PI: polyimide; N/A: not applicable.

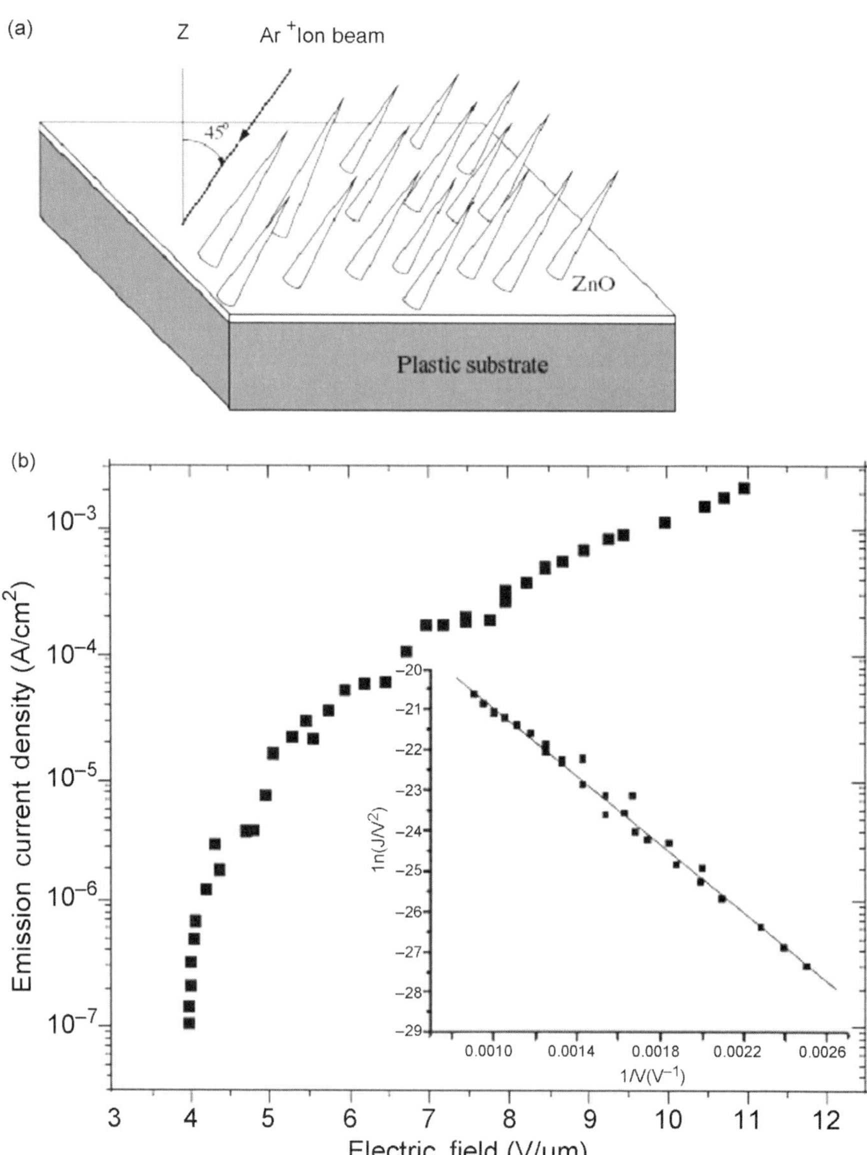

FIGURE 5.10 (a) Schematic diagram of ZnO nanoneedles with ion beam incidence angle of 45°. (b) Emission current density from ZnO nanoneedles grown on polyimide foil. The inset reveals that the field emission follows Fowler–Nordheim behavior.

(Reproduced with permission from [117] © 2005 Institute of Physics.)

Table 5.3 Preparation of other materials-based flexible field emitters and electron emission properties; all methods could be categorized into direct growth method

Preparation technique (flexible emitter)	Turn-on field	Threshold field or emission performance	Field enhancement factors	Ref.
Ion-beam irradiation (ZnO nanoneedles on PI film)	N/A	4.1 V/μm at 1 μA/cm^2 1 mA/cm^2 at 9.6 V/μm	~1134	[117]
Chemical vapor deposition (Si nanowires on carbon cloth)	0.3 V/μm at 0.01 mA/cm^2	1 mA/cm^2 at 0.7 V/μm	6.1 × 10^4	[119]
Chemical vapor deposition (ZnO nanowires on carbon cloth)	0.2 V/μm at 0.1 μA/cm^2	1 mA/cm^2 at 0.7 V/μm	4.1 × 10^4	[115]
Electrode position (ZnO nanorods/nanowalls on PET film)	1.2–2.2 V/μm at 1 μA/cm^2	2.0–4.2 V/μm at 10 μA/cm^2	5256–23 104	[118]
Evaporation of C$_{60}$ with ion beam (diamond-like carbon on PI film)	N/A	17 mA/cm^2 at 68 V/μm	N/A	[120]

PI: polyimide; PET: polyethylene terephthalate; N/A: not applicable.

Recently, ZnO nanostructures were electrochemically grown at 70°C on flexible polyester substrates. A low turn-on electric field measured at a current density of 1 μA/cm^2 for ZnO nanorods was demonstrated. Si [119] and DLC [120] nanostructures were also prepared on plastic substrates for the purpose of flexible field emitters. Table 5.3 gives an overview of combinations of emitter materials, preparation methods and emission properties from other materials.

CONCLUSION

This chapter presents the current status of development of flexible FEAs. This development has been considered in the context of preparation methods of flexible emitters and their electron emission properties, with the focus on CNT-based emitters. Three categories of preparation methods are discussed.

A lot of effort and progress has been made during the past decade in the preparation and characterization of the flexible FEAs based on CNTs and other materials. In addition, the flexible field emitters, especially CNT-based emitters, have developed into a new research field. Several strategies, such as attachment, composite/matrix and direct growth methods, have been successfully developed for CNT flexible emitters and have exhibited considerable electron emission properties. However, CNT-based flexible field emitters are still in the development stage. Therefore, there is not much information available yet regarding the lifetime and triode structures of flexible emitter arrays. The lifetime may be affected by the potential rise of the partial pressure of oxygen inside the devices. Compared to conventional field emission devices, flexible devices must be exposed to mechanical stresses, such as compression, tension and torsion. Under these conditions, there must be no leak generation to prevent oxygen gases from entering the flexible devices. For application to full-color flexible displays, flexible emitter arrays need to be prepared in triode-type configurations. As presented in this chapter, current research is focused on the preparation of flexible FEAs and field emission measurements in a diode-type arrangement. It seems that the appearance of a triode-type flexible field emitter will take more time.

Nevertheless, considering the excellent emission properties from CNTs and recent progress in this field, the authors believe that carbon-based flexible field emitters and, in particular, CNTs offer the most promising route to achieving this goal and making CNT-based flexible field emitters a commercial reality. Prototypes and widely available commercial products containing CNT-based FEAs are eagerly anticipated.

REFERENCES

[1] Temple D. Recent progress in field emitter array development for high performance applications. Mater Sci Eng R 1999;24:185–239.

[2] Brodie I, Spindt CA. Vacuum microelectronics. Adv Electron Electron Phys 1992;83:1–106.

[3] Brodie I, Schwoebel PR. Vacuum microelectronic devices [and Prolog]. Proc IEEE 1994;82:1006–34.

[4] Milne WI, Teo KBK, Amaratunga GAJ, Legagneux P, Gangloff L, Schnell JP, et al. Carbon nanotubes as field emission sources. J Mater Chem 2004;14:933–43.

[5] Spindt CA. A Thin-film field-emission cathode. J Appl Phys 1968;39:3504–5.

[6] Baker FS, Osborn AR, Williams J. Field emission from carbon fibres: a new electron source. Nature 1972;239:96–7.

[7] Lea C. Field emission from carbon fibres. J Phys D Appl Phys 1973;6:1105–14.

[8] Park KC, Moon JH, Chung SJ, Jang J, Oh MH, Milne WI. Deposition of N-type diamondlike carbon by using the layer-by-layer technique and its electron emission properties. Appl Phys Lett 1997;70:1381–3.

[9] Chuang FY, Sun CY, Chen TT, Lin IN. Local electron field emission characteristics of pulsed laser deposited diamondlike carbon films. Appl Phys Lett 1996;69:3504–6.

[10] Wang C, Bai C, Li X, Shang G, Lee I, Wang X, Qiu X, Tian F. Evidence of diffusion characteristics of field emission electrons in nanostructuring process on graphite surface. Appl Phys Lett 1996;69:348–50.

[11] Busta H, Furst D, Rakhimov AT, Samorodov VA, Seleznev BV, Suetin NV, Silzars A. Low-voltage electron emission from tipless field emitter arrays. Appl Phys Lett 2001;78:3418–20.

[12] Okano K, Hoshina K, Iida M, Koizumi S, Inuzuka T. Fabrication of a diamond field emitter array. Appl Phys Lett 1994;64:2742–4.

[13] Zhu W, Kochanski GP, Jin S, Seibles L, Jacobson DC, McCormack M, White AE. Electron field emission from ion-implanted diamond. Appl Phys Lett 1995;67:1157–9.

[14] Wang Q, Wang ZL, Li JJ, Huang Y, Li YL, Gu CZ, Cui Z. Field electron emission from individual diamond cone formed by plasma etching. Appl Phys Lett 2006;89:063105.

[15] Iijima S. Helical microtubules of graphitic carbon. Nature 1991;354:56–8.

[16] Bonard JM, Croci M, Klinke C, Kurt R, Noury O, Weiss N. Carbon nanotube films as electron field emitters. Carbon 2002;40:1715–28.

[17] Baughman RH, Zakhidov AA, de Heer WA. Carbon nanotubes – the route toward applications.. Science 2002;297:787–92.

[18] Chang NK, Su CC, Chang SH. Fabrication of single-walled carbon nanotube flexible strain sensors with high sensitivity. Appl Phys Lett 2008;92:063501.

[19] Sun Y, Wang HH. Electrodeposition of Pd nanoparticles on single-walled carbon nanotubes for flexible hydrogen sensors. Appl Phys Lett 2007;90:213107.

[20] Rowell MW, Topinka MA, McGehee MD, Prall HJ, Dennler G, Sariciftci NS, et al. Organic solar cells with carbon nanotube network electrodes. Appl Phys Lett 2006;88:233506.

[21] Kang SJ, Kocabas C, Ozel T, Shim M, Pimparkar N, Alam MA, et al. High-performance electronics using dense, perfectly aligned arrays of single-walled carbon nanotubes. Nature 2007;2:230–6.

[22] Hur SH, Park OO, Rogers JA. Extreme bendability of single-walled carbon nanotube networks transferred from high-temperature growth substrates to plastic and their use in thin-film transistors. Appl Phys Lett 2005;86:243502.

[23] Cao, Q, Hur S-H, Zhu Z-T, Sun Y, Wang, C, Meitl MA, et al. Highly bendable, transparent thin-film transistors that use carbon-nanotube-based conductors and semiconductors with elastomeric dielectrics. Adv Mater 2006;18:304–9.

[24] Nouchi R, Tomita H, Ogura A, Kataura H, Shiraishi M. Logic circuits using solution-processed single-walled carbon nanotube transistors. Appl Phys Lett 2008;92:253507.

[25] Brandon EJ, West W, Wesseling E. Carbon-based printed contacts for organic thin-film transistors. Appl Phys Lett 2003;83:3945–7.

[26] Gomer R. Field emission and field ionization. AIP Press; 1993.

[27] Fowler RH, Nordheim L. Electron emission in intense electric fields.. Proc R Soc Lond A 1928;119:173–81.

[28] Küttel OM, Gröning O, Emmenegger C, Nilsson L, Maillard E, Diederich L, Schlapbach L. Field emission from diamond, diamond-like and nanostructured carbon films. Carbon 1999;37:745–52.

[29] Sinnott SB, Andrews R. Carbon nanotubes: synthesis, properties, and applications. Crit Rev Solid State Mater Sci 2001;26:145–249.

[30] Rinzler AG, Hafner JH, Nikolaev P, Nordlander P, Colbert DT, Smalley RE, et al. Unraveling nanotubes: field emission from an atomic wire. Science 1995;269:1550–3.

[31] de Heer WA, Châtelain A, Ugarte D. A carbon nanotube field-emission electron source. Science 1995;270:1179–80.

[32] Fan S, Chapline MG, Franklin NR, Tombler TW, Cassell AM, Dai H. Self-oriented regular arrays of carbon nanotubes and their field emission properties. Science 1999;283:512–4.

[33] Saito Y, Uemura S, Hamaguchi K. Cathode ray tube lighting elements with carbon nanotube field emitters. Jpn J Appl Phys Part 2 Letters 1998;37: L346–8.

[34] Van Oostrom A. Field emission cathodes. J Appl Phys 2004;33:2917–22.

[35] Saito Y, Hamaguchi K, Mizushima R, Uemura S, Nagasako T, Yotani J, Shimojo T. Field emission from carbon nanotubes and its application to cathode ray tube lighting elements. Appl Surf Sci 1999;146:305–11.

[36] Saito Y, Uemura S. Field emission from carbon nanotubes and its application to electron sources. Carbon 2000;38:169–82.

[37] Au FCK, Wong KW, Tang YH, Zhang YF, Bello I, Lee ST. Electron field emission from silicon nanowires. Appl Phys Lett 1999;75:1700–2.

[38] Zhu W, Kochanski GP, Jin S, Seibles L. Defect-enhanced electron field emission from chemical vapor deposited diamond. J Appl Phys 1995;78:2707–11.

[39] Okano K, Koizumi S, Silva SRP, Amaratunga GAJ. Low-threshold cold cathodes made of nitrogen-doped chemical-vapour-deposited diamond. Nature 1996; 381:140–1.

[40] Wisitsora-at A, Kang WP, Davidson JL, Kerns DV. A study of diamond field emission using micro-patterned monolithic diamond tips with different Sp contents. Appl Phys Lett 1997;71:3394–6.

[41] Bonard JM, Salvetat JP, Stöckli T, Forr L, Châtelain A. Field emission from carbon nanotubes: perspectives for applications and clues to the emission mechanism. Appl Phys A 1999;69:245–54.

[42] Nilsson L, Groening O, Emmenegger C, Kuettel O, Schaller E, Schlapbach L, et al. Scanning field emission from patterned carbon nanotube films. Appl Phys Lett 2000;76:2071–3.

[43] Bonard JM, Stöckli T, Noury O, Châtelain A. Field emission from cylindrical carbon nanotube cathodes: possibilities for luminescent tubes. Appl Phys Lett 2001;78:2775–7.

[44] Thess A, Lee R, Nikolaev P, Dai H, Petit P, Robert J, et al. Crystalline ropes of metallic carbon nanotubes. Science 1996;273:483–7.

[45] Ren ZF, Huang ZP, Xu JW, Wang JH, Bush P, Siegal MP, Provencio PN. Synthesis of large arrays of well-aligned carbon nanotubes on glass. Science 1998;282: 1105–7.

[46] Chen Y, Lin Wang Z, Song Yin J, Johnson DJ, Prince RH. Well-aligned graphitic nanofibers synthesized by plasma-assisted chemical vapor deposition. Chem Phys Lett 1997;272:178–82.

[47] Lee OJ, Jeong SH, Lee KH. Field emission from single-walled carbon nanotubes aligned on a gold plate using a self-assembly monolayer. Appl Phys A 2003; 76:599–602.

[48] Lee OJ, Lee KH. Fabrication of flexible field emitter arrays of carbon nanotubes using self-assembly monolayers. Appl Phys Lett 2003;82:3770–2.

[49] Lyth SM, Hatton RA, Silva SRP. Efficient field emission from li-salt functionalized multiwall carbon nanotubes on flexible substrates. Appl Phys Lett 2007;90: 013120.

[50] Ma H, Zhang L, Zhang J, Zhang L, Yao N, Zhang B. Electron field emission properties of carbon nanotubes-deposited flexible film. Appl Surf Sci 2005; 251:258–61.

[51] Wang CY, Chen TH, Chang SC, Chin TS, Cheng SY. Flexible field emitter made of carbon nanotubes microwave welded onto polymer substrates. Appl Phys Lett 2007;90:103111.

[52] Liu Z, Shen Z, Zhu T, Hou S, Ying L, Shi Z, Gu Z. Organizing single-walled carbon nanotubes on gold using a wet chemical self-assembling technique. Langmuir 2000;16:3569–73.

[53] Yu XF, Mu T, Huang H, Liu Z, Wu N. The study of the attachment of a single-walled carbon nanotube to a self-assembled monolayer using X-ray photoelectron spectroscopy. Surf Sci 2000;461:199–207.

[54] Cui JB, Teo KBK, Tsai JTH, Robertson J, Milne WI. The role of dc current limitations in Fowler–Nordheim electron emission from carbon films. Appl Phys Lett 2000;77:1831–3.

[55] Kim DH, Lee HR, Lee MW, Lee JH, Song YH, Jee JG, Lee SY. Effect of the in situ Cs treatment on field emission of a multi-walled carbon nanotube. Chem Phys Lett 2002;355:53–8.

[56] Park N, Han S, Ihm J. Field emission properties of carbon nanotubes coated with boron nitride. J Nanosci Nanotechnol 2003;3:179–83.

[57] Dimitrijevic S, Withers JC, Mammana VP, Monteiro OR, Ager Iii JW, Brown IG. Electron emission from films of carbon nanotubes and Ta-C coated nanotubes. Appl Phys Lett 1999;75:2680–2.

[58] Wadhawan A, Ii RES, Perez JM. Effects of Cs deposition on the field-emission properties of single-walled carbon-nanotube bundles. Appl Phys Lett 2001; 78:108–10.

[59] Macaulay JM, Brodie I, Spindt CA, Holland CE. Cesiated thin-film field-emission microcathode arrays. Appl Phys Lett 1992;61:997–9.

[60] Geis MW, Twichell JC, Macaulay J, Okano KElectron field emission from diamond and other carbon materials after H, O, Cs treatment. Appl Phys Lett 1995;67:1328–30.

[61] Van der Biest OO, Vandeperre LJ. Electrophoretic deposition of materials. Ann Rev Mater Sci 1999;29:327–52.

[62] Uchikoshi T, Suzuki TS, Okuyama H, Sakka Y, Nicholson PS. Electrophoretic deposition of alumina suspension in a strong magnetic field. J Eur Ceram Soc 2004;24:225–9.

[63] Gao B, Yue GZ, Qiu Q, Cheng Y, Shimoda H, Fleming L, Zhou O. Fabrication and electron field emission properties of carbon nanotube films by electrophoretic deposition. Adv Mater 2001;13:1770–3.

[64] Choi WB, Jin YW, Kim HY, Lee SJ, Yun MJ, Kang JH, et al. Electrophoresis deposition of carbon nanotubes for triode-type field emission display. Appl Phys Lett 2001;78:1547–9.

[65] Du C, Pan N. Supercapacitors using carbon nanotubes films by electrophoretic deposition. J Power Sources 2006;160:1487–94.

[66] Imholt TJ, Dyke CA, Hasslacher B, Perez JM, Price DW, Roberts JA, et al. Nanotubes in microwave fields: light emission, intense heat, outgassing, and reconstruction. Chem Mater 2003;15:3969–70.

[67] Zhang M, Fang S, Zakhidov AA, Lee SB, Aliev AE, Williams CD, et al. Strong, transparent, multifunctional, carbon nanotube sheets. Science 2005;309: 1215–9.

[68] Panhuis M. Carbon nanotubes: enhancing the polymer building blocks for intelligent materials. J Mater Chem 2006;16:3598–605.

[69] Calvert P. Nanotube composites: a recipe for strength. Nature 1999;399:210–1.

[70] Lahiff E, Ryu CY, Curran S, Minett AI, Blau WJ, Ajayan PM. Selective positioning and density control of nanotubes within a polymer thin film. Nano Lett 2003;3:1333–8.

[71] Ahir SV, Terentjev EM. Photomechanical actuation in polymer–nanotube composites. Nat Mater 2005;4:491–5.

[72] Koerner H, Price G, Pearce NA, Alexander M, Vaia RA. Remotely actuated polymer nanocomposites – stress-recovery of carbon-nanotube-filled thermoplastic elastomers. Nat Mater 2004;3:115–20.

[73] Hinds BJ, Chopra N, Rantell T, Andrews R, Gavalas V, Bachas LG. Aligned multiwalled carbon nanotube membranes. Science 2004;303:62–5.

[74] Jung YJ, Kar S, Talapatra S, Soldano C, Viswanathan G, Li X, et al. Aligned carbon nanotube-polymer hybrid architectures for diverse flexible electronic applications. Nano Lett 2006;6:413–8.

[75] Wei BQ, Vajtai R, Jung Y, Ward J, Zhang R, Ramanath G, Ajayan PM. Microfabrication technology: organized assembly of carbon nanotubes. Nature 2002;416:495–6.

[76] Wang M, Li ZH, Shang XF, Wang XQ, Xu YB. Field-enhancement factor for carbon nanotube array. J Appl Phys 2005;98:014315.

[77] Hong NT, Yim JH, Koh KH, Lee S, Minh PN, Khoi PH. Field electron emission from free-standing flexible PDMS-supported carbon-nanotube-array films. J Vac Sci Technol B 2008;26:778–81.

[78] Kawai Y, Magario A, Noguchi T. High-brightness electron emission from flexible carbon nanotube/elastomer nanocomposite sheets. Jpn J Appl Phys 2006;45: L1186–9.

[79] Choi WB, Chung DS, Kang JH, Kim HY, Jin YW, Han IT, et al. Fully sealed, high-brightness carbon-nanotube field-emission display. Appl Phys Lett 1999;75: 3129–31.

[80] Choi JH, Park JH, Moon JS, Nam JW, Yoo JB, Park CY, et al. Fabrication of carbon nanotube emitter on the flexible substrate. Diamond Relat Mater 2006;15: 44–8.

[81] Lee CJ, Park J, Han S, Ihm J. Growth and field emission of carbon nanotubes on sodalime glass at 550°C using thermal chemical vapor deposition. Chem Phys Lett 2001;337:398–402.

[82] Hofmann S, Kleinsorge B, Ducati C, Ferrari AC, Robertson J. Low-temperature plasma enhanced chemical vapour deposition of carbon nanotubes. Diamond Relat Mater 2004;13:1171–6.

[83] Hofmann S, Ducati C, Robertson J, Kleinsorge B. Low-temperature growth of carbon nanotubes by plasma-enhanced chemical vapor deposition. Appl Phys Lett 2003;83:135–7.

[84] Min YS, Bae EJ, Oh BS, Kang D, Park W. Low-temperature growth of singlewalled carbon nanotubes by water plasma chemical vapor deposition. J Am Chem Soc 2005;127:12498–9.

[85] Cantoro M, Hofmann S, Pisana S, Scardaci V, Parvez A, Ducati C, et al. Catalytic chemical vapor deposition of single-wall carbon nanotubes at low temperatures. Nano Lett 2006;6:1107–12.

[86] Boskovic BO, Stolojan V, Khan RUA, Haq S, Silva SRP. Large-area synthesis of carbon nanofibres at room temperature. Nat Mater 2002;1:165–8.

[87] Bae EJ, Min YS, Kang D, Ko JH, Park W. Low-temperature growth of single-walled carbon nanotubes by plasma enhanced chemical vapor deposition. Chem Mater 2005;17:5141–5.

[88] Liao H, Hafner JH. Low-temperature single-wall carbon nanotube synthesis by thermal chemical vapor deposition. J Phys Chem B 2004;108:6941–3.

[89] Li WZ, Xie SS, Qian LX, Chang BH, Zou BS, Zhou WY, et al. Large-scale synthesis of aligned carbon nanotubes. Science 1996;274:1701–3.

[90] Vohs JK, Brege JJ, Raymond JE, Brown AE, Williams GL, Fahlman BD. Low-temperature growth of carbon nanotubes from the catalytic decomposition of carbon tetrachloride. J Am Chem Soc 2004;126:9936–7.

[91] Jeong SH, Ko JH, Park JB, Park W. A sonochemical route to single-walled carbon nanotubes under ambient conditions. J Am Chem Soc 2004;126: 15982–3.

[92] Hofmann S, Ducati C, Kleinsorge B, Robertson J. Direct growth of aligned carbon nanotube field emitter arrays onto plastic substrates. Appl Phys Lett 2003;83: 4661–3.

[93] Chhowalla M, Ducati C, Rupesinghe NL, Teo KBK, Amaratunga GAJ. Field emission from short and stubby vertically aligned carbon nanotubes. Appl Phys Lett 2001;79:2079–81.

[94] Teo KBK, Chhowalla M, Amaratunga GAJ, Milne WI, Pirio G, Legagneux P, et al. Field emission from dense, sparse, and patterned arrays of carbon nanofibers. Appl Phys Lett 2002;80:2011–3.

[95] Hong EH, Lee KH, Oh SH, Park CG. In-situ synthesis of carbon nanotubes on organic polymer substrates at atmospheric pressure. Adv Mater (Weinheim) 2002;14:676–9.

[96] Yoon BJ, Hong EH, Jee SE, Yoon DM, Shim DS, Son GY, et al. Fabrication of flexible carbon nanotube field emitter arrays by direct microwave irradiation on organic polymer substrate. J Am Chem Soc 2005;127:8234–5.

[97] Thostenson ET, Chou TW. Microwave processing: fundamentals and applications.. Composites A 1999;30:1055–71.

[98] Galema SA. Microwave chemistry. Chem Soc Rev 1997;26:233–8.

[99] Sim HS, Lau SP, Yang HY, Ang LK, Tanemura M, Yamaguchi K. Reliable and flexible carbon-nanofiber-based all-plastic field emission devices. Appl Phys Lett 2007;90:143103.

[100] Tanemura M, Okita T, Yamauchi H, Tanemura S, Morishima R. Room-temperature growth of a carbon nanofiber on the tip of conical carbon protrusions. Appl Phys Lett 2004;84:3831–3.

[101] Tanemura M, Okita T, Tanaka J, Yamauchi H, Miao L, Tanemura S, Morishima R. Room-temperature growth of carbon nanofibers induced by Ar$^+$-ion bombardment. Eur Phys J D 2005;34:283–6.

[102] Tanemura M, Hatano H, Kitazawa M, Tanaka J, Okita T, Lau SP, et al. Room-temperature growth of carbon nanofibers on plastic substrates. Surf Sci 2006;600:3663–7.

[103] Tan TT, Sim HS, Lau SP, Yang HY, Tanemura M, Tanaka J. X-ray generation using carbon-nanofiber-based flexible field emitters. Appl Phys Lett 2006; 88:103105.

[104] Jo SH, Wang DZ, Huang JY, Li WZ, Kempa K, Ren ZF. Field emission of carbon nanotubes grown on carbon cloth. Appl Phys Lett 2004;85:810–2.

[105] Suzuki K, Matsumoto H, Minagawa M, Tanioka A, Hayashi Y, Fukuzono K, Amaratunga GAJ. Carbon nanotubes on carbon fabrics for flexible field emitter arrays. Appl Phys Lett 2008;93:053107.

[106] Dong L, Jiao J, Tuggle DW, Petty JM, Elliff SA, Coulter M. ZnO nanowires formed on tungsten substrates and their electron field emission properties. Appl Phys Lett 2003;82:1096–8.

[107] Wan Q, Yu K, Wang TH, Lin. CL. Low-field electron emission from tetrapod-like ZnO nanostructures synthesized by rapid evaporation. Appl Phys Lett 2003; 83:2253–5.

[108] Tseng YK, Huang CJ, Cheng HM, Lin IN, Liu KS, Chen IC. Characterization and field-emission properties of needle-like zinc oxide nanowires grown vertically on conductive zinc oxide films. Adv Funct Mater 2003;13:811–4.

[109] Jo SH, Lao JY, Ren ZF, Farrer RA, Baldacchini T, Fourkas JT. Field-emission studies on thin films of zinc oxide nanowires. Appl Phys Lett 2003;83: 4821–3.

[110] Li SY, Lin P, Lee CY, Tseng TY. Field emission and photofluorescent characteristics of zinc oxide nanowires synthesized by a metal catalyzed vapor–liquid–solid process. J Appl Phys 2004;95:3711–6.

[111] Xu CX, Sun XW, Fang SN, Yang XH, Yu MB, Zhu GP, Cui YP. Electrochemically deposited zinc oxide arrays for field emission. Appl Phys Lett 2006;88:161921.

[112] Yang CJ, Wang SM, Liang SW, Chang YH, Chen C, Shieh JM. Low-temperature growth of ZnO nanorods in anodic aluminum oxide on Si substrate by atomic layer deposition. Appl Phys Lett 2007;90:033104.

[113] Cao B, Teng X, Heo SH, Li Y, Cho SO, Li G, Cai W. Different ZnO Nanostructures fabricated by a seed-layer assisted electrochemical route and their photoluminescence and field emission properties. J Phys Chem C 2007;111: 2470–6.

[114] Wei A, Sun XW, Xu CX, Dong ZL, Yu MB, Huang W. Stable field emission from hydrothermally grown ZnO nanotubes. Appl Phys Lett 2006;88:213102.

[115] Jo SH, Banerjee D, Ren ZF. Field emission of zinc oxide nanowires grown on carbon cloth. Appl Phys Lett 2004;85:1407–9.

[116] Lee CJ, Lee TJ, Lyu SC, Zhang Y, Ruh H, Lee HJ. Field emission from well-aligned zinc oxide nanowires grown at low temperature. Appl Phys Lett 2002;81: 3648–50.

[117] Yang HY, Lau SP, Yu SF, Huang L, Tanemura M, Tanaka J, et al. Field emission from zinc oxide nanoneedles on plastic substrates. Nanotechnology 2005; 16:1300–3.

[118] Pradhan D, Kumar M, Ando Y, Leung KT. One-dimensional and two-dimensional ZnO nanostructured materials on a plastic substrate and their field emission properties. J Phys Chem C 2008;112:7093–6.

[119] Zeng B, Xiong G, Chen S, Wang W, Wang DZ, Ren ZF. Field emission of silicon nanowires grown on carbon cloth. Appl Phys Lett 2007;90:033112.

[120] Chen H, Iliev MN, Liu JR, Ma KB, Chu WK, Badi N, et al. Room-temperature deposition of diamond-like carbon field emitter on flexible substrates. Nucl Inst Method Phys Res B 2006;243:75–8.

Flexible Solar Cells Made of Nanowires/Microwires

Jongseung Yoon[1], Yugang Sun[2] and John A. Rogers[3]

[1]*Department of Materials Science and Engineering, Beckman Institute,
University of Illinois at Urbana–Champaign, Urbana, Illinois, USA*
[2]*Center for Nanoscale Materials, Argonne National Laboratory, Argonne, Illinois, USA*
[3]*Department of Materials Science and Engineering, Frederick Seitz Materials Research Laboratory,
University of Illinois at Urbana–Champaign, Urbana, Illinois, USA*

INTRODUCTION

Global energy demand is projected to more than double by 2050 owing to the growth in population and economies [1, 2]. More than 80% of current primary energy consumption is obtained from fossil fuels. The reserve of fossil fuels will be depleted in the next century. Finding sufficient supplies of alternative energy for the future is, therefore, one of society's most daunting challenges. Fortunately, sunlight provides the most abundant carbon-neutral energy source, which far exceeds human needs even in the most aggressive scenarios. For example, the sun delivers 120 000 terawatts of photon radiation on the surface of the Earth. If technologies are able to convert the solar energy deposited on 0.2% of the land with an efficiency of 10%, 24 terawatts of power can be generated to support societal needs (i.e. 13 terawatts). As a result, developing new strategies to harvest solar energy, efficiently with low cost, represents an important challenge. Solar energy conversion approaches can be classified into three categories: solar electricity, solar fuel and solar thermal. Although systems based on each of these three can exploit the solar resource at levels that exceed demand, this chapter focuses on solar electricity (i.e. photovoltaics).

The photovoltaic effect was discovered by A. E. Becquerel in 1839 [3], but the first photovoltaic device for converting solar energy into electricity (i.e. solar cell) was not built until 1883, by Charles Fritts [4]. This first system was constructed with semiconductor selenium coated with a thin layer of gold, resulting in a junction to induce the photovoltaic effect with an energy

CONTENTS

conversion efficiency of $\sim$1%. The first practical photovoltaic device was invented in 1954 by Bell Laboratory scientists, Daryl Chapin, Calvin Fuller and Gerald Pearson [5]. This group used doped silicon (Si) to create p–n junctions and increase light absorption, resulting in efficiencies up to $\sim$6%. This breakthrough invention led to the first solar battery on April 25, 1954. The US satellite Vanguard 1, launched in March 1958, was the first spacecraft with solar panels (arrays of solar cells) to provide power. Since then, solar cells with improved energy conversion efficiencies have been developed through the use of new materials (e.g. GaAs) and multiple junctions. Record efficiencies are higher than 40%, although different research teams reported different numbers with slight differences. For example, the National Renewable Energy Laboratory (NREL), the Fraunhofer Institute in Germany and Spectrolab Inc. have announced the energy conversion efficiencies of their cells as 41.6%, 41.1% and 40.8%, respectively, which were certified by NREL. Uncertified records of 42.8% and 43% were also reported by research groups at the University of Delaware and the University of New South Wales, respectively.

Current, commercially available solar cells are electrically connected and encapsulated as a module that often incorporates a sheet of glass on the front (sun up) side, allowing light to pass through while protecting the semiconductor wafers. These cells, therefore, are mechanically rigid, making them difficult to transport and to integrate with curvilinear surfaces, for applications such as solar-powered electronic textiles, rollable solar cell panels for portable devices, etc. This chapter focuses on progress in the development of flexible solar cells to reduce the weight and cost, to improve shock resistance and to facilitate transportation, storage and installation, compared to conventional systems. Although many studies report flexible devices using thin films of organics, amorphous and polycrystalline Si and other inorganic semiconductors, systems made with one-dimensional (1D) nano/microstructures of inorganic semiconductors with high crystallinity (even monocrystallinity) provide significant enhancement in energy conversion efficiency. The first section summarizes the operating principles of two major classes of photovoltaic devices (i.e. p–n junction solar cells and photoelectrochemical cells) and their feasibility in flexible devices. A following section presents some perspectives on trends for future work.

BASIC OPERATING PRINCIPLES OF SOLAR CELLS

The most important component of a solar cell (or photovoltaic device) is a semiconductor that can strongly absorb photons to excite electrons in the valence band to the conduction band, thereby leaving holes (i.e. positive charges) behind. The separated electrons and holes must diffuse in different directions to avoid electron–hole recombination. The charge separation is usually driven by a potential difference induced by band bending at an interface in the semiconductor. Depending on the characteristics of interface, photovoltaic devices are classified into two major categories: p–n junction cells and photoelectrochemical (PEC) cells.

p–n Junction solar cells

Since the birth of the first practical solar cell, which was based on a p–n junction formed in Si, extensive efforts have been devoted to exploring p–n junctions with new materials and integrating multiple p–n junctions to increase the photon-to-electricity conversion efficiency of the resulting cells. Figure 6.1 presents the configuration of a p–n junction solar cell and the mechanism for charge separation and migration under illumination. A typical cell consists of a layer of p-type semiconductor material (i.e. p-type base) intimately overlaid with a layer of n-type semiconductor (i.e. n-type emitter), leading to the formation of a p–n junction (Figure 6.1a). Formation of the junction causes band bending of both n- and p-semiconductor at the interface, i.e. the conduction and valence bands bend downward in the p-type side and upward in the n-type side, owing to matching of Fermi levels (E_f) of both sides (Figure 6.1b). A back contact and front finger contact serve as both charge collectors and electrical connection to external load. The use of arrayed finger contacts in the front (i.e. the side of n-type emitter) of the cell minimizes the absorption of sunlight by the contacts. When a photon reaches the surface of the n-type emitter, there are several possibilities: (1) the photon passes through both n- and p-type semiconductor layers, usually when its energy is lower than the bandgap (E_g) of the material; (2) the photon is absorbed by the semiconductor, when its energy is higher than

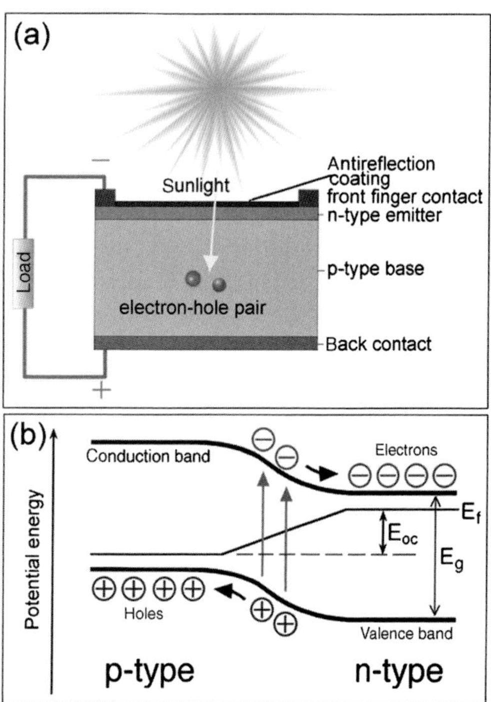

FIGURE 6.1 *(a) Schematic illustration of the operating principles of a p–n junction solar cell for converting photon energy to electricity. (b) Energy levels and charge flow at the p–n junction.*

the bandgap of the semiconductor; and (3) the photon reflects from the semiconductor surface. An antireflection coating layer can minimize the probability of this third outcome.

Photogenerated electrons concentrated in the conduction band and the corresponding holes in the valence band tend to move across the interface of the p–n junction of a solar cell, owing to the action of the electrical field associated with band bending. The diffusion of electrons and holes to different directions, i.e. to the n-type and p-type side, respectively, leads shifts of the Fermi levels of both n- and p-type regions (i.e. upward for n-type and downward for p-type). These shifts result in a built-in potential, which equals the open circuit voltage (V_{oc}) if the cell is ideally equivalent to a current source in parallel with a diode (Figure 6.1b). This potential drives all the mobile charge carriers out of the region adjacent to the junction, which is called the depletion region (or space charge region), under constant light illumination. When ohmic metal–semiconductor contacts are made to both the n-type and p-type sides of the cells and electrodes connect to an external load, electrons created on the n-type side and those that migrate from the p-type side power the load (Figure 6.1a). The electrons continue to move in the wire until they reach the p-type semiconductor, where they recombine with holes created on the p-type side and those that migrate from the n-type side. The performance of a photovoltaic cell is usually characterized by the energy conversion efficiency (η), which corresponds to the percentage of optical energy that is converted to electrical energy when the cell is connected to an electrical circuit. This parameter is determined by the ratio of the maximum power point $(P_m,$ in W) divided by the product of the input light irradiance $(E,$ in W/m^2) under standard test conditions (STC) and the surface area of the cell $(A_c$ in m^2):

$$\eta = \frac{P_m}{E \times A_c}$$

For solar cells, STC specifies a temperature of 25°C and an irradiance of 1000 W/m^2 with an air mass 1.5 (AM 1.5) spectrum, which corresponds to the irradiance and spectrum of sunlight incident on a clear day on a surface at 37 degrees with respect to the sunlight and the sun at an angle of 41.81 degrees above the horizon. P_m is determined by the product of current (I) and voltage (V) at the maximum power point. The overall behavior of a solar cell is also often characterized with a fill factor (FF) that corresponds to the ratio of P_m divided by V_{oc} and the short circuit current (I_{sc}):

$$FF = \frac{P_m}{V_{oc} \times I_{sc}} = \frac{\eta \times A_c \times E}{V_{oc} \times I_{sc}}$$

Conventional p–n junction solar cells are usually manufactured with thick, rigid supports consisting of semiconductor wafers surrounded by glass and metal packaging materials. For example, monocrystalline Si solar cells and polycrystalline Si solar cells consist of thick p-type base Si wafers and glass substrates to support the p–n junctions, respectively. The rigidity associated with these designs restricts application possibilities, and increases the cost and difficulty of transport and installation. Exploiting thin plastic sheets as supports can change this scenario, thereby enabling the possibility of mechanically flexible solar cells. In such a scenario, the p–n junctions (i.e. the critical components of the solar cells), of course, must also be flexible. As discussed in Chapter 10, elementary bending mechanics dictates that even monocrystalline inorganic semiconductors become flexible when their thicknesses are sufficiently small, e.g. on the scale from tens of nanometers to tens of micrometers. In this chapter, flexible solar cells made of transfer-printed microribbons on plastic substrates are discussed, and several examples of modules produced in this manner are presented (see Flexible p–n junction solar cells, below).

Photoelectrochemical solar cells

Electric fields for driving separation of photogenerated electron–hole pairs can also be generated at semiconductor/electrolyte interfaces. Photovoltaic devices based on this class of junctions are referred to as photoelectrochemical (PEC) cells or electrochemical photovoltaic (EPV) cells. Figure 6.2(a) illustrates the

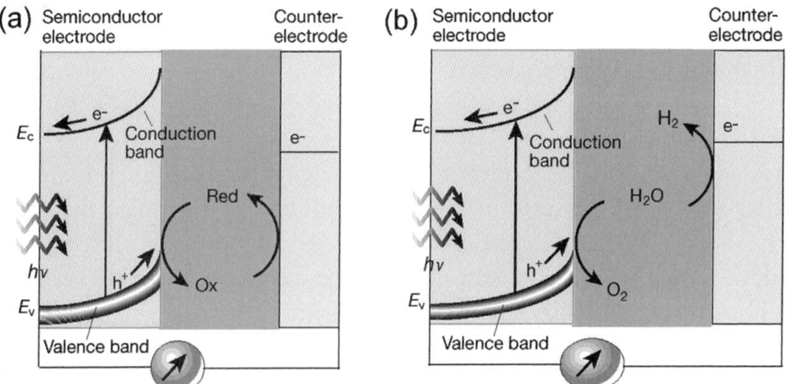

FIGURE 6.2 *Fundamental operating principles of photoelectrochemical cells consisting of n-type semiconductor photoanodes under different energy conversion pathways: (a) photon-to-electricity conversion in a regenerative cell; and (b) photon-to-chemical fuel (e.g. hydrogen) through photon-splitting of water. (Reprinted with permission from [6] © 2001 Macmillan Magazines Ltd.)*

operating principles of a PEC cell constructed with an n-type semiconductor/ electrolyte interface, where the conduction and valence bands of the semiconductor bend upward [6]. The band bending provides the driving force to separate the electron–hole pairs generated in the semiconductor when the cell is under illumination with photons having energies higher than the bandgap of the semiconductor. Electrons in the conduction band flow from the junction to the bulk of the semiconductor, while leftover holes in the valence band migrate from the bulk to the junction. When the cell is connected to an external load through an ohmic metal–semiconductor contact and a metal counter-electrode, the electrons move out of the semiconductor to the wire to drive the external load. The electrons continue to flow into the counter-electrode and jump into the electrolyte. In parallel, the holes pass through the semiconductor/electrolyte interface into electrolyte to oxidize appropriate species in the electrolyte. The oxidized product (Ox) recovers to its original state (Red) through reduction by the electrons created in the conduction band that travel into the electrolyte. As a result, no net change occurs in the cell during operation. PEC cells might offer advantages in cost and ease of fabrication compared with p–n junction solar cells. For example, the junction for efficient charge separation can be easily formed in a PEC cell by simply immersing a semiconductor electrode in an appropriate electrolyte. This process avoids the need to form p–n junctions associated with cells of the type described in the previous section. The liquid electrolyte provides a natural, readily conformable and strain-free junction.

If the band structure of a semiconductor photoanode is appropriate, the photogenerated electrons and holes can drive reactions to produce chemical fuels (Figure 6.2b). For example, water molecules can be directly split in oxygen and hydrogen when the edge energy levels of the conduction and valence bands of the photoanode are lower than the reduction potential of H^+/H_2 and higher than the reduction potential of O_2/H_2O, respectively. Hydrogen is a clean energy carrier and can be converted into electricity in fuel cells. The conversion of solar energy directly into fuels in this manner can eliminate the need for external wires and a separate electrolyzer. More important, if a third electrode is added to a PEC cell, in situ chemical storage can be achieved, thereby providing power 24 hours a day.

The direct contact of a semiconductor photoanode and electrolyte always leads to oxidation and corrosion of the anode (e.g. Si and GaAs) due to the migration of holes created in the valence band of the semiconductor across the photoanode/electrolyte interface. The corrosion process results in gradual degradation of performance. Semiconductor oxides with large bandgaps are promising as electrodes because of their resistance to corrosion. However, the large bandgaps prevent such electrodes from efficiently absorbing sunlight.

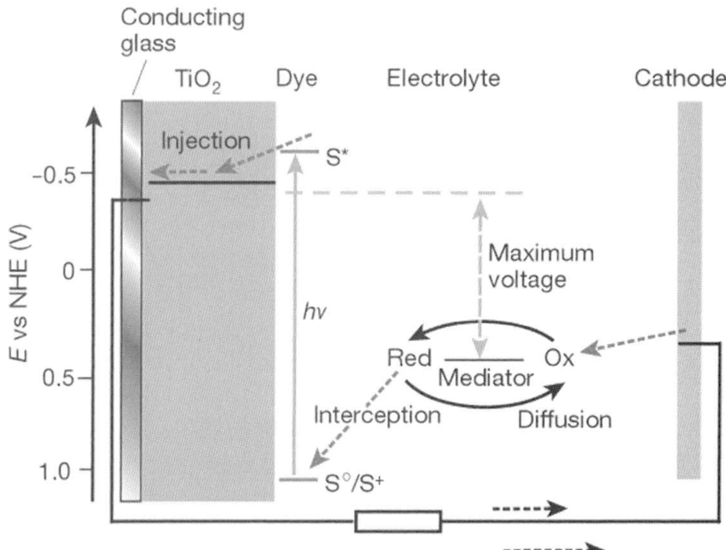

FIGURE 6.3 *Operating principle of a dye-sensitized photoelectrochemical cell with a photoanode of large bandgap semiconductor, such as TiO₂. In a typical case, photoillumination excites the dye molecules from ground state to an excited state with an energy level higher than the conduction band of TiO₂, resulting in the injection of electrons from the dye molecules to the conduction band of TiO₂. This charge separation leads to the oxidation of the dye molecules. The oxidized dye molecules in turn oxidize the mediators, redox species dissolved in the electrolyte, and they are reduced back to original dye molecules for sunlight harvesting. The electrons migrate from the TiO₂/electrolyte interface to the electron collector electrode, and then travel through the wires and load to the counter-electrode for reinjection into electrolyte to reduce the oxidized mediator molecules. (Reprinted with permission from [6] © 2001 Macmillan Magazines Ltd.)*

Advances led by M. Grätzel, beginning in 1991, circumvent this challenge through the addition of strongly absorbing dye molecules to the electrode surfaces. In one example, the surfaces of large-bandgap TiO_2 photoanodes are decorated with dye molecules to absorb sunlight while retaining the inherent chemical stability of TiO_2 [7]. Figure 6.3 shows the principle of such a dye-sensitized PEC cell, where the processes of absorption of sunlight and generation of charges are separated [6]. Photons excite the dye molecules from their ground state ($S°$) to an excited state (S^*), the energy level of which is higher than the energy of the conduction band edge of TiO_2. This energy difference results in the injection of an electron from an excited dye molecule into the conduction band of the oxide. The electrons collected in the conduction band travel through an external load to the counter-electrode by following the same path as shown in Figure 6.2. The oxidized dye molecules (S^+) are regenerated by accepting electrons from electron donors in the electrolyte, which usually consists of an organic solvent containing a redox system, such as I_3^-/I^-. The resulting triiodide ions can be reduced back to iodide ions at the counter-electrode, where photoexcited electrons are collected after traveling through the complete circuit. In this manner, the dye-sensitized solar cell (DSC) efficiently converts sunlight to electricity without suffering any permanent chemical transformation. The voltage generated under illumination corresponds to the difference between the Fermi level of the electron in the oxide and the reducing potential of the redox pair in the electrolyte solution.

Photoanodes of classic DSCs consist of mesoporous TiO_2 films on transparent conductive supports, formed by various approaches including sol-gel process [8–10], paste screen printing [11], dip-coating combined with layer-by-layer assembly [12, 13] and atomic layer deposition (ALD) [14–17]. Figure 6.4(a) shows a scanning electron microscopy (SEM) image of a mesoscopic TiO_2 film prepared by depositing hydrothermally synthesized TiO_2 colloids followed with high-temperature annealing [18]. The sintering process at high temperature converts TiO_2 to the anatase phase and fuses neighboring nanoparticles. The fusion generates crystalline interconnections between nanoparticles, resulting in a transport pathway for electrons from the TiO_2 to the conductive support, which serves the electron collector (Figure 6.4b). Glass slides or flexible plastic sheets coated with transparent conductive layers of fluorine-doped tin dioxide (FTO) or tin-doped indium oxide (ITO) usually serve as the photoanode supports. By considering the flexibility associated with thin TiO_2 films (with thickness of $< 20\ \mu m$) and electrolyte layers (usually in liquid phase), flexible DSCs can be fabricated by using transparent plastic substrates covered with transparent conductive layers as electron collectors and counter-electrodes. In this configuration, excitation of dye molecules can be carried out from both sides, i.e. front illumination from the side of photoanodes and back illumination from the side of counter-electrodes. Front illumination is more efficient because unwanted absorption of sunlight by electrolytes and reflection of counter-electrodes can be eliminated (Figure 6.4b). Back illumination is required when the photoanode support is opaque. Figure 6.4(c) provides the configuration of a flexible DSC with a Ti foil (0.2 mm in thickness) as a support for a mesoporous TiO_2 thin film. A polyethylene naphthalate (PEN) sheet coated with an ITO layer serves as the counter-electrode after it is decorated with Pt catalyst through electrochemical deposition [19]. The modification with Pt catalyst does not significantly decrease the transparency of the ITO/PEN sheet. As a result, back illumination of a DSC shown in Figure 6.4(c) can still generate high energy conversion efficiency (7.2%), in particular, when the thickness of the electrolyte layer is minimized. The use of Ti foils rather than plastic sheets as supports for the

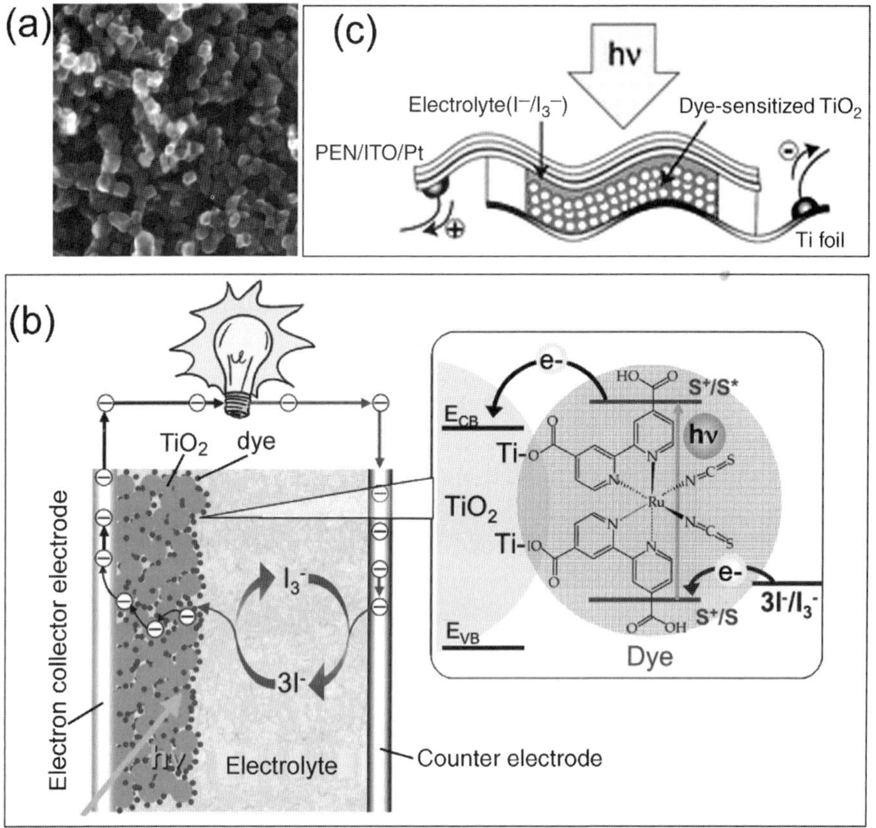

FIGURE 6.4   *(a) SEM image of an annealed mesoscopic film of colloidal TiO₂ nanoparticles supported on a fluorine-doped tin dioxide (FTO) glass substrate. The average size of the nanoparticles is ∼20 nm. (b) Schematic illustration of the typical configuration of the most popular dye-sensitized photoelectrochemical (PEC) cell, which uses dye-derivatized TiO₂ nanoparticles to harvest light. The sensitizing dye molecules are cis-Ru(SCN)₂L₂, where L represents 2,2′-bipyridyl-4,4′-dicarboxylate. The mediator redox couple is iodide/triiodide dissolved in electrolyte. (Modified with permission from [18] © 2005 American Chemical Society.) (c) Schematic configuration of a flexible dye-sensitized PEC cell with photoanode as shown in (a) on a flexible Ti foil. PEN: polyethylene naphthalate; ITO: indium–tin oxide.*
(Reprinted with permission from [19] © 2006 The Royal Society of Chemistry.)

TiO_2 film allows high-temperature annealing to convert TiO_2 to pure anatase phase, which is critical for efficient charge separation under illumination. The electrolyte in the cell of Figure 6.4(c) is a solution of 0.60 M butylmethylimidazolium iodide (BMII), 0.03 M I_2, 0.10 M guanidinium thiocyanate (GuNCS) and 0.50 M 4-*tert*-butylpyridine in a mixture of acetonitrile and valeronitrile (85:15 v/v) and is sealed between the photoanode and counter-electrode with a hot-melt gasket with 25 μm thickness, made of ionomer Surlyn 1702 (DuPont). Using organic solvents increases difficulties in encapsulation and operation, in particular, when DSCs have large areas. Recently, electrolytes in ionic liquids [20–23] and polymeric matrices [24, 25] have been evaluated for solvent-free DSCs. Such devices exhibit comparable efficiencies (∼9% versus ∼11%) to those with solvent electrolyte at optimized conditions.

DSCs made of mesoporous TiO_2 photoanodes are possibly flexible, with special configurations as shown in Figure 6.4(c), but no results have been reported on how bending influences the performance of the cells. Recent studies conclude that increasing the organization of mesoporous TiO_2 films and/or using arrayed tubular nanowires of TiO_2 can enhance the performance of resulting DSCs [26–30]. Most important, arrays of vertically oriented nanowires on a plastic substrate exhibit much higher resistance to fracture or deformation than porous films when the substrate is bent (Figure 6.5) [31]. As highlighted in Figure 6.5(a), bending the plastic substrate generates compressive or tensile stress in the mesoporous network of colloidal TiO_2 nanoparticles. When this stress exceeds a critical value, the network deforms via cracks and/or delamination from the substrate. By comparison, arrayed nanowires can efficiently release the bending stress in the film through adjustments of gaps between the nanowires (Figure 6.5b). The next section focuses on flexible PEC cells with such designs.

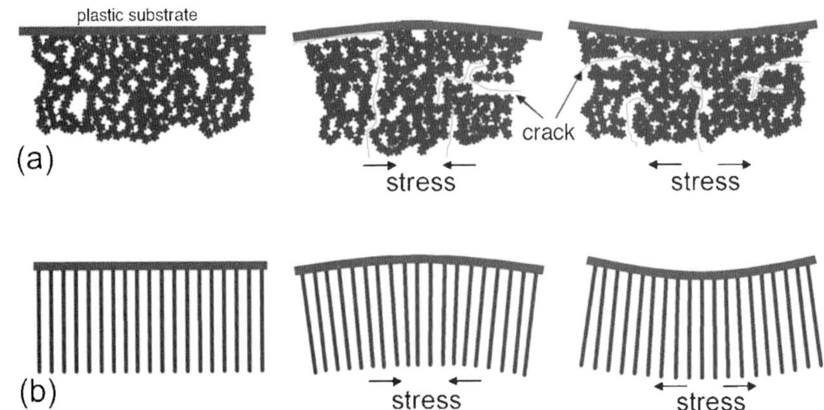

FIGURE 6.5 *Schematic illustration of structural responses of (a) mesoscopic porous network and (b) aligned nanowire array under tensile and compressive strains. (a) Stresses in the mesoporous network film are released through the formation of cracks in the film or delamination from the supporting substrate. (b) Stresses in nanowire array are released through adjustment of the wire/wire gaps.*
(Reprinted with permission from [31] © 2008 American Institute of Physics.)

FLEXIBLE P–N JUNCTION SOLAR CELLS

In this section, a type of flexible p–n junction solar cell is discussed that uses printed assemblies of ultrathin Si cells (i.e. microcells) derived from bulk monocrystalline wafers. This approach, pioneered by Rogers and co-workers at the University of Illinois at Urbana–Champaign, represents a clear example of how inherently rigid and fragile monocrystalline semiconductor materials can be used as active building blocks for high-performance and mechanically flexible p–n junction solar cells and modules. The outcomes rely critically on anisotropic etching methods to create the microcells, dry transfer printing techniques to assemble them into modules, and mechanical analysis for optimized layouts.

Fabrication of arrays of ultrathin monocrystalline silicon microsolar cells

As discussed in Chapter 3, monocrystalline nanoscale and microscale structures such as wires, ribbons, bars and membranes with thicknesses in the range of a few hundred nanometers to tens of micrometers can be generated from bulk Si (111) wafers [32]. The fabrication scheme exploits anisotropic (i.e. orientation dependent) etching of Si with wet chemical etchants, where the etching front in one crystallographic direction advances much faster than in other directions owing to different chemical reactivities of certain crystallographic planes. For example, potassium hydroxide (KOH), a well-known anisotropic chemical etchant for Si, etches the (110) plane of Si about 600 times more rapidly than the (111) plane [33, 34]. In this etching process, Si atoms at the surface are oxidized to silicon dioxide (SiO_2) by reducing water molecules (H_2O) to hydrogen (H_2) (Eqn 6.1). The SiO_2 formed at the Si surface complexes with OH^-, which can then be dissolved in the aqueous solution for the continued etching of underlying Si (Eqn 6.2):

$$Si + 2H_2O \rightarrow SiO_2 + 2H_2 \tag{6.1}$$

$$SiO_2 + 2OH^- \rightarrow H_2SiO_4{}^{2-} \tag{6.2}$$

Such anisotropic wet chemical etching processes combined with top–down lithographic techniques can create printable forms of microstructures and nanostructures of single-crystalline Si with precise control over the geometries and spatial layouts. Figure 6.6 depicts fabrication steps to produce monocrystalline Si nanoribbons/microribbons or bars with rectangular cross-sections. The process begins with patterning lines of mask layers (e.g. photoresist or

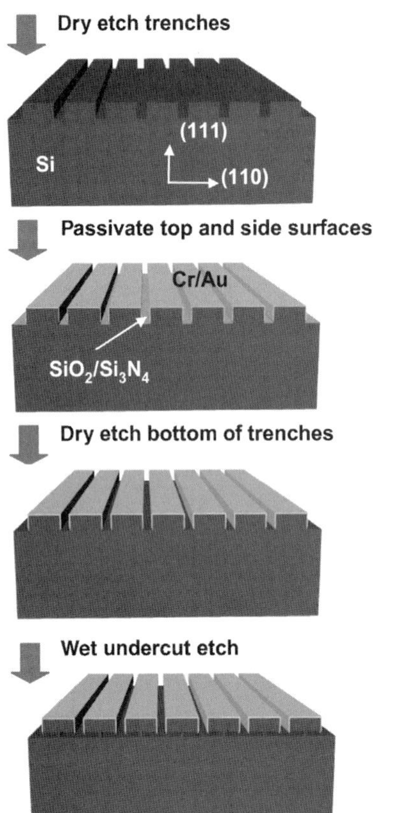

FIGURE 6.6 *Schematic illustration of the fabrication steps to create microscale and nanoscale ribbons or bars of monocrystalline Si from bulk wafers by use of anisotropic wet-chemical etching techniques.*

SiO_2) perpendicular to the Si (110) planes, followed by dry etching to form trenches on the Si (111) wafer. In the next step, SiO_2 and Si_3N_4 are deposited as protective layers over the entire surface. Electron-beam evaporation of metals (e.g. Cr/Au) at an oblique angle and dry plasma etching of unprotected SiO_2 and Si_3N_4 exposes Si (110) planes at the bottom edges of the trenches. Etching by KOH then proceeds along the [110] direction until the two etching fronts meet each other to complete the undercut process.

Arrays of nanoribbons/microribbons or bars of monocrystalline Si are unique platforms for high-performance Si solar cells, partialy because they enable module designs that might be difficult to achieve using conventional wafer-based technologies. The resulting systems may offer new opportunities for monocrystalline Si photovoltaics. The fabrication of microbar solar cells (or microcells) begins with trench formation by top–down photolithography and inductively coupled plasma reactive ion etching (ICP RIE, STS) on a p-type Si (111) Czochralski wafer (resistivity: 10–20 Ωcm), where the long axes of trenches are aligned perpendicular to the Si [110] direction of the wafer as described in Figure 6.6. The spacing between trenches defines the width of microcells, while their depth, controlled by RIE etching time, determines the thickness of microcells. Figure 6.7(a) shows an SEM image of the resulting trench structures after the deep RIE process and removal of masking materials. After the trench formation, electronic junctions and electrical contacts are implemented into the structure by doping processes. Thermal diffusion of solid-state phosphorus (PH-1000N, Saint-Gobain) and boron sources (BN-1250, Saint-Gobain) as n- and p-type dopants creates rectifying p–n junctions as well as p+ and n+ ohmic contacts on the surface of trenches. SiO_2 (900 nm) layers grown by plasma-enhanced chemical vapor deposition (PECVD) serve as patterned diffusion barriers for selective area doping, with assistance of photolithography and etching with buffered

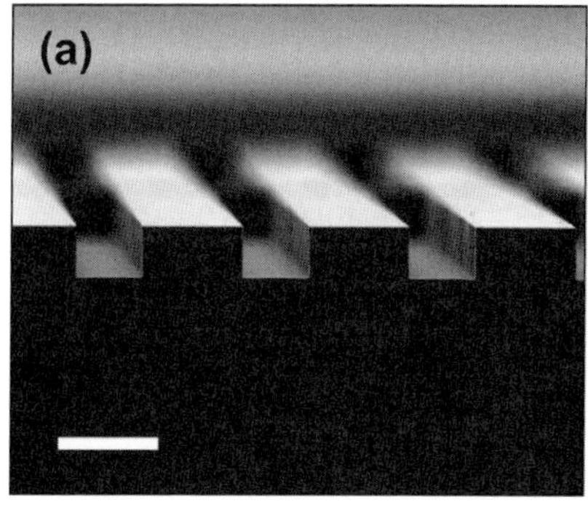

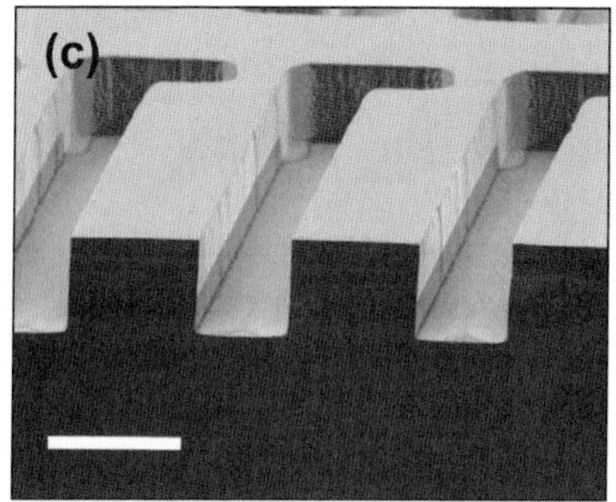
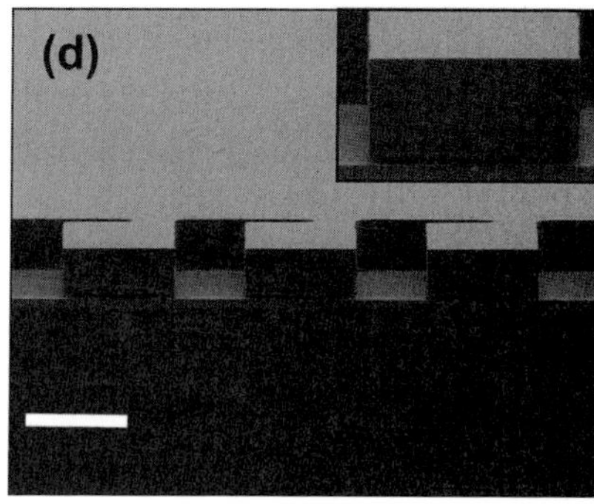

FIGURE 6.7 *SEM images of (a) trenches formed on a Si (111) wafer after dry etching, (b) trenches with patterned SiO₂ window (bright region) for boron doping, (c) trenches after deposition of passivation layer and angled evaporation of metals, and (d) microcells after KOH undercut etching. All scale bars are 50 μm.*
(Reprinted with permission from [37], © 2008 Macmillan Magazines Ltd.)

oxide etchant (BOE). Figure 6.7(b) shows an SEM image of the trenches with patterned SiO_2 masking layer for boron doping. Once the doping processes are completed, the entire surfaces of trenches are coated with PECVD-grown SiO_2 (100 nm) and Si_3N_4 (300 nm) as passivation layers to protect the Si at the top surfaces and sidewalls of trenches from KOH undercut etching. Angled electron-beam evaporation of metals [e.g. Cr/Au (8/80 nm)] provides partial coverage of the top and sidewalls of trenches (Figure 6.7c). CHF_3/O_2 plasma etching (40/2 sccm, 50 mTorr, 150 W, 10 min) removes unprotected SiO_2 and Si_3N_4 at the bottom surface of trenches, and thus selectively exposes Si (110) planes. A hot KOH solution (e.g. 80°C) then undercuts the trench to create the three-dimensional shape of the microcells, which remain tethered to the source wafer at both ends (Figure 6.7d). The newly created bottom surfaces of the microbars resulting from the undercut etching are then doped with boron to form a p+ back surface field (BSF), yielding fully functioning and print-ready microcells on the donor substrate.

Transfer printing, planarization and interconnect formation

Arrays of Si microcells fabricated in the manner described above can be retrieved from the source wafer, and delivered onto foreign substrates (i.e. receiving substrate) using deterministic, dry transfer assembly techniques. A key advantage of this printing-based approach is that it separates fabrication steps that are optimized for and compatible with the source wafer from receiving substrates. In this way, monocrystalline Si photovoltaic modules can be formed on various types of cost-effective and unconventional substrates including mechanically flexible, thin sheets of plastics. In a typical transfer-printing process, an elastomeric poly(dimethylsiloxane) (PDMS) stamp is laminated against the top surfaces of Si microcells on a source wafer, where conformal contact with the stamp occurs spontaneously owing to van der Waals' interactions between PDMS and Si. Relief features on the stamp can be designed to retrieve a selected fraction of cells, thereby providing a route to arbitrary layouts of cells on the receiving substrate. Because of the viscoelastic nature of PDMS, the adhesion between the stamp and Si is strongly affected by the rate of separation [35]. For example, peeling back the stamp from the donor substrate with a sufficiently high speed leads to adhesion between Si and the surface of the PDMS stamp that is strong enough to separate microcells efficiently from the donor substrate with their lithographically defined order and alignment maintained. As described in the previous section, the fabricated microcells are freestanding along their lengths, and partially connected through narrow and sharp 'anchor' regions to the donor substrate at both ends. Such anchoring elements, which maintain the lithographically defined cell layout even after complete undercut etching, are designed to fracture during the retrieval action of the stamp owing to stress concentration [36]. The stamp 'inked' with microcells can then be used to transfer them onto virtually any type of target substrate including low-cost, lightweight and mechanically flexible plastics with the aid of thin adhesive layers. Given their rectangular cross-section and thickness (e.g. ~15–20 μm), planarization of microcells on the receiving substrate becomes critically important for their interconnection to form modules. Among many different kinds of thin film adhesives available, photocurable polymers (e.g. epoxy and polyurethane) are particularly useful because they can serve not only as adhesives for printing but also as planarizing media. Figure 6.8(a) illustrates transfer printing where a flat stamp is used, to enable printing and planarization in a single step with a photocurable polyurethane (e.g. NOA, Norland products). As the 'inked' flat stamp (Figure 6.8b) is gently pressed down on the receiving substrate, the NOA material spontaneously fills the volumes between adjacent microcells through capillary action. Curing under ultraviolet (UV) light while the stamp is in

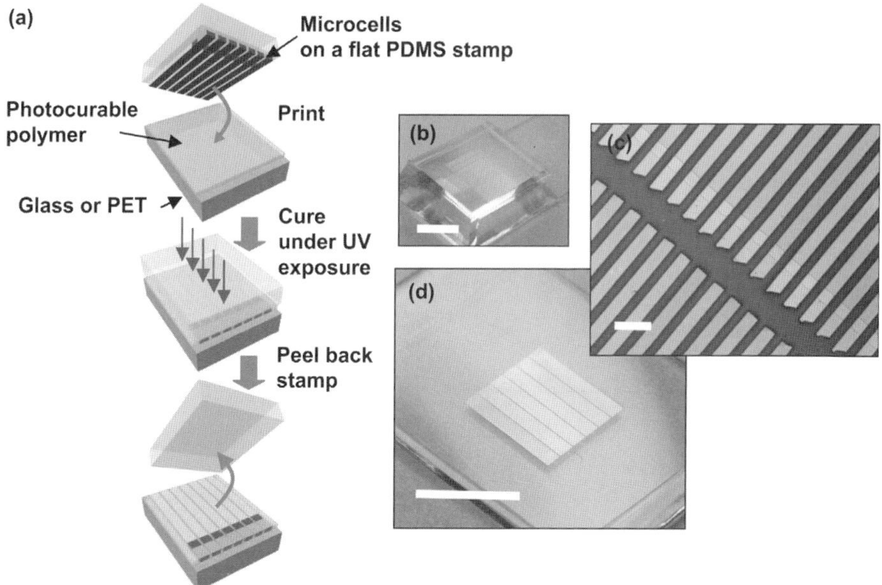

FIGURE 6.8   *(a) Schematic illustration of processing steps for printing and planarizing Si microcells in a single step using a flat poly(dimethylsiloxane) (PDMS) stamp, where the photocurable polymer (NOA 61, ~30 μm in thickness) is used as both an adhesive and a planarizing medium. (b) Optical image of an array of microcells on a flat elastomeric PDMS stamp, immediately after retrieval from a source wafer. (c) Optical microscope image of an array of microcells on a PDMS stamp. (d) Optical image of a printed array of microcells on a glass substrate.*
(Reprinted with permission from [37], © 2008 Macmillan Magazines Ltd.)

contact, and then peeling back the stamp, yields printed and planarized arrays of microcells on the receiving substrate (Figure 6.8c, d).

Stamps with appropriately designed relief features can selectively retrieve cells and print them in layouts that are different from those in the source wafer, in a step and repeat manner. For example, arrays of microcells with large spacings can be fabricated via this approach for modules with definable levels of transparency or for ultrathin form-factor concentrator modules that use microlens arrays [37]. Automated printing systems can be used for high-throughput, high-efficiency and high-accuracy processing. A typical system consists of x-, y-, z-axis translation, tilt and rotation stages to control the position and orientation of a stamp relative to source and target substrates with precise control. Alignment and contact forces between the stamp and Si can be manipulated by load cells and integrated vision systems, all under computer control. Figure 6.9(a) shows the tool used for printing microcells described here. In a repetitive printing process, partially cured and surface-activated

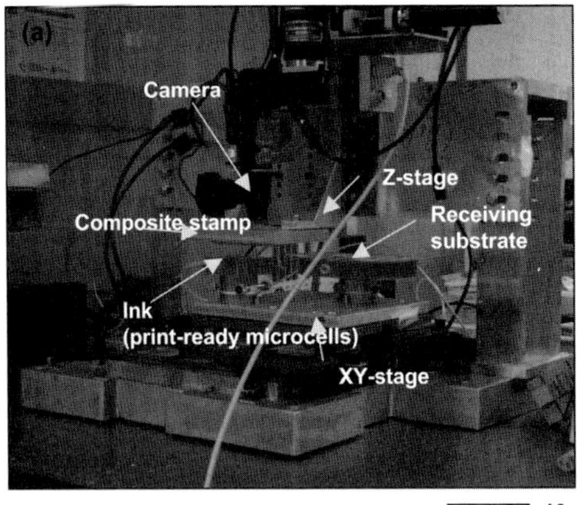

FIGURE 6.9 *(a) Optical image of automated printing machine. (b) Schematic illustration of processing steps for planarizing printed Si microcells, where microcells are printed at selected cell spacings on a poly(dimethylsiloxane) (PDMS)-coated substrate. (c) Optical image of printed microcells on a polyethylene terephthalate (PET) substrate with intercell spacing of ~400 μm.*
(Reprinted with permission from [37], © 2008 Macmillan Magazines Ltd.)

(e.g. by exposure to UV ozone) thin PDMS layers can be used as substrate adhesives. In this case, planarization using photocurable polymers is performed as a final step after multiple printing cycles. The process involves spin-coating the polymer onto the printed arrays of microcells followed by contact of a flat PDMS stamp, UV curing and removing the stamp. Figure 6.9(b, c) shows schematic illustrations of this planarization approach, and an image of sparse arrays of microcells printed on a poly(ethyleneterephthalate) (PET) substrate. The distance between microcells is ~400 μm, while the spacing on the source wafer was 26 μm. Overall yields of the entire processes including cell fabrication, printing and planarization are >99.5%.

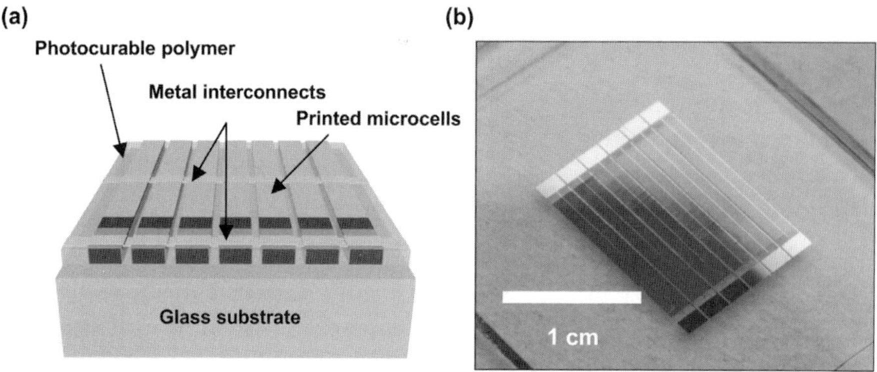

FIGURE 6.10   *(a) Schematic illustration and (b) optical image of the completed module. (Reprinted with permission from [37], © 2008 Macmillan Magazines Ltd.)*

After printing and planarization, metal interconnects can be implemented by various methods, including etch-back, lift-off or shadow mask patterning of vacuum-evaporated metals, or direct ink writing of conductive inks [38]. Figure 6.10(a, b) shows a schematic illustration and a photographic image of a completed module, where approximately 130 microcells are interconnected with each metal line (Cr/Au: 10 nm/600 nm). The substrate is glass, and the photocurable polyurethane material serves as the printing and planarizing medium. Figure 6.11 illustrates the entire process of cell fabrication and

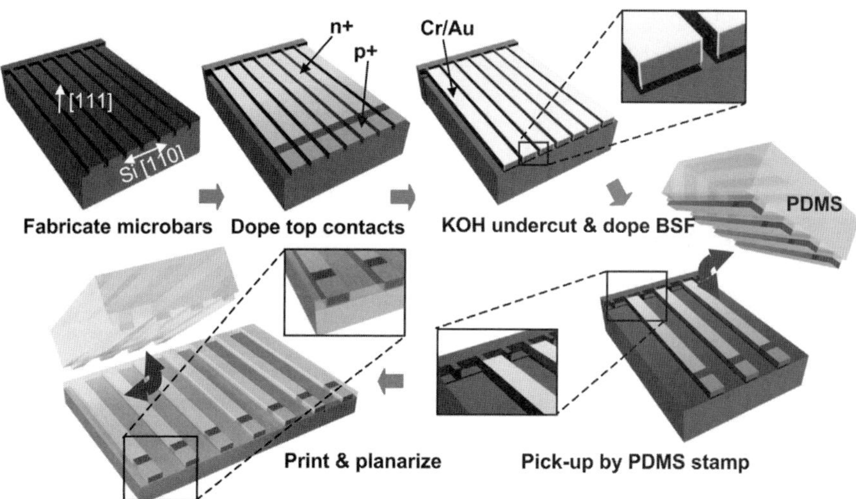

FIGURE 6.11   *Schematic illustration of steps for fabricating ultrathin microcells from a bulk wafer, and printing them onto a target substrate. KOH: potassium hydroxide; BSF: back surface field; PDMS: poly(dimethylsiloxane).*

transfer printing. Overall, this printing-based approach offers many advantages over conventional Si solar cell technologies, including the use of low-cost and flexible substrates, large-area coverage, area multiplication, unusual module fabrication, and so on.

Photovoltaic characteristics of printed microcells

Figure 6.12(a) shows a representative doping design for a microcell [length (L) = 1.55 mm] with phosphorus-doped (L_{n+} = 1.4 mm), boron-doped (L_{p+} = 0.1 mm) and undoped (L_p = 0.05 mm) regions. Surface doping concentrations of the n+ (phosphorus), p+ (boron) and back surface field (boron) regions are 1.2×10^{20} cm^{-3}, 1.8×10^{20} cm^{-3} and 5.8×10^{19} cm^{-3}, respectively, as

(a)

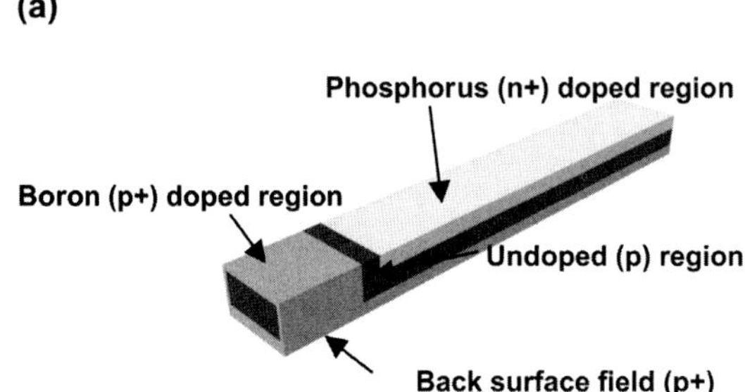

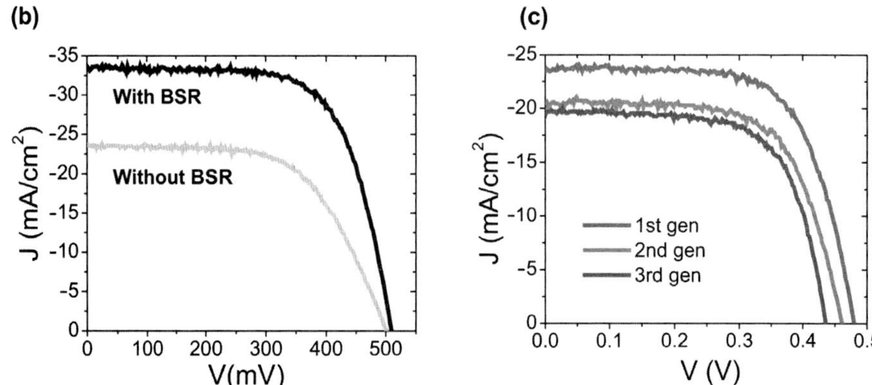

FIGURE 6.12 *(a) Schematic illustration of microcells' doping layout. (b) Representative current density (J) and voltage (V) characteristics of an individual microcell. BSR: back-side reflector. (Reprinted with permission from [37] © 2008 Macmillan Magazines Ltd.) (c) J–V curves of individual microcells corresponding to the first, second and third generations from a single source wafer, with thickness of ~15 μm.*

measured by secondary ion mass spectrometry (SIMS). There are two critically important aspects in these doping configurations. First, the boron region on the top surface is connected to back surface field on the bottom through doping of the side surfaces. As a result, both emitter (n+) and base (p+) contacts of microcells are accessible from the same surface (i.e. top), thereby greatly simplifying the printing and interconnection steps. For example, microcells can be printed on electrically non-conductive, plastic substrates using polymeric glue layers without any complication in access to the base contact. Second, the phosphorus-doped area of the top surface is partially extended down the side-walls to roughly one-third of the cell thickness (Figure 6.1b). Such extended emitter doping prevents electrical shorts that would otherwise be caused by metal interconnects in the n+ region, due to slight imperfections in the planarization process.

Photovoltaic characteristics of the printed microcells and modules were measured under simulated AM 1.5 illumination (1000 W/m^2) at room temperature. Figure 6.12(b) shows representative current density (J)–voltage (V) curves of individual microcells on an NOA-coated glass substrate with and without a diffuse back-side reflector (BSR). These microcells are not covered with antireflection coatings or treated with surface texturization. Without a BSR, a typical microcell with $\sim$15 μm thickness exhibits a short-circuit current density, J_{sc}, of $\sim$23.6 mA/cm^2, an open-circuit voltage, V_{oc}, of $\sim$503 mV, a fill factor of $\sim$0.61 and overall solar-energy conversion efficiency (η) of $\sim$7.2%. The efficiency calculation uses the spatial dimensions (i.e. the top surface area) of the microcells; contributions of light incident on the edges are not included. The average efficiency of microcells of 15–20 μm thickness is in the range of 6–8%. This value increases to 10–13% with a diffuse BSR on the back side of the glass substrate.

The fabrication processes allow the source wafer to be reused multiple times for nearly full utilization of the material. After a given set of microcells is transferred, the surface of the source wafer can be chemically polished in wet chemical etchants of Si [e.g. KOH or tetramethylammonium hydroxide (TMAH)] and cleaned, to prepare the wafer for another cycle of cell fabrication steps. This process, in principle, can be repeated until the entire wafer is consumed. For example, Figure 6.12(c) shows J–V curves of the first, second and third generation microcells produced from the same source wafer. The strategy for fabricating microcells with thinner Si for large-area modules significantly increases the efficiency in utilizing Si without scarification of energy conversion efficiency in comparison with commercial monocrystalline Si solar cell modules fabricated with conventional Si wafers with thickness of $\sim$300 μm (or even thicker). The dependence of performance on the thickness of Si has been evaluated by both experimental measurements and simulation on the microcells shown in Figure 6.12(a), leading to a consistent conclusion: energy conversion efficiency sharply increases with thickness of the Si cells up to $\sim$15 µm, followed by a gradual saturation from 20 to 30 µm to a plateau above $\sim$40 µm. The increase in efficiency with larger thickness is mainly ascribed to increased light absorption associated with the longer optical path lengths. The total absorption does not significantly increase when the thickness of Si is above $\sim$40 µm. However, the bulk recombination of minority carriers in thicker Si increases, resulting in worse tolerance on impurity. As a result, using thin monocrystalline Si down to $\sim$40 µm for solar cells can increase material utilization efficiency as well as reduce the cost associated with growth of highly pure materials. Current techniques in industry, however, cannot afford cost-effective capability to slice monocrystalline Si to wafers with thickness less than 300 µm. Therefore, the approach described for fabrication of microcells in this chapter represents a promising strategy for efficient utilization of Si and improvement in performance with low-quality materials.

Integration of mechanically flexible modules and photovoltaic characteristics

Ultrathin microcells can also be further exploited for module designs that may be difficult to achieve by conventional wafer-based Si solar cell technologies (Figure 6.13a). In the design of mechanically flexible modules with inherently rigid and fragile materials, optimized mechanical designs are needed to maintain maximum strains below the intrinsic elastic deformation limit of constituting materials. A heterogeneous composite structure (Figure 6.13b) composed of Si microcells, metal interconnects and polymer encapsulating layers can be configured to isolate the most critical and mechanically fragile parts (i.e. Si and metal interconnects) from bending induced strain by implementing what is called a neutral mechanical plane. This plane corresponds to the position where there is zero strain under bending. For a homogeneous sheet, this plane is located exactly halfway between the top and bottom surfaces. The position of the neutral mechanical plane of the composite structure can be determined by analytical calculation, where the system is modeled as a composite beam as shown in Figure 6.13(c), where W, W_{Si} and W_{NOA} are the widths of the beam, the Si microcell and the distance between adjacent microcells, respectively, and t, t_m, b and $a-t$ are the thicknesses of the microcell, metal interconnect layer and polymer (NOA) layers above and below the microcell. The strain is then given by $\varepsilon_{yy} = (z - z_0)/R$, where R is the bending radius and z_0 is the position of the neutral mechanical plane. The quantity of

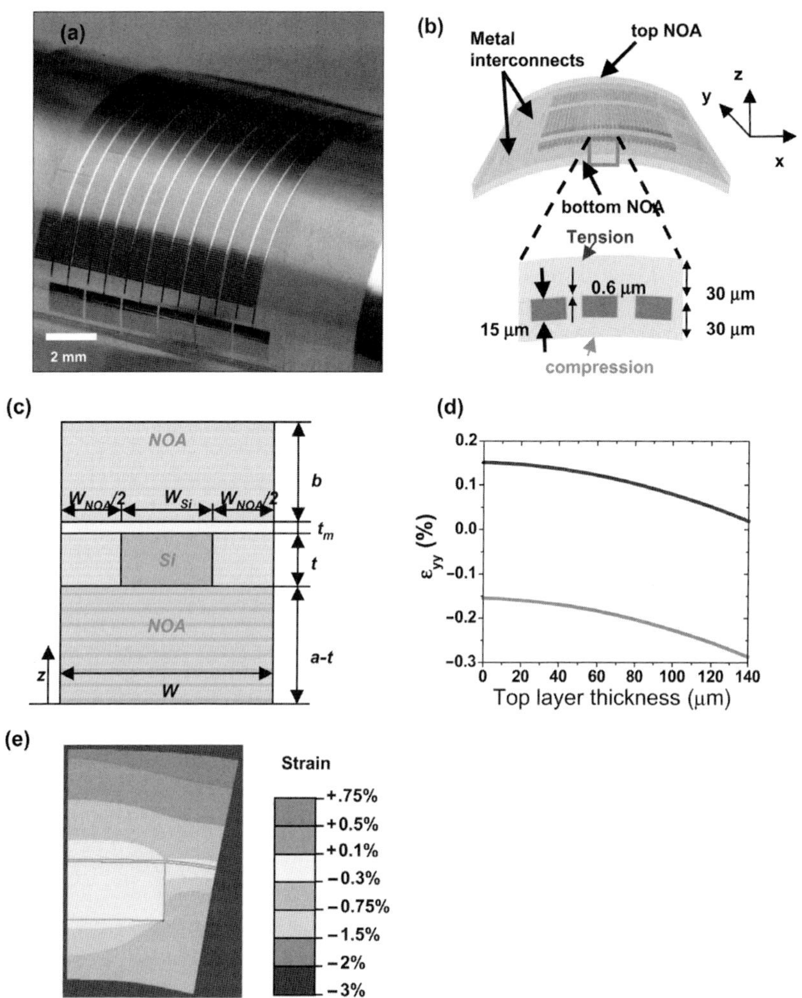

FIGURE 6.13 *(a) Optical image of a module bent along a direction parallel to the widths of the microcells, to a bending radius (R) of 4.9 mm. (Reprinted with permission from [37] © 2008 Macmillan Magazines Ltd.) (b) Schematic illustration of an optimized design in which the neutral mechanical plane is positioned near the center of the microcells. (c) Cross-sectional schematic illustration of a model composite structure composed of Si microcell, metal layer and polymer encapsulation layer, with key parameters. (d) Analytically calculated bending strains (ε_{yy}) at the top (top line) and bottom (bottom line) surface of Si microcell bent along the cell length direction as a function of the top polymer layer thickness. (e) Color contour plot of calculated bending strain (ε_{xx}) through the cross-section of a mechanically flexible microcell module, bent along the cell width direction at R = 4.9 mm.*

z_0 is given by:

$$z_0 = \frac{a-t}{2} \times$$

$$\frac{\left(1+\frac{b}{a-t}\right)^2 + 2\frac{b}{a-t}\frac{t+t_m}{a-t} + \frac{W_{NOA}}{W}\frac{t}{a-t}\left(2+\frac{t}{a-t}\right) + \frac{E_{Si}t}{E_{NOA}(a-t)}\frac{W_{Si}}{W}\left(2+\frac{t}{a-t}\right) + \frac{E_{Au}t_m}{E_{NOA}(a-t)}\left(2+\frac{2t+t_m}{a-t}\right)}{1 + \frac{b}{a-t} + \frac{W_{NOA}}{W}\frac{t}{a-t} + \frac{E_{Si}t}{E_{NOA}(a-t)}\frac{W_{Si}}{W} + \frac{E_{Au}t_m}{E_{NOA}(a-t)}}$$

Figure 6.13(d) depicts the strain (ε_{yy}) at the top and bottom surfaces of a Si microcell with $R = 4.9$ mm using the analytical expressions described above. These predictions are supported by finite element modeling (Figure 6.13e). The maximum strains in the Si and the metal interconnects are less than 0.3% for the bending radius of 4.9 mm. The fabrication of mechanically flexible modules with these optimized designs begins with transfer-printing of arrays of ultrathin microcells onto a photocurable polyurethane (NOA, thickness: ~30 μm)-coated PET substrate in a similar manner described in the previous section. After metal interconnects (Cr/Au, 10 nm/600 nm) are formed on the printed and planarized arrays of microcells, an additional polyurethane layer (thickness: ~30 μm) is spin-coated on top of it as an encapsulation layer. After curing under UV exposure, the PET substrate is removed, resulting in arrays of interconnected microcells embedded in a polyurethane matrix. Figure 6.14 illustrates these processes.

Photovoltaic properties of such flexible modules can be evaluated under various bending configurations. For example, the modules can be bent along or perpendicular to the length direction of microcells, and in inward or outward directions as schematically illustrated in Figure 6.15. Figure 6.16 shows *J–V* characteristics of a module at bending radii of 12.6, 8.9, 6.3 and 4.9 mm under AM 1.5 illumination, tested in outward bending, and along or perpendicular to the length direction of microcells. As predicted by mechanics analysis, the module performance is unaltered down to a bending radius of 4.9 mm, where the efficiency and fill factor of the module were ~6% and 0.60, respectively. The slightly reduced performance compared to that of individual flat cells can be partially attributed to the shadowing effect and resistive losses arising from metal interconnects. Fatigue tests, with bending cycles up to 200 cycles, showed little change in performance (Figure 6.17).

Future directions for transfer-printed flexible p–n junction solar cells

The module designs described in the preceding sections may create new possibilities for monocrystalline Si photovoltaics, particularly in applications

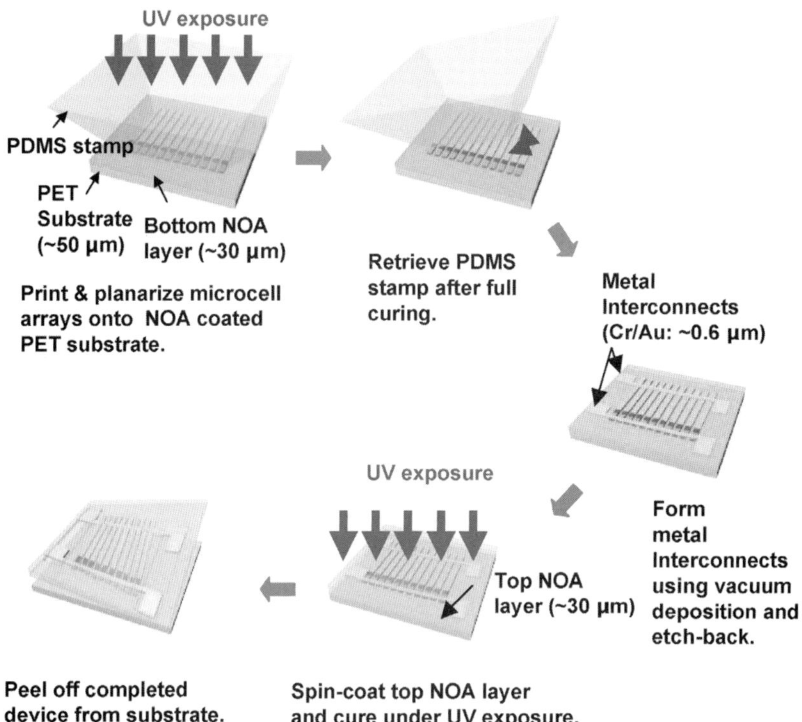

FIGURE 6.14 *Schematic illustration of fabrication steps for a mechanically flexible microcell module, with key dimensions indicated. PDMS: poly(dimethylsiloxane); PET: polyethylene terephthalate.*
(Reprinted with permission from [37], © 2008 Macmillan Magazines Ltd.)

that benefit from mechanical flexibility, thin, lightweight and large-area construction, together with potential to further reduce the module cost by the use of low-purity Si in ultrathin geometries. The approaches described here provide examples for how inherently rigid semiconductor materials, not restricted to Si, can be used as key components in mechanically flexible photovoltaic devices with properties comparable to those of similarly designed, conventional rigid systems. Similar concepts and other recently developed approaches in stretchable electronics [39, 40] can be readily applied to monocrystalline Si and other semiconductor materials (e.g. GaAs) to form solar modules that can operate under even more extreme mechanical deformations. Low-cost printing and fabrication techniques suitable for interconnection, together with other approaches to reduce cost or increase performance, are also important areas for further work.

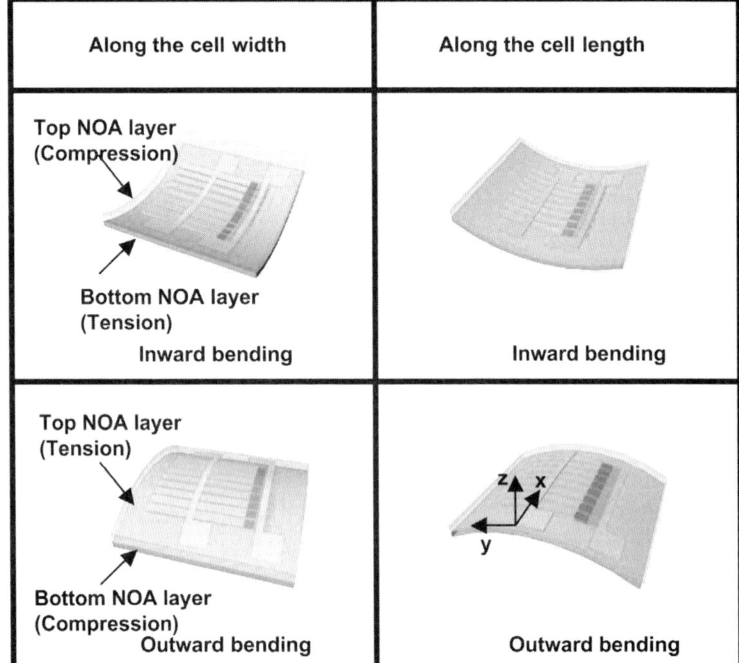

FIGURE 6.15 *Schematic illustration of bending geometries used for electrical characterization of mechanically flexible microcell modules. (Reprinted with permission from [37] © 2008 Macmillan Magazines Ltd.)*

FLEXIBLE PHOTOELECTROCHEMICAL CELLS

As discussed earlier in the chapter (Photoelectrochemical solar cells), use of well-aligned 1D nanostructures perpendicular to the supporting electrodes yields better mechanical flexibility than mesoporous films of colloidal nanoparticles. In addition, the aligned porosity, crystallinity and orientation of the arrayed nanotubes/nanowires are beneficial to formation of electron transport pathways for vectorial charge transfer between interfaces. This section focuses on flexible PEC cells made of arrays of aligned TiO_2 nanotubes and ZnO nanowires based on the dye-sensitized configuration (Figure 6.4b). Representative examples are discussed with details. Systems that use aligned Si nanowires, without dye molecules, for highly efficient PEC cells are also discussed. A final section reviews future challenges in developing low-cost, high-performance, flexible PEC cells with the use of aligned 1D nanostructures.

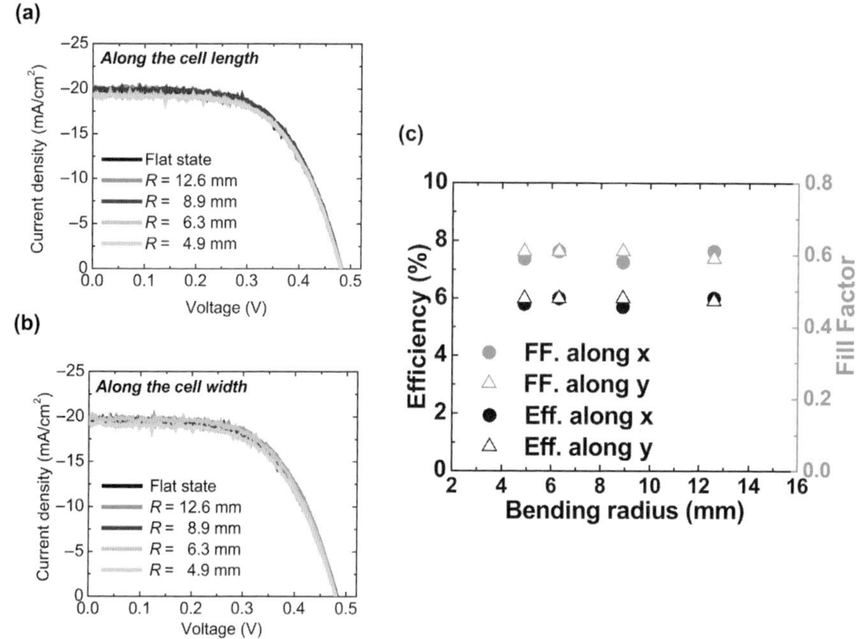

FIGURE 6.16 *J–V data from a module under AM 1.5 illumination in a flat configuration and bent along (a) the cell length and (b) width directions, at R = 12.6, 8.9, 6.3 and 4.9 mm. (c) Plot of efficiency and fill factor at various bending radii. Directions of x and y are denoted in Figure 6.15.*
(Reprinted with permission from [37] © 2008 Macmillan Magazines Ltd.)

Arrays of aligned TiO$_2$ nanotubes

Ordered, vertically oriented TiO$_2$ nanotube arrays are usually synthesized through potentiostatic anodization of titanium foils or titanium films deposited on conductive glass substrates [41]. This method was pioneered by Gong et al. with a first demonstration in the synthesis of titania nanotubes with lengths less than 500 nm by electrochemically anodizing a titanium foil in an aqueous hydrofluoric acid (HF) solution [42]. The growth conditions, including electrolytes, solvents and pH of solutions, have since been optimized to enable the growth of nanotubes with high levels of control. For example, systematic work conducted in Grimes' group shows that TiO$_2$ nanotubes can be grown to lengths > 200 μm with growth rates of ~15 μm/h [29]. In a typical synthesis, a Ti foil with thickness of ~250 μm, precleaned with ethanol, is immersed in a 0.30 wt% NH$_4$F ethylene glycol solution, with 2% H$_2$O in volume, followed by anodization at 60 V for 17 h. The oxidation results in the

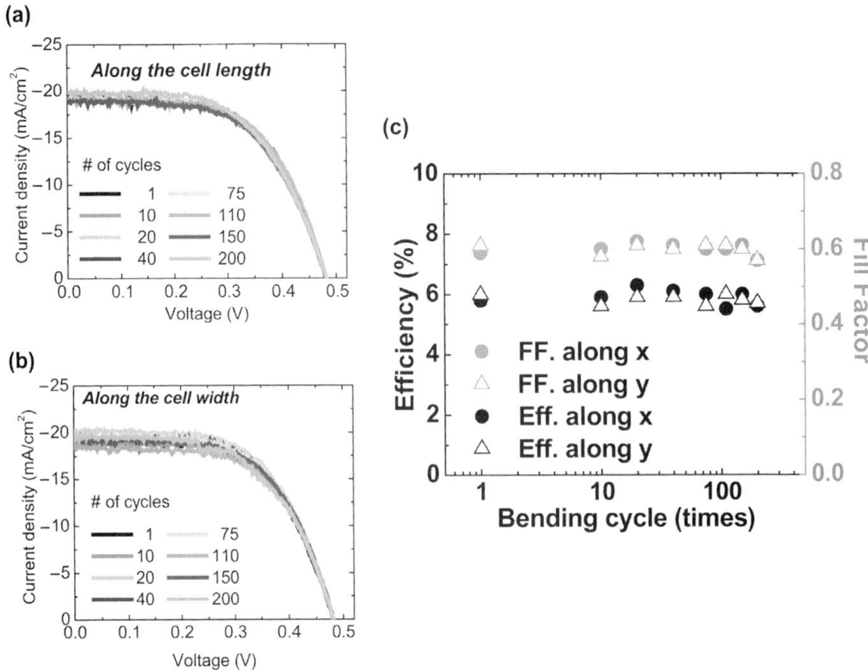

FIGURE 6.17 *J–V curves from a mechanically flexible module measured under AM 1.5 illumination bent along (a) the cell length and (b) width directions, after selected numbers of bending cycles up to 200 times. (c) Plot of efficiency and fill factor as a function of bending cycles. Directions of x and y are denoted in Figure 6.15.*
(Reprinted with permission from [37] © 2008 Macmillan Magazines Ltd.)

formation of nanotubes with lengths of 220 μm, average inner diameters of 110 nm and outer diameters of 160 nm. Long nanotubes can also be synthesized in other electrolytes, such as 1–2 wt% H_2O and 0.3–0.6 wt% NH_4F in formamide and/or *N*-methylformamide, 2–4% HF (48% aqueous) in dimethyl sulfoxide. Maintaining low concentrations of water (< 5%) in polar organic electrolytes is critical for growing arrays of very long nanotubes [43]. Excessive water leads to thick oxide layers that block the growth, resulting in short tubes. By contrast, oxidation of Ti is challenging in non-aqueous electrolytes owing to a lack of OH^-. In general, the formation of nanotubes in a fluoride electrolyte results from the three simultaneous processes: (1) oxidation of metallic Ti into TiO_2 driven with high positive voltage; (2) dissolution of metallic Ti under an electrical field; and (3) anisotropic dissolution of Ti and TiO_2 through chemical etching by fluoride ions with the assistance of an electric field. As a result, the lengths of TiO_2 nanotubes can be tuned by controlling the growth time, composition of electrolyte and anodization potential. As-synthesized amorphous TiO_2 nanotubes require high-temperature annealing in air (or oxygen) to increase their crystallinity. Increasing the annealing temperature increases the crystallinity. In fact, when annealing between 500 and 620°C, higher crystalline conversion efficiencies are achieved for nanotubes with higher temperatures. Excessive temperatures lead to the growth of the oxide layer underneath the nanotubes, which eventually distorts and destroys the array [44]. As a result, the conditions for synthesizing and annealing TiO_2 nanotubes must be optimized carefully.

The inset of Figure 6.18(a) shows an SEM image of an array of TiO_2 nanotubes with pore sizes of ~70 nm and wall thicknesses of ~8 nm [45]. Their lengths are ~7 μm, resulting from an anodization time of 4 h. Tuning the anodization time from 0 to 20 h generates nanotubes with varying lengths up to 14 μm. In the synthesis, a Ti foil is oxidized at 35 V in a two-electrode cell containing a Pt counter-electrode and at room temperature. The electrolyte includes 0.25 wt% NH_4F and 0.75 wt% H_2O in ethylene glycol. To make the as-grown nanotubes suitable for DSCs, they are soaked in a 40 mM aqueous solution of $TiCl_4$ for 30 min at 70°C. This process improves photocurrent and photovoltaic performance of the resulting DSCs, for reasons that are not clear yet. The $TiCl_4$-treated TiO_2 nanotubes are then rinsed with water and ethanol, followed by annealing in air at 500°C for 3 h. X-ray diffraction (XRD) patterns of the annealed nanotubes shown in the SEM image exhibit pure anatase phase of TiO_2. Figure 6.18(a) sketches the configuration of a flexible DSC fabricated with crystalline nanotubes on the original Ti foil. The structure and operation are similar to the flexible, mesoporous cell

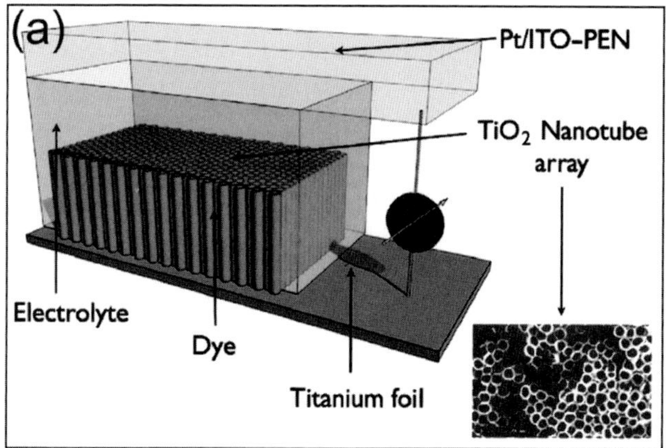

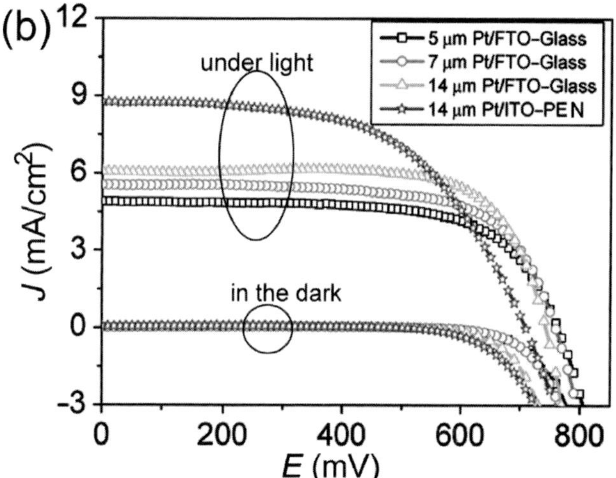

FIGURE 6.18 *(a) Schematic configuration of a flexible dye-sensitized photoelectrochemical (PEC) cell, which employs dye-derivatized TiO_2 nanotubes on a Ti foil to harvest light. The inset presents a typical SEM image of the aligned TiO_2 nanotubes. (b) Comparison of current–voltage characteristics of different PEC cells. These cells consist of TiO_2 nanotubes with different lengths and different counter-electrodes, i.e. rigid Pt/fluorine-doped tin dioxide (FTO)–glass substrates and flexible Pt/indium–tin oxide (ITO)–polyethylene naphthalate (PEN) substrates. The numbers in the box represent the lengths of the TiO_2 nanotubes. (Reprinted with permission from [45] © 2008 American Chemical Society.)*

shown in Figure 6.4(c). The crystallized TiO_2 nanotubes are sensitized by immersing them in a 0.3 mM solution of an indoline organic dye coded as D205 in acetonitrile and *tert*-butyl alcohol (1:1 v/v) for 16 h. The electrolyte employs non-solvent, room-temperature, binary ionic liquid electrolyte containing 0.05 M I_2, 0.5 M *N*-butyl benzimidazole (NBB), 0.1 M GuNCS in a mixture of 65 vol% 1-propyl-3-methylimidazolium iodide (PMII) and 35 vol% 1-ethyl-3-methyl-imidazolium tetracyanoborate $[EMIB(CN)_4]$, resulting in easy fabrication and operation compared to the cell of Figure 6.4(c). An ITO-PEN film coated with Pt catalyst provides the counter-electrode. Owing to the opacity of the Ti foil, the cell shown in Figure 6.18(a) is excited through back illumination, i.e. sunlight passes through the Pt/ITO-PEN counter-electrode to excite dye molecules. Figure 6.18(b) compares the current–voltage curves of DSCs made of nanotubes with different lengths and different counter-electrodes when they are exposed to the AM 1.5 illumination. The curves clearly show that illumination significantly increases current flows in the cells. The short circuit photocurrent densities are 4.95, 5.58 and 6.11 mA/cm^2 for the DSCs (with rigid Pt/FTO-glass counter-electrodes) made of nanotubes with lengths of 5, 7 and 14 μm, respectively. The V_{oc} only slightly decreases with the lengths of the nanotubes (i.e. 763, 759 and 743 mV, respectively). As a result, the DSCs with longer TiO_2 nanotubes have higher photovoltaic performance. For example, the energy conversion efficiency is 3.29% for the cell with 14 μm nanotubes; the efficiency decreases to 2.52% for the cell with 5 μm nanotubes. Replacing counter-electrodes with a Pt/ITO–PEN electrode results in a flexible DSC with comparable photovoltaic performance. Figure 6.18(b) compares the current–voltage curves of DSCs made with 14 μm nanotubes and different counter-electrodes, clearly showing that the flexible cell exhibits higher J_{sc} (8.99 vs 6.11 mA/cm^2) and lower V_{oc} (709 vs 743 mV) than the rigid cell. The energy conversion efficiency slightly increases to 3.58% (vs 3.29%).

Arrays of aligned ZnO nanowires

Aligned TiO_2 nanotubes offer improved electron transport pathways compared to mesoporous networks of colloidal TiO_2 nanoparticles, but polycrystalline nanotubes still suffer strong electron scattering at the grain boundaries. Arrays of nanowires/nanotubes with single crystallinity can avoid these limitations. One example is aligned nanowires of ZnO, a large bandgap (> 3 eV) semiconductor, which can be configured into DSCs otherwise similar to those made with TiO_2 nanotubes (Figure 6.18a).

Growth of aligned ZnO nanowires has been pioneered in Yang's group using the vapor–liquid–solid (VLS) process, in which gold nanoparticles serve as catalysts to promote the growth of ZnO nanowires perpendicular to the surfaces of (110) sapphire substrates due to epitaxial correlation between the lattice of (110) sapphire and that of (0001) ZnO (i.e. the longitudinal axis of the ZnO wires) [46]. Since the original work, researchers have developed effective approaches to grow aligned ZnO nanowires on various substrates including flexible plastic sheets [47, 48]. A comprehensive overview of the growth of ZnO nanowires appears in Chapter 7. Figure 6.19(a) shows an SEM image of an array of ZnO nanowires grown on FTO-coated glass through a seed-mediated process in aqueous solution at relatively low temperature ($< 100°C$) [49]. In the first step, the conductive glass substrate is coated with a 10–15 nm thick film of ZnO quantum dots with diameters of 3–4 nm by dipping in a concentrated ethanol solution. In the second step, immersing the glass substrate with ZnO seeds in an aqueous solution of 25 mM zinc nitrate hydrate, 25 mM hexamethylenetetramine and 5–7 mM polyethylenimine drives the growth of ZnO nanowires from the ZnO dots at 92°C. The growth tends to slow after ~2.5 h. Replacing the growth bath with fresh solution every 2.5 h enables the continuous growth of ZnO nanowires with very large lengths. The cationic polyelectrolyte, i.e. polyethylenimine, plays an important role in preventing lateral growth of the nanowires, enabling relatively high nanowire densities. The high aspect ratios and densities (up to 35 billion wires/cm^2) that can be achieved are well suited for DCSs because of the increased surface areas for dye adsorption. In a typical ZnO nanowire-based DSC, sensitizing occurs with

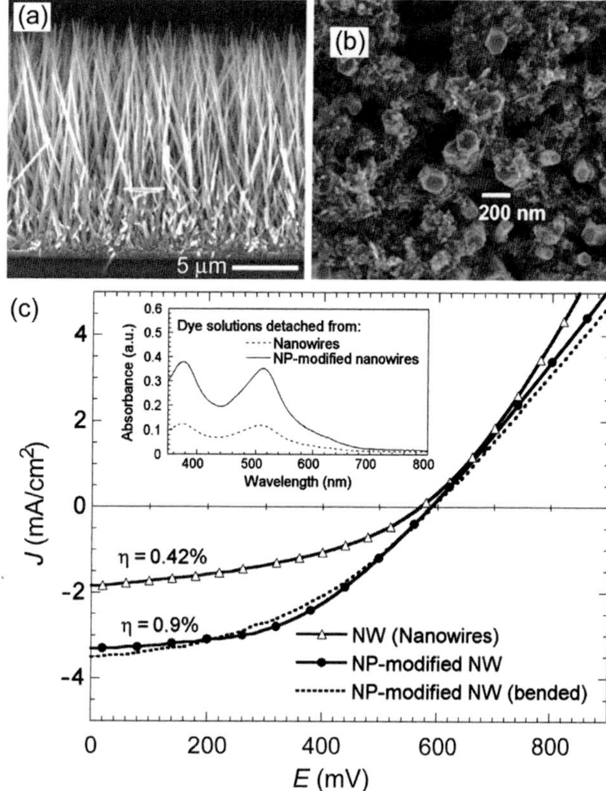

FIGURE 6.19 *(a) Side-view SEM image of a ZnO nanowire array grown on a fluorine-doped tin dioxide (FTO)-coated glass substrate. (Reprinted with permission from [49] © 2005 Nature Publishing Group.) (b) Top-view SEM image of arrayed ZnO nanowires decorated with ZnO nanoparticles that filled in the gaps between nanowires and attach to the surfaces of the nanowires. (c) Current–voltage characteristics of dye-sensitized solar cells (DSCs) based on dye-derivatized ZnO nanowires (NW, triangles) and nanoparticle (NP)-modified ZnO nanowires before (circles) and after (dashed) the cell has undergone five cycles of bending/relaxing. The inset compares the difference of absorption of dye molecules detached from pristine ZnO nanowires (dashed) and nanoparticle-modified ZnO nanowires (solid), clearly showing that the modification with ZnO nanoparticles increases the loading of the dye molecules. The dye molecules were detached from different ZnO films with the same area (i.e. 1 cm^2) in 8 ml KOH aqueous solution (with a concentration of 0.1 mM). (Reprinted with permission from [31] © 2008 American Institute of Physics.)*

an ethanolic (dry) solution of 0.5 mM *cis*-bis(isothiocyanato)bis(2,2′-bipyridyl-4,4′-dicarboxylato)-ruthenium(II) bis-tetrabutylammonium [(Bu$_4$N)$_2$Ru(dcbpyH)$_2$(NCS)$_2$, coded as N719 dye]. The resulting sensitized ZnO nanowires on FTO glass serve as the photoanode of the cell, while an FTO-coated glass substrate modified with Pt catalysts serves as counter-electrode. The electrolyte contains 0.5 M LiI, 50 mM I$_2$ and 0.5 M 4-tertbutylpyridine in 3-methoxypropionitrile. Preliminary results show that devices with this design can achieve energy conversion efficiencies of 1.5%, which is lower than those made with TiO$_2$ nanotubes and porous films. This limitation is mainly ascribed to the low surface area of the single-crystalline ZnO nanowires and the correspondingly low amount of dye molecules adsorbed on their surfaces.

Surface area and dye load can be significantly improved by filling the gaps between ZnO nanowires with ZnO nanoparticles. Figure 6.19(b) shows an SEM image of aligned ZnO nanowires grown on a plastic PET substrate coated with ITO followed by filling with ZnO nanoparticles with a diameter of 20 nm [31]. The ZnO nanowires are grown through a seed-mediated hydrothermal reaction at low temperature (85°C), which is compatible with the plastic substrate. The average diameters and lengths of the ZnO nanowires are 190 nm and 7 μm, respectively. Spin-coating a ZnO particle solution, prepared by dispersing 1.6 g ZnO nanoparticles and 2% titanium isoproxide in 8 ml methanol solution containing 0.02 M acetic acid, fills the vacancies between the nanowires. Titanium isoproxide and acetic acid facilitate the attachment of ZnO nanoparticles to the surfaces of the nanowires, thereby forming crystalline connections between particles and nanowires. The filling of nanoparticles indeed increases the loading of dye N719. The inset of Figure 6.7(c) compares the absorption spectra of two solutions of dye molecules detached from a film of pure nanowires and a film of nanoparticle-decorated nanowires. The difference in absorbance indicates that the addition of nanoparticles increases the surface adsorption by more than two times compared to nanowires alone. As shown in Figure 6.19(c), the higher dye loading in the film of nanoparticle-modified nanowires induces an improvement in energy conversion efficiency by ~110% (i.e. 0.9% vs 0.42%). The nanowire/nanoparticle composite film on plastic substrate exhibits excellent mechanical bendability. For example, the film shown in Figure 6.19(b) does not show any visible cracking after it has been bent for 1000 cycles to a bending radius of 5 mm and the resulting DSC retains its performance even after it is bent to a radius of 5 mm. Although addition of ZnO nanoparticles increases the surface areas for higher dye loading, the energy conversion efficiency of the resulting DSCs is still far lower than that of TiO$_2$-based DSCs. A possible reason for this outcome is that the energy levels of the dye molecules adopted from the TiO$_2$-based DSCs may not be compatible with the energy levels of

contained the same solution as the PEC cell. The short circuit current density is 1.43 mA/cm^2, which is lower than DSCs as shown in Figures 6.18 and 6.19. The low photocurrent density is ascribed to the low coverage (6.5%) of Si wires on the Si wafer. Therefore, increasing the coverage can potentially further increase the photocurrent density of PEC cells made from Si wire arrays.

Extremely high temperatures for VLS growth make it challenging (or even impossible) to grow Si wires directly on flexible plastic substrates. Although chemical etching methods under mild conditions have been developed to fabricate Si nanowires from bulk Si wafers [51], their alignment is modest and the bulk mother wafers are still too rigid for flexible solar cells. On the other hand, the arrays can be transferred to a flexible support, retaining their aligned orientations after they are grown on rigid Si wafers at high temperatures [55]. Figure 6.20(c) shows an SEM image of an array of Si wires partially embedded in a PDMS film fabricated by drop-casting diluted PDMS precursors on the arrayed wires shown in Figure 6.20(a), followed by curing and peeling off. The Si wires embedded in the PDMS film retain their ordered configurations even when the sample is bent. Figure 6.20(d, e) presents large-area samples of aligned Si wires embedded in PDMS films, clearly showing their mechanical flexibility and strong absorption of light, indicated by the black color. The wire arrays supported with PDMS films are suitable for flexible PEC cells. The remaining challenge is the formation of contacts to the wires to collect photogenerated electrons.

Future directions for flexible photoelectrochemical cells

Advantages associated with PEC cells make them a potential replacement for p–n junction solar cells because of the potential for lower fabrication costs. The relatively low energy conversion efficiency of PEC cells, however, currently prevents them from being widely used in solar economy. Achieving flexible PEC cells with high energy conversion efficiency and large areas still faces many challenges. For instance, the development of effective methodologies for growing aligned 1D nanowire arrays on large-area flexible substrates (in particular, transparent plastic sheets coated with transparent conductive layers) represents the most critical step to realize high-performance, flexible PEC cells. Low-temperature hydrothermal processes that have been used for growing ZnO nanowire arrays may be promising. Eliminating the use of organic solvent-based electrolytes is important to facilitate fabrication of PEC cells. Solvent-free electrolytes with ionic liquids and polymeric matrices as hosting media, in which diffusion coefficients of electrolyte species are comparable to or even higher than those in organic solvents, may replace the electrolytes widely used in current PEC cells, resulting in quasi-solid or fully solid PEC cells. Regardless of materials (in particular, low bandgap semiconductors) for the photoanodes, these elements almost always suffer from degradation during operation due to corrosion caused by oxidation. One possible solution for suppressing corrosion is to decorate surfaces of traditional semiconductors (e.g. Si) with very small precious metal nanoparticles (with sizes <10 nm), resulting in modulation of the band structures near the semiconductor/electrolyte interfaces. The variation of band configurations may force photogenerated holes (for photoanodes) to dissipate into the electrolyte through the metal nanoparticles rather than the semiconductor surfaces, leading to avoidance of oxidation [56, 57]. Other challenges include improvement of absorption efficiency and stability of dye molecules used in DSCs. High quantum-yield quantum dots and plasmonic metal nanoparticles have been demonstrated as potential replacements for the current dye molecules [58]. The use of these nanoparticles could also enable operation over an improved spectral range to collect solar energy more efficiently [59, 60].

CONCLUSION

In summary, this chapter provides an overview of mechanically flexible solar cells made of nanostructures/microstructures of inorganic semiconductors, with a special emphasis on p–n junction and PEC systems. By employing dry transfer printing techniques and appropriate mechanical designs, intrinsically rigid and fragile semiconducting materials in bulk can be readily used as building blocks for flexible p–n junction solar cells. As an example, monocrystalline Si-based flexible solar cells and modules were described. In the second part, various recent approaches towards flexible PEC cells employing nanotubes and nanowires are reviewed. Nanotubes/nanowires made of TiO_2, ZnO and Si either grown on or transferred onto flexible substrates represent promising photoanode materials for mechanically flexible PEC solar cells.

ACKNOWLEDGMENT

Y. Sun acknowledges the support by the US Department of Energy, Office of Science, Office of Basic Energy Sciences, under contract DE-AC02-06CH11357. J. Yoon acknowledges support from a Beckman postdoctoral fellowship.

REFERENCES

[1] Lewis NS, Nocera DG. Powering the planet: chemical challenges in solar energy utilization. Proc Natl Acad Sci USA 2006;103:15729–35.

[2] Whitesides GM, Crabtree GW. Don't forget long-term fundamental research in energy. Science 2007;315:796–8.

[3] http://en.wikipedia.org/wiki/A_E_Becquerel.

[4] http://en.wikipedia.org/wiki/Charles_Fritts.

[5] Meet the 2008 National Inventors Hall of Fame Inductees. http://www.invent.org/2008induction/1_3_08_induction_chapin.asp.

[6] Gratzel M. Photoelectrochemical cells. Nature 2001;414:338–44.

[7] O'Regan B, Gratzel M. A low-cost, high-efficiency solar cell based on dye-sensitized colloidal TiO_2 films. Nature 1991;353:737–40.

[8] Hossain MF, Biswas S, Shahjahan M, Majumder A, Takahashi T. Fabrication of dye-sensitized solar cells with TiO_2 photoelectrode prepared by sol-gel technique with low annealing temperature. J Vac Sci Technol A 2009;27:1042–6.

[9] Chen Y, Stathatos E, Dionysiou DD. Sol-gel modified TiO_2 powder films for high performance dye-sensitized solar cells. J Photochem Photobiol A 2009;203:192–8.

[10] Hocevar M, Krasovec UO, Berginc M, Drazic G, Hauptman N, Topic M. Development of TiO_2 pastes modified with Pechini sol-gel method for high efficiency dye-sensitized solar cell. J Sol Gel Sci Technol 2008;48:156–62.

[11] Ramasamy E, Lee WJ, Lee DY, Song JS. Portable, parallel grid dye-sensitized solar cell module prepared by screen printing. J Power Sources 2007;165:446–9.

[12] Chi B, Zhao L, Li J, Pu J, Chen Y, Wu CC, Jin T. TiO_2 mesoporous thick films with large-pore structure for dye-sensitized solar cell. J Nanosci Nanotechnol 2008;8:3877–82.

[13] Fan Q, McQuillin B, Bradley DDC, Whitelegg S, Seddon AB. A solid state solar cell using sol-gel processed material and a polymer. Chem Phys Lett 2001;347:325–30.

[14] Martinson ABF, Hamann TW, Pellin MJ, Hupp JT. New architectures for dye-sensitized solar cells. Chem Eur J 2008;14:458–67.

[15] Hamann TW, Martinson ABF, Elam JW, Pellin MJ, Hupp JT. Atomic layer deposition of TiO_2 on aerogel templates: new photoanodes for dye-sensitized solar cells. J Phys Chem C 2008;112:10303–7.

[16] Standridge SD, Schatz GC, Hupp JT. Toward plasmonic solar cells: protection of silver nanoparticles via atomic layer deposition of TiO_2. Langmuir 2009;25:2596–600.

[17] Standridge SD, Schatz GC, Hupp JT. Distance dependence of plasmon-enhanced photocurrent in dye-sensitized solar cells. J Am Chem Soc 2009;131:8407–9.

[18] Gratzel M. Solar energy conversion by dye-sensitized photovoltaic cells. Inorg Chem 2005;44:6841–5.

[19] Ito S, Ha NLC, Rothenberger G, Liska P, Comte P, Zakeeruddin SM, Pechy P, et al. High-efficiency (7.2%) flexible dye-sensitized solar cells with Ti-metal substrate for nanocrystalline-TiO_2 photoanode. Chem Commun 2006;42:4004–6.

[20] Ito S, Zakeeruddin SM, Comte P, Liska P, Kuang DB, Gratzel M. Bifacial dye-sensitized solar cells based on an ionic liquid electrolyte. Nat Photonics 2008;2:693–8.

[21] Kuang D, Uchida S, Humphry-Baker R, Zakeeruddin SM, Gratzel M. Organic dye-sensitized ionic liquid based solar cells: remarkable enhancement in performance through molecular design of indoline sensitizers. Angew Chem Int Ed 2008;47:1923–7.

[22] Cao Y, Zhang J, Bai Y, Li R, Zakeeruddin SM, Gratzel M, Wang P. Dye-sensitized solar cells with solvent-free ionic liquid electrolytes. J Phys Chem C 2008; 112:13775–81.

[23] Shi D, Pootrakulchote N, Li R, Gui J, Wang Y, Zakeeruddin SM, et al. New efficiency records for stable dye-sensitized solar cells with low-volatility and ionic liquid electrolytes. J Phys Chem C 2008;112:17046–50.

[24] Wang P, Zakeeruddin SM, Moser JE, Nazeeruddin MK, Sekiguchi T, Gratzel M. A stable quasi-solid-state dye-sensitized solar cell with an amphiphilic ruthenium sensitizer and polymer gel electrolyte. Nat Mater 2003;2:402–7.

[25] Bach U, Lupo D, Comte P, Moser JE, Weissortel F, Salbeck J, et al. Solid-state dye-sensitized mesoporous TiO_2 solar cells with high photon-to-electron conversion efficiencies. Nature 1998;395:583–5.

[26] Adachi M, Murata Y, Takao J, Sakamoto M, Wang F. Highly efficient dye-sensitized solar cells with a titania thin-film electrode composed of a network structure of single-crystal-like TiO_2 nanowires made by the 'oriented attachment' mechanism. J Am Chem Soc 2004;126:14943–9.

[27] Zukalova M, Zukal A, Kavan L, Nazeeruddin MK, Liska P, Gratzel M. Organized mesoporous TiO_2 films exhibiting greatly enhanced performance in dye-sensitized solar cells. Nano Lett 2005;5:1789–92.

[28] Mor GK, Shankar K, Paulose M, Varghese OK, Grimes CA. Use of highly-ordered TiO_2 nanotube arrays in dye-sensitized solar cells. Nano Lett 2006;6:215–8.

[29] Shankar K, Mor GK, Prakasam HE, Yoriya S, Paulose M, Varghese OK, Grimes CA. Highly-ordered TiO_2 nanotube arrays up to 220 μm in length: use in water photoelectrolysis and dye-sensitized solar cells. Nanotechnology 2007;18: 065707.

[30] Zhu K, Neale NR, Miedaner A, Frank AJ. Enhanced charge-collection efficiencies and light scattering in dye-sensitized solar cells using oriented TiO_2 nanotubes arrays. Nano Lett 2007;7:69–74.

[31] Jiang CY, Sun XW, Tan KW, Lo GQ, Kyaw AKK, Kwong DL. High-bendability flexible dye-sensitized solar cell with a nanoparticle-modified ZnO-nanowire electrode. Appl Phys Lett 2008;92:143101.

[32] Mack S, Meitl MA, Baca AJ, Zhu ZT, Rogers JA. Mechanically flexible thin-film transistors that use ultrathin ribbons of silicon derived from bulk wafers. Appl Phys Lett 2006;88:213101.

[33] Kovacs GTA, Maluf NI, Petersen KE. Bulk micromachining of silicon. Proc IEEE 1998;86:1536–51.

[34] Bean KE. Anisotropic etching of silicon. IEEE Trans Electron Dev 1978;25:1185–93.

[35] Meitl MA, Zhu ZT, Kumar V, Lee KJ, Feng X, Huang YY, et al. Transfer printing by kinetic control of adhesion to an elastomeric stamp. Nat Mater 2006;5: 33–8.

[36] Meitl MA, Feng X, Dong J, Menard E, Ferreira PM, Huang Y, Rogers JA. Stress focusing for controlled fracture in microelectromechanical systems. Appl Phys Lett 2007;90:083110.

[37] Yoon J, Baca AJ, Park SI, Elvikis P, Geddes JB III, Li L, et al. Ultrathin silicon solar microcells for semitransparent, mechanically flexible and microconcentrator module designs. Nat Mater 2008;7:907–15.

[38] Ahn BY, Duoss EB, Motala MJ, Guo X, Park SI, Xiong Y, et al. Omnidirectional printing of flexible, stretchable, and spanning silver microelectrodes. Science 2009;323:1590–3.

[39] Kim DH, Song J, Choi WM, Kim HS, Kim RH, Liu Z, et al. Materials and noncoplanar mesh designs for integrated circuits with linear elastic responses to extreme mechanical deformations. Proc Natl Acad Sci USA 2008; 105:18675–80.

[40] Kim DH, Ahn JH, Choi WM, Kim HS, Kim TH, Song J, et al. Stretchable and foldable silicon integrated circuits. Science 2008;320:507–11.

[41] Grimes CA. Synthesis and application of highly ordered arrays of TiO_2 nanotubes. J Mater Chem 2007;17:1451–7.

[42] Gong D, Grimes CA, Varghese OK, Hu WC, Singh RS, Chen Z, Dickey EC. Titanium oxide nanotube arrays prepared by anodic oxidation. J Mater Res 2001;16:3331–4.

[43] Paulose M, Shankar K, Yoriya S, Prakasam HE, Varghese OK, Mor GK, et alAnodic growth of highly ordered TiO_2 nanotube arrays to 134 μm in length. J Phys Chem B 2006;110:16179–84.

[44] Varghese OK, Gong D, Paulose M, Grimes CA, Dickey EC. Crystallization and high-temperature structural stability of titanium oxide nanotube arrays. J Mater Res 2003;18:156–65.

[45] Kuang D, Brillet J, Chen P, Takata M, Uchida S, Miura H, et al. Application of highly ordered TiO_2 nanotube arrays in flexible dye-sensitized solar cells. ACS Nano 2008;2:1113–6.

[46] Huang MH, Mao S, Feick H, Yan H, Wu Y, Kind H, et al. Room-temperature ultraviolet nanowire nanolasers. Science 2001;292:1897–9.

[47] Wang JX, Sun XW, Yang Y, Huang H, Lee YC, Tan OK, Vayssieres L. Hydrothermally grown oriented ZnO nanorod arrays for gas sensing applications. Nanotechnology 2006;17:4995–8.

[48] Xu S, Wei Y, Kirkham M, Liu J, Mai W, Davidovic D, et al. Patterned growth of vertically aligned ZnO nanowire arrays on inorganic substrates at low temperature without catalyst. J Am Chem Soc 2008;130:14958–9.

[49] Law M, Greene LE, Johnson JC, Saykally R, Yang P. Nanowire dye-sensitized solar cells. Nat Mater 2005;4:455–9.

[50] Kayes BM, Atwater HA, Lewis NS. Comparison of the device physics principles of planar and radial p–n junction nanorod solar cells. J Appl Phys 2005;97: 114302.

[51] Peng K, Wang X, Lee S-T. Silicon nanowire array photoelectrochemical solar cells. Appl Phys Lett 2008;92:163103.

[52] Goodey AP, Eichfeld SM, Lew K-K, Redwing JM, Mallouk TE. Silicon nanowire array photoelectrochemical cells. J Am Chem Soc 2007;129:12344–5.

[53] Struthers JD. Solubility and diffusivity of gold, iron, and copper in silicon. J Appl Phys 1956;27:1560.

[54] Maiolo JRI, Kayes BM, Filler MA, Putnam MC, Kelzenberg MD, Atwater HA, Lewis NS. High aspect ratio silicon wire array photoelectrochemical cells. J Am Chem Soc 2007;129:12346–7.

[55] Plass KE, Filler MA, Spurgeon JM, Kayes BM, Maldonado S, Brunschwig BS, et al. Flexible polymer-embedded Si wire arrays. Adv Mater 2009;21:325–8.

[56] Nakato Y, Ueda K, Yano H, Tsubomura H. Effect of microscopic discontinuity of metal overlayers on the photovoltages in metal-coated semiconductor-liquid junction photoelectrochemical cells for efficient solar energy conversion. J Phys Chem 1988;92:2316–24.

[57] Takabayashi S, Ohashi M, Mashima K, Liu Y, Yamazaki S, Nakato Y. Surface structures, photovoltages, and stability of n-Si(111) electrodes surface modified with metal nanodots and various organic groups. Langmuir 2005;21:8832–8.

[58] Lopez-Luke T, Wolcott A, Xu LP, Chen S, Wen Z, Li J, et al. Nitrogen-doped and CdSe quantum-dot-sensitized nanocrystalline TiO_2 films for solar energy conversion applications. J Phys Chem C 2008;112:1282–9.

[59] Tian Y, Tatsuma T. Mechanisms and applications of plasmon-induced charge separation at TiO_2 films loaded with gold nanoparticles. J Am Chem Soc 2005;127:7632–7.

[60] Chen ZH, Tang YB, Liu CP, Leung YH, Yuan GD, Chen LM, et al. Vertically aligned ZnO nanorod arrays sensitized with gold nanoparticles for Schottky barrier photovoltaic cells. J Phys Chem C 2009;113:13433–7.

Zinc Oxide Nanowire Arrays on Flexible Substrates: Wet Chemical Growth and Applications in Energy Conversion

Sheng Xu[*], Benjamin Weintraub[*] and Zhong Lin Wang

School of Materials Science and Engineering, Georgia Institute of Technology, Atlanta, Georgia, USA

INTRODUCTION

This chapter provides an overview of zinc oxide (ZnO) nanowire synthesis strategies pertinent to flexible organic substrates. A thorough evaluation of the current status of ZnO nanowire growth in the literature is provided, with an emphasis on process control in terms of patterned and unpatterned substrates. In addition to synthesis, energy harvesting applications based on a nanogenerator are reviewed. Lastly, current processing obstacles and future research directions are discussed.

History of nanowires

The first report of one-dimensional (1D) semiconducting single crystal structures dates back to 1964, when Wagner and Ellis successfully synthesized silicon (Si) whiskers with diameters up to the micrometer scale through the proposed vapor–liquid–solid (VLS) mechanism at 950°C [1]. Westwater et al. [2] and Morales and Lieber [3] went on to further develop the VLS growth mechanism in the late 1990s and synthesize nanometer-scale Si wires. High-temperature catalyst-assisted vapor transport growth of 1D nanowires usually follows the VLS mechanism. A metal catalyst forms a liquid alloy with the

CONTENTS

[*]The authors contributed equally to this work.

nanowire component under the reaction conditions. The liquid droplet serves as a preferential site for absorption of the gas-phase reactant and the nucleation site for crystallization when supersaturated. Nanowire growth begins after the liquid becomes supersaturated in reactant materials and continues as long as the catalyst alloy remains in a liquid state and the reactant is available.

History of ZnO nanostructures
Synthesis methods

ZnO is traditionally synthesized at high temperatures (450–900°C) in the gas phase. Huang et al. first reported the VLS catalytic growth of ZnO nanowires by vapor transport using an Au catalyst on Si substrates in 2001 [4].

Others used catalyst-free vapor–solid (VS) processes such as metal–organic chemical vapor deposition (MOCVD) [5, 6]. These methods can produce high-quality, single-crystalline nanowires with high aspect ratios, but are limited in terms of sample uniformity, low product yield and, most notably, substrate choice.

Wet chemical methods are an attractive alternative to vapor transport processes since they can be carried out at temperatures as low as 50°C. In 1990, Vergés et al. reported the first ZnO microrods grown in an aqueous solution at temperatures of less than 80°C [7]. About 10 years later, others successfully decreased the size of the structures and reported well-oriented ZnO nanorod arrays [8, 9]. Typically, an aqueous solution of $Zn(NO_3)_2$ and $C_6H_{12}N_4$ [hexamethylenetetramine (HMTA)] is heated to temperatures between 50 and 95°C and ZnO forms on a substrate placed in the solution. In general, HMTA decomposes into ammonia and formaldehyde. The ammonia provides a steady source of hydroxide ions to form zinc hydroxide. Finally, the hydroxide compound undergoes a condensation reaction to form ZnO.

Applications

Because it is difficult to manipulate such small structures, aligned growth is an efficient approach to self-assemble nanowires into useful devices such as nanowire lasers, light-emitting diodes (LEDs), solar cells, water-splitting photocatalysts, nanogenerators, piezotriggers and field-effect transistors (FETs) [10–13]. The substrate is a crucial factor in determining the nanowire orientation. In order to achieve vertical alignment, an epitaxial relationship must be satisfied between ZnO and the underlying substrate. As a result, substrate candidates must have wurtzite crystal structures similar to ZnO, such as sapphire, GaN, AlGaN or AlN. The substrate choices needed for high-temperature vapor transport synthesis are quite limited and unsuitable for polymers.

GROWTH OF ZINC OXIDE NANOWIRES ON FLEXIBLE SUBSTRATES

Growth of ZnO nanowires on flat substrates

The low temperatures and substrate independence are key prerequisites for integrating ZnO into flexible electronic devices. Although more difficult to achieve, there have been reports of ZnO synthesis on polymer substrates such as poly(dimethylsiloxane) (PDMS), polystyrene, polyurethane, Au-coated Kapton, and polystyrene bead-coated polycarbonate [14–16]. In the case of PDMS, a 5 cm diameter substrate was fully coated with ZnO nanorods by employing seeds, but microcracks formed after flexing. Growth on Au-coated Kapton yielded high-density, size-controlled ZnO nanorods without using seeds. The alignment of ZnO nanorods grown directly on polycarbonate was enhanced by a self-assembled polystyrene sphere monolayer.

Unpatterned growth

ZnO nanorods can be synthesized on organic substrates in a variety of fashions. ZnO seeds can by sputtered at room temperature on virtually any substrate, leading to high-density nanorod growth. For example, Liu et al. successfully carried out synthesis on a 4-inch thermoplastic polyurethane substrate (Figure 7.1) [17]. The technique is general, and polyimide, polystyrene, polyethylene terephthalate and polyethylene naphthalate are just a few examples of substrates seen in the literature. In addition, ZnO nanorods can be electrochemically deposited provided that the flexible substrate is either conductive or precoated with a metal or a transparent conducting oxide such as indium–tin oxide (ITO). A typical experimental step would consist of a $ZnCl_2/KCl$ electrolyte with O_2 bubbled into the solution to maintain an oxygen-saturated electrolyte. A standard three-electrode set-up with a saturated Ag/AgCl reference electrode and Pt counter electrode is often used. Typical potentials are approximately -1 V. ZnO nanorods electrochemically deposited on an Au film with a curvature of radius less than 10 μm have been achieved [18].

Patterned growth

Essential for the fabrication of novel nanoelectronics and integration into pre-existing microelectromechanical system (MEMS) and nanoelectromechanical system (NEMS) devices is the position-controlled growth of nanowires. In 1964, Wagner et al. proposed that controlled growth could be obtained through appropriate use of impurities in patterns or films on substrate surfaces [1]. Recent efforts at high-temperature vapor transport growth have shown that the Au catalyst size and position dictate the dimensions and spatial registration of the nanowires. Several techniques, including photolithography [19],

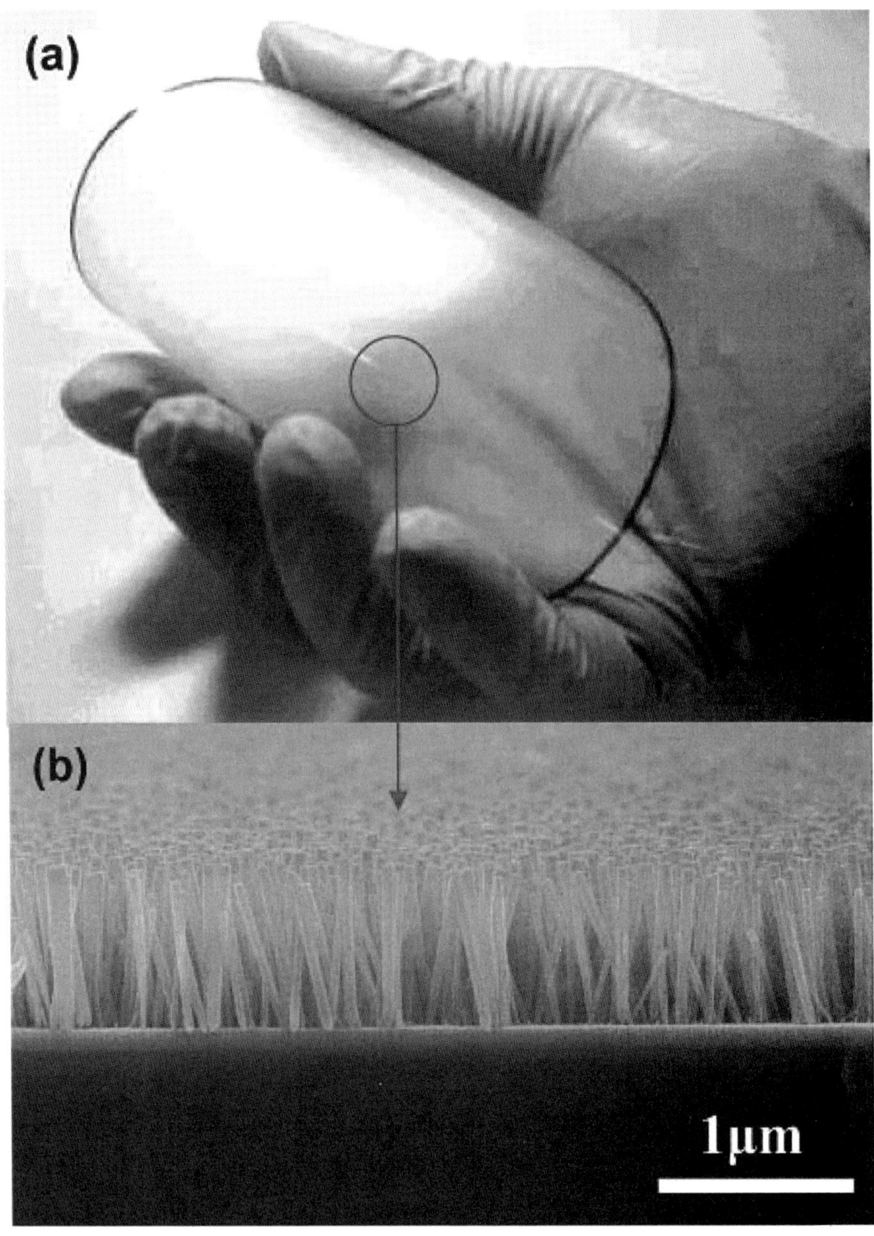

FIGURE 7.1 *ZnO unpatterned nanorod growth on a thermoplastic polyurethane (TPU) substrate using ZnO seeds. (a) Photograph demonstrating large-scale ZnO nanorod synthesis on 4-inch TPU substrate. (b) Profile SEM image showing uniform nanorod length on TPU substrate.*
(Images courtesy of Liu et al. [17].)

electron-beam lithography (EBL) [20] and nanosphere lithography [21], have been used to form Au islands for this purpose. In most cases, multiple nanowires grow at each Au island location because at high temperatures (450–900°C for vapor transport) metal mobility and diffusion rates are high. However, using ultrathin Au islands and hexagonal array microsphere shadowing, individual nanowires have been achieved at each growth site using vapor transport synthesis [22]. Using low temperature (50–95°C) solution-based methods, individual ZnO nanowires have been defined using patterned organic masks on Si substrates. At these temperatures, the organic mask remains intact during the synthesis, which allows for other types of processing methods. Kim et al. used EBL and a poly(methylmethacrylate) (PMMA) mask above a ZnO thin film seeding layer to achieve 250 nm diameter nanorods, although the vertical alignment was poor [23]. Cui and Gibson followed a similar approach, but used an electrochemical synthesis method to circumvent the seed requirement and achieved 120 nm diameter nanowires [24]. Others have relied on microstamped self-assembled monolayers (SAMs) to define micrometer-size ZnO nanowires [25].

Site-selective growth of individual ZnO nanorods in patterned arrays by a wet chemical approach on organic substrates was achieved by Weintraub et al. in 2007 [26]. In this experiment, a polyimide substrate was coated with an Au thin film to serve as a catalyst for nanorod synthesis. The substrate did not require a ZnO thin film or nanoparticle seeds and the synthesis was carried out at low temperatures ($< 70°C$), making it an important synthesis technique for flexible electronic applications. After defining the growth regions by EBL and developing the features (Figure 7.2), it was found that nucleation could be achieved within larger pattern sizes ($> 1\ \mu m$) without an applied potential, but it would result in low growth density. On the other hand, at submicrometer feature sizes the nucleation process was hampered since there were fewer nucleation sites exposed on the Au. It is known that an applied potential increases the nucleation density of ZnO nanorods by up to four orders of magnitude on polished surfaces. The applied potential was important for achieving high-density nucleation within the micrometer-sized patterns and single nanorod nucleation in the 200 nm diameter patterns.

The synthesis technique allows for the position control of ZnO nanorods of micrometer- and submicrometer-sized patterns. Figure 7.3 demonstrates micrometer scale patterning of 5×5 arrays with circular patterns ranging in diameter from 1 to 9 μm. The nanorods range in diameter from 50 to 150 nm and have a uniform length of 1.4 μm. The nanorods exhibited a high degree of vertical alignment and growth density while fully occupying the patterned regions (Figure 7.3c). A small amount of non-vertical growth was found at the resist mask boundaries.

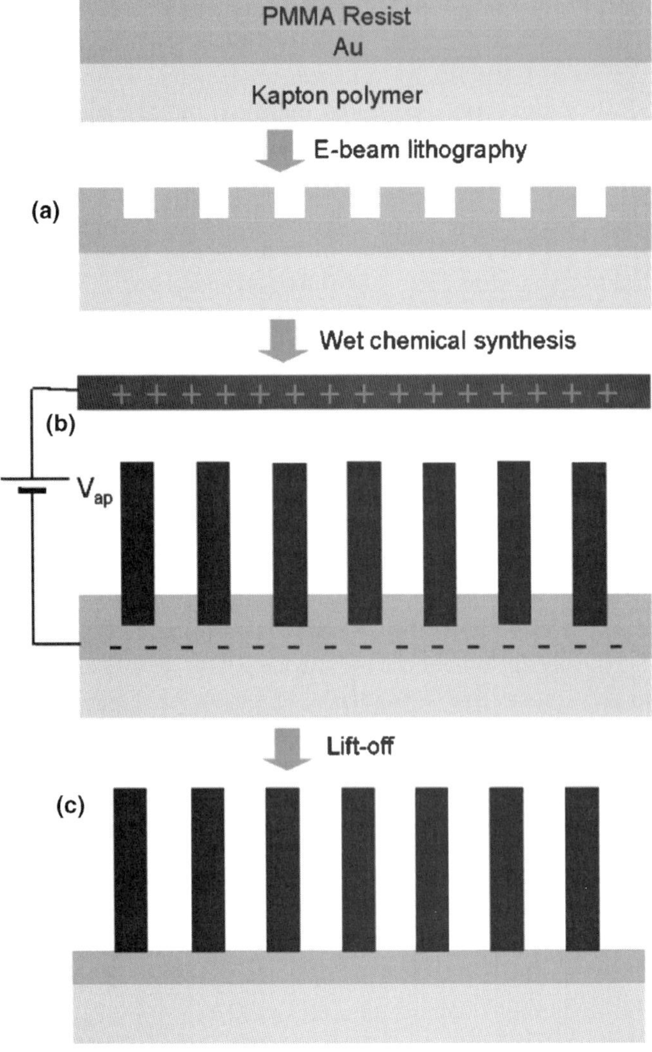

FIGURE 7.2 *Process flow for lithography-based density-controlled ZnO nanorod array growth on a plastic substrate. Flexible Kapton polymer substrate is coated in a Au thin film and poly(methylmethacrylate) (PMMA) resist. (a) Desired density is defined using electron-beam lithography, and the Au pattern is subsequently exposed during the developing step. (b) The patterned substrate is immersed in a nutrient solution bath and held at a −500 mV bias relative to the reference electrode. (c) The PMMA resist is finally removed in the lift-off step.*

In addition, submicrometer-sized patterns were defined for growth of single ZnO nanorod arrays. Figure 7.3(d) shows a low-magnification scanning electron microscope (SEM) image of a 25 × 25 array of individual ZnO nanorods separated by 1.25 μm. Circular patterns of 200 nm were defined in the resist to

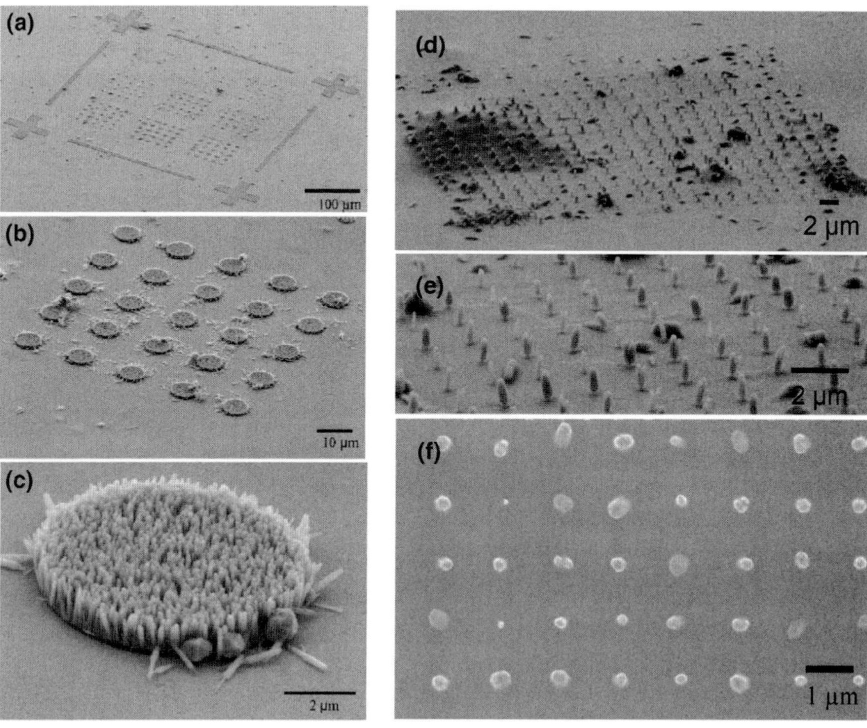

FIGURE 7.3 *(a–c) SEM images of micrometer-scale ZnO nanorod patterns of 5 × 5 arrays. (a) Low-magnification tilted SEM image of ZnO nanorod arrays ranging in diameter from 1 to 9 µm. (b) Tilted SEM image of a patterned ZnO nanorod array with a 9 µm circular diameter. (c) High-magnification tilted SEM image of a single patterned ZnO nanorod array with a 9 µm circular diameter. (d–f) SEM image of a 25 × 25 array of patterned single ZnO nanorods with a spacing of 1.25 µm. (d) Low-magnification SEM image at a 70° stage tilt of a ZnO nanorod array. (e) Higher magnification SEM image of the array. (f) Top-view SEM image of the array.*

nucleate a single ZnO nanorod. ZnO nanorods did not grow on the PMMA resist regions owing to a lack of nucleation sites. It was found that 200 nm was the optimum circular diameter to nucleate a single nanorod under the given experimental conditions. The nanorod length was controlled essentially by the synthesis time. Figure 7.3(e) shows a high-magnification SEM image of the ZnO nanorod array profile. Figure 7.3(f) shows a top SEM image of the array and demonstrates that the nanorods exhibit a high degree of vertical alignment. The base diameter was defined by the resist mask. Above the resist, lateral growth occurred along the non-polar planes, resulting in nanorods with a diameter of 200 ± 50 nm. This lateral growth caused some of the nanorods to detach from the substrate. In general, nanorod/substrate adhesion is a major obstacle for the practical realization of flexible electronics based on inorganic nanostructures such as ZnO.

Growth of ZnO nanowire arrays on microfibers

A unique advantage for ZnO nanowire arrays is that they can be grown at ~80°C on substrates of any shape and any material. ZnO can nucleate and grow along the polar axis as long as there is ZnO seed. ZnO nanowires can grow not only on flat surfaces, like the inorganic substrates (e.g. Si, GaN) and polymer films (e.g. polycarbonate, Kapton film), but also on curved surfaces, like flexible microfibers. Such versatile assembly of well-designed nanowire arrays on flexible microfibers can enable applications such as functional textiles and wearable microelectronics.

Unpatterned growth

In 2008, Qin et al. demonstrated a method of growing ZnO nanowire arrays on a kind of flexible microfiber [27]. The fibers used in their experiments were Kevlar 129 fibers, which had high strength, modulus, toughness and thermal stability. Basically, they used ZnO seed film to initiate the growth of ZnO nanowires. Otherwise, it would be very difficult to nucleate onto the flexible microfibers. So before the hydrothermal growth of ZnO nanowire arrays, they deposited ZnO seed film onto the surface of the fibers with a magnetron sputtering machine. Then, the fiber was put into the nutrient solution for ZnO nanowire growth. The nutrient solution was composed of a 1:1 ratio of zinc nitrate hexahydrate and HMTA. ZnO nanowires were then able to grow radially out from the fiber surface.

A typical SEM image of a Kevlar fiber covered by the grown ZnO nanowires is shown in Figure 7.4(a). Along the entire length of the fiber, ZnO nanowires grew radially and exhibited a very uniform coverage and well-preserved cylindrical shape of the fiber. Some splits in the nanowire arrays can be identified (Figure 7.4b), which were produced probably owing to the growth-induced surface tension in the seeding layer. All of the ZnO nanowires are single crystalline,

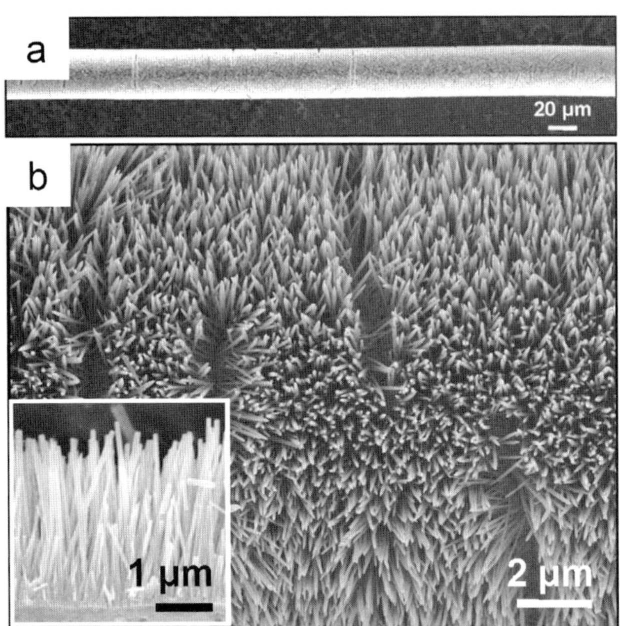

FIGURE 7.4 *Unpatterned growth of ZnO nanowire arrays on Kevlar fibers. (a) SEM image of a Kevlar fiber covered with ZnO nanowire arrays along the radial direction. (b) Higher magnification SEM image and a cross-section image (inset) of the fiber, showing the distribution of nanowires.*

and each nanowire has a hexagonal cross-section with a diameter in the range of 50–200 nm and a typical length of 3.5 μm.

Patterned growth

Patterned growth of ZnO nanowire arrays has been achieved using patterned gold catalysts [28], seed layers [17, 29] and SAMs [30]. In 2007, Morin et al. developed a simple catalyst-free process that could be used not only on flat polymer films but also on microfibers [31]. Their main working principle is that by employing ultraviolet (UV)-irradiated polymer surfaces, the region that has been irradiated generates a carboxylic group on top of the polymer surface, while the unexposed region remains unchanged. Astonishingly, ZnO nanowire arrays could only nucleate and grow selectively on the regions where there were no carboxylic groups. This is because protonated HMTA species combined with the surface carboxylic acid groups, thus blocking the nucleation of ZnO in these UV-radiated regions. Based on the reported pKa of HMTA (5.4–5.5) [32], and the measured pH of 6.6 under their reaction conditions, approximately 7% of the HMTA species present should be protonated.

In their experiments, single filaments of polyester [polyethylene tere-phthalate (PET)], the material on which ZnO nanowire arrays were grown, were cleaned by a standard cleaning procedure. First, they were sonicated in acetone for 20 min, then rinsed with distilled water, and then baked at 60°C for at least 24 h to get rid of any absorbed moisture. For surface functionalization, a single filament was strung through slits in a polycarbonate backing film and then photo-oxidation was carried out using a UV lamp with wavelength at 254 nm at a nominal intensity of 18.5 mW/cm^2 for 50 min through a common photo mask (Figure 7.5a). Then, the photo-oxidized substrate was immersed into a 6 ml vial containing 3.5 ml of nutrient solution, composed of a 1:1 ratio of zinc nitrate and HMTA. The morphology of the ZnO nanostructures in their experiments could be varied by using different compositions of the nutrient solution. To grow ZnO nanorods, a solution of 12 mM zinc nitrate hexahydrate and 12 mM HMTA was used, while a mixed solution of 25 mM zinc nitrate hexahydrate, 25 mM HMTA and 0.18 mM citric acid was used to produce ZnO plates, where citric acid acted as a capping agent to the basal planes of the nanowires to inhibit the axial growth of the nanoplates, as will be discussed in a later section. To avoid random deposition of the heterogeneously formed precipitates from the solution body, the substrate was positioned with the UV-irradiated face facing downward [33]. After that, the sample was removed from the vial, rinsed with ethanol and dried with N$_2$ gas (Figure 7.5b).

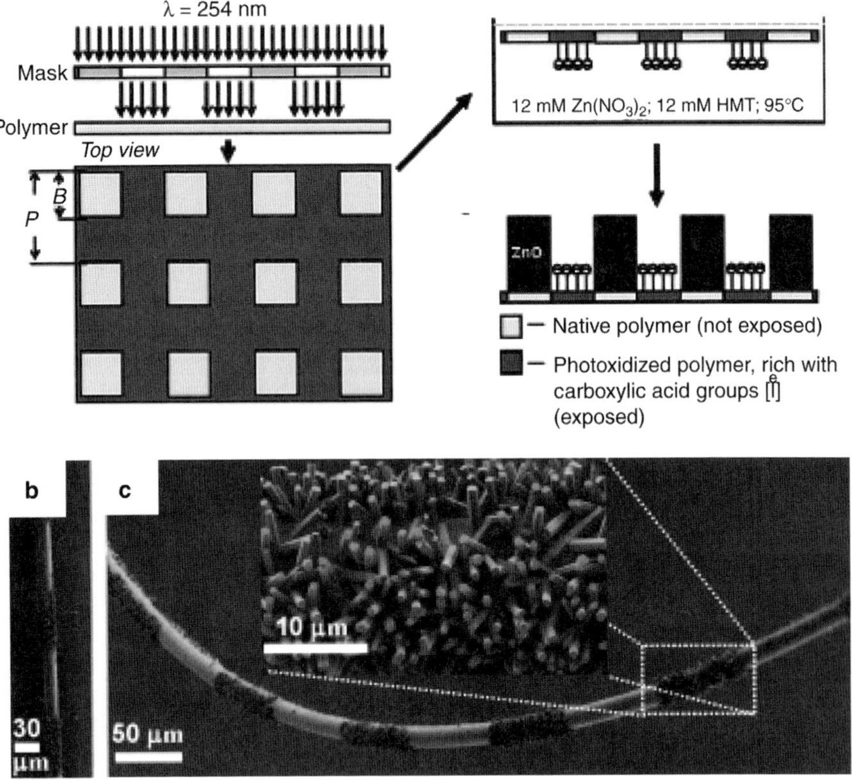

FIGURE 7.5 *(a) Schematic illustration of the process employed to produce ordered microarrays of ZnO. (b) SEM images of patterned ZnO nanorods assembled on flexible polyethylene terephthalate (PET) filaments.*
(Image courtesy of Morin et al. [31].)

The strategy is anticipated to expand to other inorganic and polymer materials with complex substrate geometries, which enables many macroelectronic or microelectronic applications. In the future, the fundamental concepts demonstrated in this work may be further applied to self-assembled block copolymer nanostructures to achieve truly bottom–up nanoscale assembly of nanomaterials.

Growth mechanism

In general, ZnO nanostructures are formed by the hydrolysis of zinc nitrate in water in the presence of HMTA. The detailed role of HMTA is still under investigation. HMTA is a nonionic cyclic tertiary amine that can act as a Lewis base with metal ions and a bidentate ligand capable of bridging two zinc (II) ions in solution [34]. HMTA also hydrolyzes to produce formaldehyde and ammonia under the given reaction conditions; the latter of which can form ammonium hydroxide and complex with zinc (II) to form $Zn(NH_3)_4^{2+}$ [35]. Zinc (II) is known to coordinate in tetrahedral complexes. Under the given pH and temperature, zinc (II) is thought to exist primarily as $Zn(OH)_4^{2-}$ and $Zn(NH_3)_4^{2+}$. The ZnO is formed by the dehydration of these intermediates.

Seeded growth

One main advantage of wet chemical growth is that it can be achieved independently of the substrate by using ZnO seeds in the form of thin films or nanoparticles. In this way, the nucleation step is bypassed and only the necessary conditions for growth are considered. Alignment of the ZnO nanocrystals occurs on flat surfaces regardless of their crystallinity or surface chemistry, including ZnO and Al_2O_3 single crystals, transparent conducting oxides such as indium–tin oxide (ITO) and fluorine-doped tin oxide (FTO), amorphous oxides including glass and Si with its native oxide, and the oxide-free metals such as Au and Ti. However, this approach typically requires higher temperature substrate processing. ZnO seeds must be annealed at 150°C to improve particle adhesion to the substrate, and ZnO nanorod vertical alignment is improved by textured ZnO seeds annealed at 350°C [15, 36]. Through the use of seeds, wafer-scale synthesis has been achieved [15].

Au/polyimide

ZnO synthesis can also be carried out on catalysts such as Au, Ag and Pt. Such metals can be deposited onto polymeric substrates as thin films to achieve nanowire growth. In this case, the surface roughness of the thin film must be carefully controlled to promote nucleation. In addition, the nanowire orientation is largely a function of the surface topography. In order to achieve vertically aligned nanowires, an appropriate process flow must be employed for smooth substrate surfaces. Figure 7.6 shows an example of ZnO nanorod growth on an Au-coated polyimide substrate at 70°C. The nanorods have an aspect ratio of approximately 15:1. The 50 nm Au thin film was deposited by DC magnetron sputtering. Vertical alignment was achieved by carefully controlling the surface roughness of the substrate.

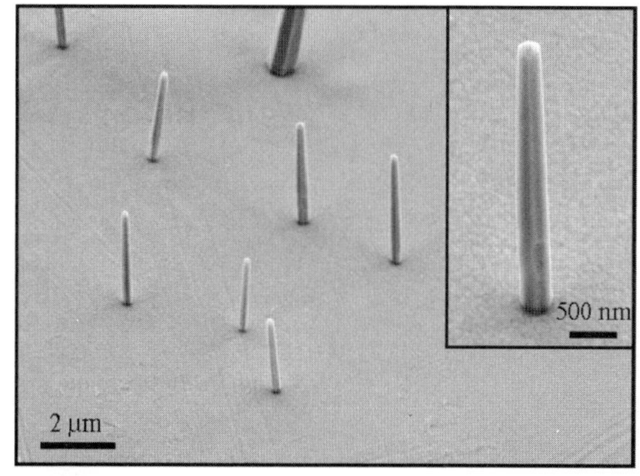

FIGURE 7.6 *SEM image of vertically aligned ZnO nanowires grown on an Au-coated polyimide substrate.*

Electrochemical deposition

Electrochemical deposition of ZnO is a powerful technique for achieving large-area, uniform nanorod synthesis. However, it requires a conductive substrate and materials that can withstand acid reaction environments. Typically, a $ZnCl_2$/KCl electrolyte is bubbled with O_2 to maintain an oxygen-saturated electrolyte. A standard three-electrode set-up with a saturated Ag/AgCl reference electrode and Pt counter-electrode is often used in conjunction with a potentiostat to control the current–voltage characteristics throughout the process.

In addition, the nucleation and growth rate of ZnO can be enhanced by electrochemical methods in a solution of zinc nitrate and HMTA. As mentioned previously, Weintraub et al. used an electrochemical technique combined with a patterned substrate to achieve individual ZnO nanorod growth [26]. In this case, the applied potential causes an increased OH^- concentration by the reduction of NO_3^- in an electrochemical reaction at the negative electrode. As such, the reaction would proceed as follows:

$$NO_3^- + H_2O + 2e^- \rightarrow NO_2^- + 2OH^- \tag{7.1}$$

$$Zn^{2+} + 2OH^- \rightarrow ZnO + H_2O \tag{7.2}$$

The HMTA still promotes 1D growth by capping the non-polar facets and providing a hydroxide source. This method is particularly useful because enhanced nucleation can be achieved without the use of seeds. Seeding often requires a high-temperature annealing step ($> 200°C$), making it difficult to use polymer seeded substrates. Therefore, the electrochemical approach, which can be carried out on any conductive substrate, is useful for growth on conductive polymers.

Zinc nitrate hexahydrate and hexamethylenetetramine

The most commonly used chemical agents for the hydrothermal synthesis of ZnO nanowires are probably zinc nitrate hexahydrate and HMTA [9, 33]. The chemistry of the growth has already been well documented. Zinc nitrate hexahydrate salt provides Zn^{2+} ions required for building up ZnO nanowires. Water molecules in the solution provide O^{2-} ions. Even though the exact function of HMTA during the ZnO nanowire growth is still unclear, it is believed to act as a weak base, which would slowly hydrolyze in the water solution and gradually produce OH^-. This is critical in the synthesis process because, if the HMTA hydrolyzes very quickly and produces a lot of OH^- in a short period, the Zn^{2+} ions in solution would precipitate out very quickly owing to the high pH environment, which would have to make a contribution to the ZnO nanowire-oriented growth, and eventually result in fast consumption of the nutrient and prohibit further growth of ZnO nanowires.

$$(CH_2)_6N_4 + 6\ H_2O \leftrightarrow 4\ NH_3 + 6HCHO \tag{7.3}$$

$$NH_3 + H_2O \leftrightarrow NH_3 \cdot H_2O \tag{7.4}$$

$$NH_3 \cdot H_2O \leftrightarrow NH_4^+ + OH^- \tag{7.5}$$

$$Zn^{2+} + 2OH^- \leftrightarrow Zn(OH)_2 \tag{7.6}$$

$$Zn(OH)_2 \leftrightarrow ZnO + H_2O \tag{7.7}$$

The growth process of ZnO nanowires can be controlled through the five chemical reactions listed above. All five reactions are actually in equilibrium and can be controlled by adjusting the reaction parameters, such as precursor concentration, growth temperature and growth time, to push the reaction equilibrium forward or backward. In general, precursor concentration determines the nanowire density. Growth time and temperature control the ZnO nanowire morphology and aspect ratio.

Nanowire density controlled by concentration

The density of ZnO nanowires on the substrate, which is Kapton film coated with gold in this case, could be controlled by the initial concentration of the zinc salt and HMTA. To explore the relationship between the precursor concentration and the density of the ZnO nanowire arrays, a series of experiments was performed by varying the precursor concentration but keeping the ratio constant between the zinc salt and HMTA [33]. Experimental results show that the density of the nanowire arrays is closely related to the precursor concentration. Detailed analysis of the measured data is shown in Figure 7.7 (red line). From 0.1 mM to 5 mM, the density of the ZnO nanowires, defined as number of nanowires per 100 μm^2, increases dramatically, possibly for the following reasons. The zinc chemical potential inside the solution body increases with zinc concentration. To balance the increased zinc chemical potential in solution, more nucleation sites on the substrate surface will be generated. So, the density of ZnO nanowires increases. When further increasing the zinc concentration, the density of ZnO nanowires remains approximately steady with a slight tendency toward a decrease.

The steady/saturated density may be understood from the nucleation and growth process. The nanowire density is decided by the number of nuclei

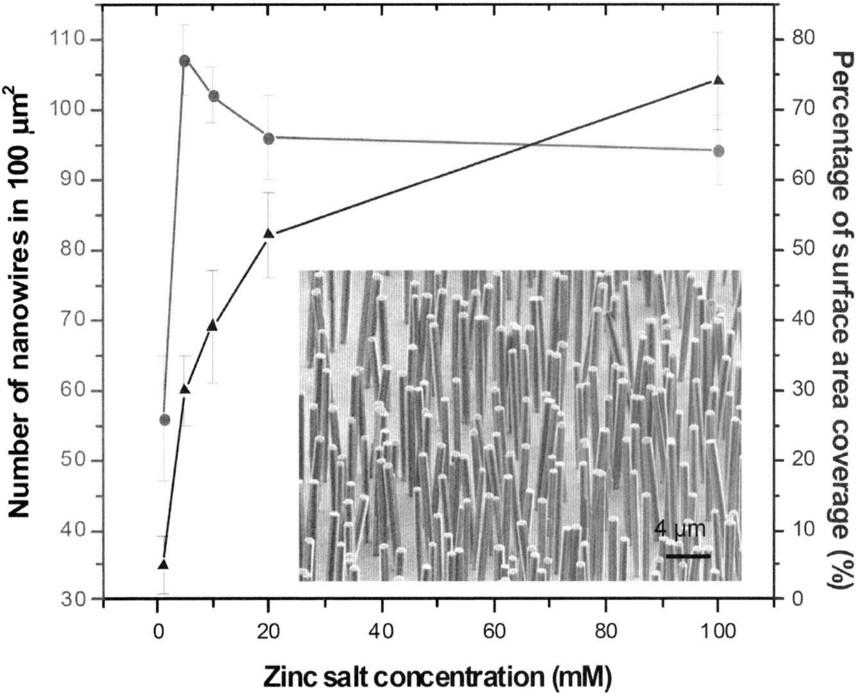

FIGURE 7.7 *Nanowire density controlled by concentration. Plot of ZnO nanowire density in a 100 µm² area (line with filled circle data points) and plot of area percentage covered by ZnO nanowires (line with filled triangle data points). Each data point was obtained from four different areas. Inset is a typical image of ZnO nanowires grown at 5 mM.*

formed at the very beginning of the growth, which continued to grow and form nanorods (shorter nanowires). The arrival of more ions on the substrate may not initiate new nuclei at a later stage, for two possible reasons. First, considering the critical size required for a nucleus to grow into a crystal, no new nanorods would form if the nuclei are smaller than the critical size. Second, owing to the existence of the first group of grown nanorods, the newly arrived ions on the substrate have a higher probability of reaching the existing nanowires than the newly formed nuclei; thus, the nuclei may not exceed the critical size and they will eventually dissolve into the solution body. In such cases, a continuous increase in solution concentration may not increase the density of the nanowires when its density is larger than the saturation density. This also explains why the nanowires grown in these experiments have fairly uniform height. Even though the density of the ZnO nanowires remains steady at high precursor concentration level, the surface coverage percentage increases slightly owing to the lateral growth of the nanowires (Figure 7.7, blue line).

Morphology evolution with growth time

Wurtzite ZnO has polar surfaces, such as (0001) and (000-1), and non-polar surfaces such as {01-10} and {2-1-10}. Three different growth stages were defined during the nanowire growth: at the beginning, in the middle and at the end [33]. At different growth stages, the axial growth at polar surfaces is found to have a different relative growth rate to the lateral growth at non-polar surfaces. At the beginning of the growth period, which is defined as from the very start to around 30 min, lateral growth seems to be more significant than axial growth (Figure 7.8a). When the growth time exceeds 30 min, the width of ZnO nanowires is almost independent of time, indicating little lateral growth (Figure 7.8b). This indicates that, at the beginning growth stage, lateral growth is more significant than axial growth. In the middle growth stage, from 30 min to around 6 h, axial growth is dominant. At the final growth stage, from 6 h to 48 h in these experiments, lateral growth and axial growth seem to be equally significant, as can be seen from Figure 7.8(b, c). In this period, the length of the ZnO nanowires almost doubles. At the same time, their width is also doubled. The aspect ratio remains nearly constant.

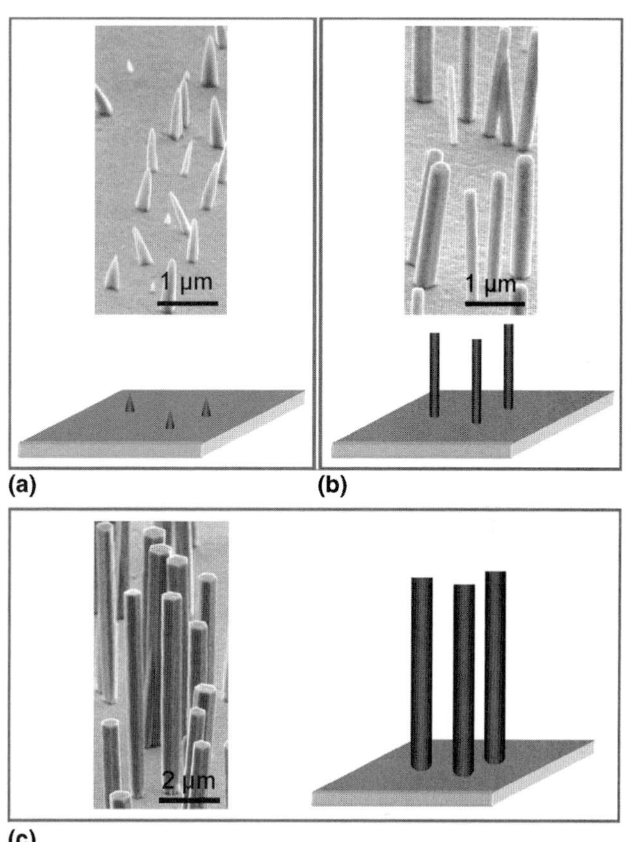

FIGURE 7.8 *Morphology of ZnO nanowire evolution with growth time: (a) 0.5 h, (b) 6 h, and (c) 48 h.*

Morphology and density controlled by growth temperature

Temperature is an important factor in maintaining a high aspect ratio and the morphology of ZnO nanostructures, and 70°C is the optimum temperature for attaining a high aspect ratio and well-defined hexagonal prism shape of the ZnO nanowires (Figure 7.9b) [33]. At a lower temperature, e.g. 60°C (Figure 7.9c), the aspect ratio becomes smaller than that at 70°C. The aspect ratio of the ZnO nanowires is determined by the relative growth rate of the polar and non-polar surfaces. Therefore, the relative growth rate of polar surface to non-polar surface at low temperatures is smaller than that at high

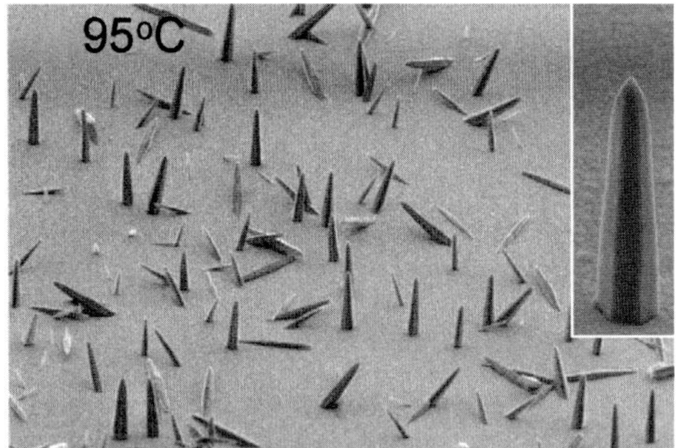

(a)

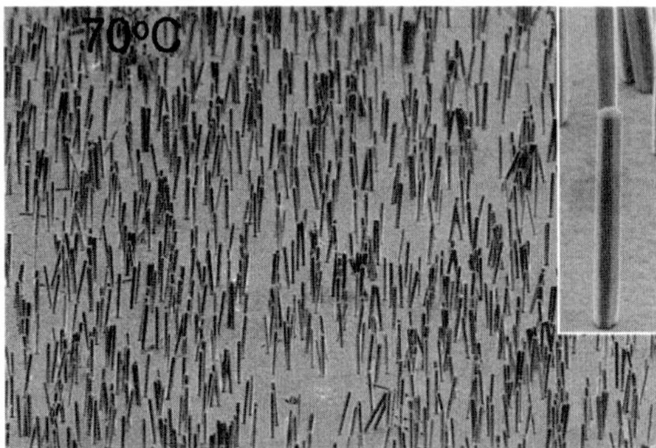

(b)

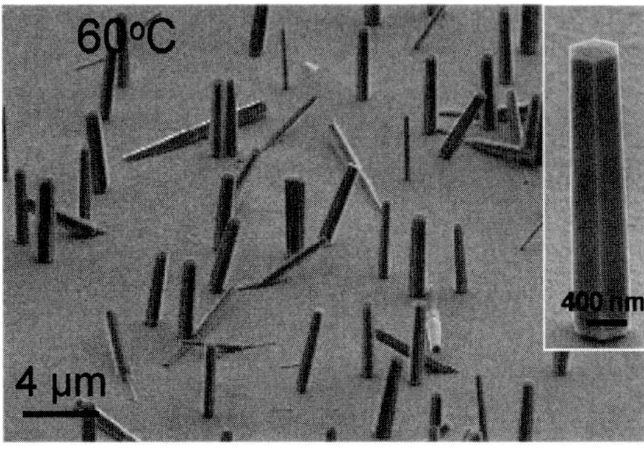

(c)

FIGURE 7.9 *Morphology and density change with growth temperature: 60° tilt degree view of ZnO nanowires synthesized at (a) 95°C, (b) 70°C, and (c) 60°C. The three images are recorded at the same magnification and the three inset images are at the same magnification.*

temperatures. If the growth temperature is increased further, e.g. to 95°C, pyramid-shaped ZnO nanorods are grown (Figure 7.9a), which means that the axial growth rate is so fast that the lateral growth cannot keep up with the axial growth.

From another point of view, as discussed previously, HMTA functions as a slowly decomposing weak base, which maintains the weak basic environment in the solution and successively provides Zn^{2+} ions with OH^- ions. From reaction (7.3), it is found that 7 moles of reactant produce 10 moles of product. So, by increasing the reaction temperature, reaction (7.3) would move forward by virtue of entropy increase, which means that HMTA decomposes more quickly at high temperature than at low temperature. That is to say, at the very beginning, HMTA has already decomposed to a relatively large degree and has produced enough OH^-, resulting in a sufficient and thorough growth of ZnO at the beginning. Therefore, the base of the ZnO nanorods is thicker than at lower temperature. Furthermore, both the axial and lateral growth rates are greatly increased at higher temperature. As time goes by, the supply of Zn^{2+} is limited in the reaction container and gradually becomes exhausted, which leads to the formation of incompletely shaped nanowires (pyramids or tapered nanorods).

The temperature-dependent growth behavior of the ZnO nanowire density may be understood from the basic theory of crystal nucleation and growth. When the temperature is low, the mobility and the diffusion length of the ions on the substrate are rather limited, which prevent the ions from diffusing around, and therefore large nuclei are formed. Thus, the nanowire density is low. At high temperatures, the mobility and diffusion length of the ions are large enough to reach the sites of the first grown nanowires; thus, no new nuclei are formed because the precursor ions have a higher affinity to the already formed seeds than bare substrate, resulting in a lower density of nanowires on the substrate. At an optimum temperature, such as 70°C, the mobility of the ions is moderate and their diffusion length is within a small range in the vicinity of the substrate; thus, the accumulation of ions in local regions results in a high density of nanowires. But their size is small because of the conservation of total ions that are able to migrate near the substrate.

FIGURE 7.10 *ZnO nanowires grown at 5 mM, 80°C for 48 h.*

As discussed above, the concentration controls the nucleation density, the growth time controls the aspect ratio, and the temperature controls the morphology and the aspect ratio. By adjusting these three parameters, ZnO nanowires with a high aspect ratio can be achieved (Figure 7.10).

Additives to modify the growth

Additives could be added to the growth system to modify the growth habits of the ZnO nanowires. Commonly used additives for hydrothermally synthesizing ZnO nanowires may be placed in two categories. The first is additives that absorb onto the side surfaces and enhance the vertical growth, including polyethylenimine (PEI) [37] and ethylenediamine (EN) [38] (Figure 7.11a). The other category is additives that can cap onto the basal plane of the ZnO nanowire and promote lateral growth, a typical example of which is the citrate ions (Figure 7.11b) [39].

Growth orientation

Wang et al. and Lee et al. have both indicated that the Zn-terminated (0001) surface is much more catalytically active than the O-terminated (000-1) surface for the growth of ZnO nanowires by physical vapor deposition methods [40–42]. It is reasonable to infer that in the wet chemical growth of ZnO nanowire arrays the Zn-terminated ZnO (0001) polar surface is chemically active and the oxygen-terminated (000-1) polar surface is inert. So, the wet chemically grown ZnO nanowires are all growing along the <0001> direction,

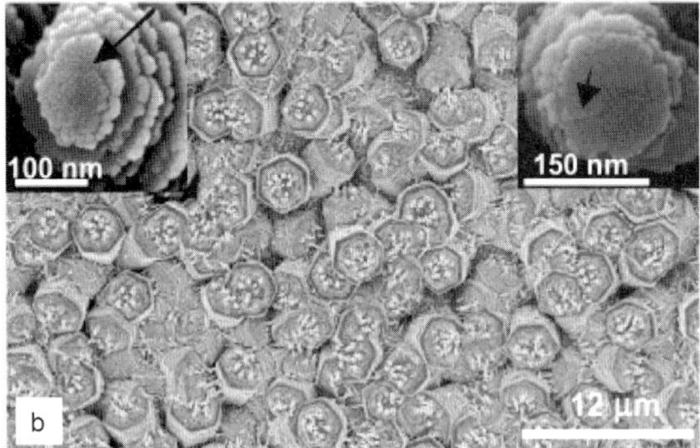

FIGURE 7.11 *(a) Typical scanning electron microscopy cross-section of a cleaved nanowire array on fluorine-doped tin dioxide (FTO) with polyethylenimine (PEI) as additive. (b) Large arrays of well-aligned helical ZnO nanorods on top of base ZnO rods grown with citric acid as additive.*
(Images courtesy of Law et al. [37] and Tian et al. [39].)

which is also the fast growth direction. As indicated in the transmission electron microscope (TEM) images [14], the corresponding electron diffraction pattern illustrates that the nanowire growth direction is <0001> and its side surfaces are {2-1-10}.

APPLICATIONS IN ENERGY CONVERSION: NANOGENERATOR

Fundamentals of piezoelectricity

The non-central symmetry of the wurtzite crystal structure leads to piezoelectricity in ZnO. Piezoelectricity arises from atomic scale polarization. The ZnO unit cell in Figure 7.12(a) shows a Zn^{2+} cation situated inside a tetrahedral cage surrounded by O^{2-} anions. Under non-stress conditions, the positive and negative center of mass coincide so all ionic dipoles cancel (Figure 7.12b). However, when the crystal is subjected to a mechanical force, F, the positive and negative centers of mass become displaced with respect to each other, leading to a net polarization, P (Figure 7.12b). In single-crystal ZnO, this results in a macroscopic dipole whereby positive and negative electric charge build up on opposite crystal faces. The polarization and applied stress are linearly related, as shown by the following equation:

$$P_i = d_{ij}T_j$$

where T_j, a mechanical stress exerted along direction j, and P_i, the induced polarization along direction i, are related by the piezoelectric coefficient, d_{ij}. The strongest piezoelectric coefficient (d_{33}) in ZnO arises from a stress applied along the [0001] direction resulting in polarization across the (0001) and (000$\bar{1}$) planes. The effective piezoelectric coefficient (d_{33}) of a ZnO nanobelt was experimentally measured by piezo-response force microscopy ranging from 14.3 pm/V to 26.7 pm/V, depending on the frequency [43]. It is worth noting that these are higher than the ZnO bulk value of 9.93 pm/V. The converse piezoelectric effect also exists, whereby an applied electric field, E_i, induces strain, S_j, in the crystal represented by the following equation:

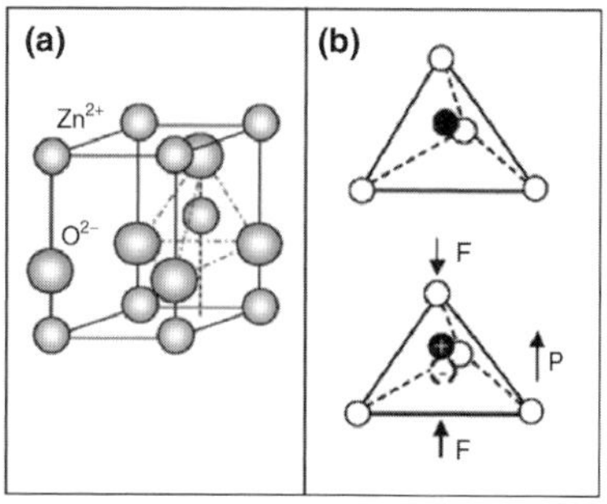

$$S_j = d_{ij} E_i$$

FIGURE 7.12 *(a) Wurtzite structure model of ZnO emphasizing the tetrahedral coordination. (b) Schematic showing the piezoelectric effect in a tetrahedrally coordinated cation–anion unit.*

The piezoelectric coefficient, d_{ij}, is defined in the same way as above. In general, ZnO piezoelectric crystals are electromechanical transducers that can convert electric signals to mechanical signals, and vice versa.

Nanogenerator based on atomic force microscope

Theory behind the nanogenerator

ZnO is unique in that it is both piezoelectric and semiconducting, which allows for a variety of electromechanically coupled devices [44–47]. The theory behind such devices is best explained by considering a single aligned nanowire (Figure 7.13a). Upon being mechanically deflected by an external force, a strain field is established in the nanowire with the right side under tension ($\varepsilon > 0$) and the left side under compression ($\varepsilon < 0$) (Figure 7.13b). The strain creates an electric field inside the nanowire. As a result, a piezoelectric potential of V_s^+ to V_s^- runs across the nanowire from the tensile to the compressed side (Figure 7.13d). The potential is due to the relative displacement of Zn^{2+} and O^{2-} ions in the wurtzite crystal structure and remains as long as the strain field exists. The bottom of the nanowire is grounded ($V_s = 0$). These principles will be used to describe electromechanically coupled devices based on ZnO nanowires such as a nanogenerator.

Piezoelectric nanogenerator

The nanogenerator is an electromechanical transducer which converts mechanical energy into electrical energy using aligned ZnO nanowires. The nanogenerator can potentially harness energy from the environment in the form of mechanical, vibrational and hydraulic energy for electric power generation. To demonstrate the principle, a single nanowire on a sapphire substrate has been deflected by a conductive atomic force microscope (AFM) tip in tapping mode, leading to a voltage output of 8 mV [47].

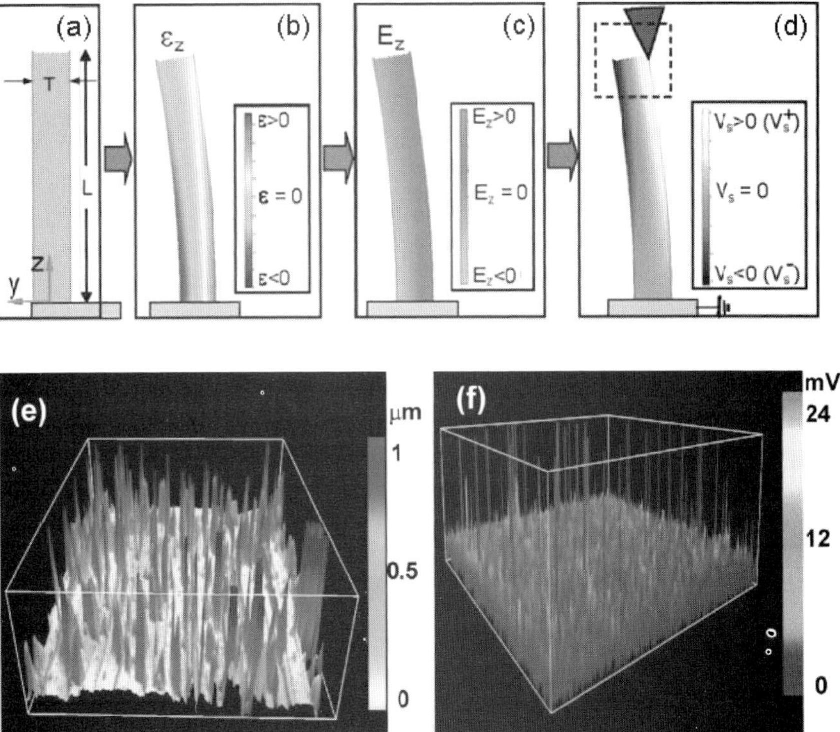

FIGURE 7.13 *Principle of coupled electromechanical devices. (a) Vertically aligned ZnO nanowire. (b) Strain field distribution after mechanical deflection. (c) Piezoelectric induced electric field along z-direction. (d) Potential distribution across nanowire due to piezoelectric effect. (e) 3D plot of the atomic force microscope (AFM) topographic image. (f) Voltage output profile obtained by scanning the AFM tip over a 40 × 40 µm area of aligned ZnO nanowires on a plastic substrate.*

The operating principles behind the nanogenerator are as follows [44–47]. Ionic charges are separated as shown in Figure 7.13(a–d); this is the initial nanogenerator charging process. The final step is the discharge process which involves the coupling of piezoelectric and semiconducting properties. The platinum-coated AFM tip or zigzag electrode deflects the nanowire. ZnO has an electron affinity of 4.5 eV and Pt has a work function of 6.1 eV, so at the nanowire tip the Pt/ZnO junction forms a Schottky barrier. At first, the tip is in contact with the tensile side (V_s^+). The potential of the metal tip is nearly zero $(V_m = 0)$. A metal/n-type semiconductor junction is formed with $\Delta V = V_m - V_s^+ < 0$, resulting in a reverse-biased Schottky junction. Thus, little current flows during this initial process of accumulating and preserving charge. In the discharge step, the tip or electrode is in contact with the compressed side of the nanowire ($V_s{}^-$). A metal/semiconductor junction forms with $\Delta V = V_m - V_s{}^- > 0$, resulting in a forward-biased Schottky junction, and current is allowed to flow. The driving force for free electrons to flow from the ZnO nanowire to the metal is ΔV, but the output voltage is the difference in Fermi level between the tip and the grounded side. The above-proposed mechanism can also be well explained by a band diagram, which has been described in detail elsewhere [48, 49].

Nanogenerator based on flexible substrate

In addition to rigid sapphire substrates, experiments have been carried out on flexible Au-coated polyimide substrates stimulated by an AFM tip [14]. In these experiments, the highest reported voltage output is 45 mV. The output power of a single nanowire is approximated as 10–20 pW. Figure 7.13(e) shows an AFM surface topography plot of the nanowire arrays on a polyimide substrate. A nanowire position versus piezoelectric voltage output plot acquired by AFM can also be seen in Figure 7.13(f).

One problem when mechanically straining nanorods on a flexible substrate is nanorod detachment from the substrate. Steps must be taken to improve the nanorod bond to the substrate. In the above experiment, a 200 nm PMMA film encapsulates the nanorod array. The nanorod tips are then exposed to UV light to develop a portion of the PMMA. In this way, the nanorods are fastened more securely to the substrate.

Nanogenerator based on microfibers

The ceramic or semiconductor substrates used for growing ZnO nanowires are hard and brittle, and cannot be used in areas that require a foldable or flexible power source, such as implantable biosensors in muscles or joints, and power generators built into walking shoes. It is necessary to use conductive polymers, which are likely to be biocompatible and safe, as substrates. Two advantages are offered by this approach. One is the cost-effective, large-scale solution of growing ZnO nanowire arrays at a temperature below 100°C. The other is the large degree of choice of flexible plastic substrates for growing aligned ZnO nanowire arrays, which could play an important role in flexible and portable electronics for harvesting low-frequency ($\sim$10 Hz) energy from the environment, such as from body movement (e.g. gestures, respiration or locomotion).

Fabrication of the fiber nanogenerator

Qin et al. recently demonstrated fiber-based nanogenerators [27]. The design of the direct current nanogenerator is based on the zigzag-electrode mechanism [44], but in this case they have replaced the zigzag electrode by an array of metal wires, which are actually metal-coated ZnO nanowire arrays.

The top and side surfaces of the as-grown ZnO nanowire on microfibers are smooth and clean, indicating that they are able to form the reliable metal semiconductor junctions needed for fabricating electronic nanodevices. The spacing between the nanowires is in the order of a few hundred nanometers, which is large enough for them to be bent to generate the piezoelectric potential. The tips of the nanowires are separated from each other owing to their small tilting angles ($< \pm 10°$), but their bottom ends are tightly connected (Figure 7.4b inset). As a result, a continuous ZnO film at the root of the nanowires forms, which could serve as a common electrode for the nanogenerator. In reality, the metal nanowire arrays were made by metal coating of ZnO nanowire arrays grown on Kevlar fibers (Figure 7.14b, c). In practice, any fiber should work as long as it has good electrical conductivity. The metal to be coated is required to form a Schottky contact with ZnO.

After growing the ZnO nanowires on the fibers, the nanowires need to be fastened to make them robust and resistive to the mechanical forces. To maintain the high flexibility of the fiber after growing a crystalline film and nanowires, a surface coating strategy was introduced to improve the mechanical performance of the fiber and the binding of nanowires. Two layers of tetraethoxysilane (TEOS) were infiltrated – one above and one below the ZnO seed layer – as binding agents. The Si–O bonds in TEOS are highly reactive with the OH groups on the ZnO surface, and its organic chains bind firmly to the body of the aromatic polyamide fiber. As a result, the ZnO seed layer and the fiber core were tightly bound together by a thin layer of TEOS. Furthermore, since TEOS could easily form cross-linked chains, the nanowires were firmly bundled and bound together at their roots and fixed on the ZnO seeding layer, successfully preventing them from scratching or stripping off during mechanical brushing and sliding.

To demonstrate the power generation ability of the ZnO nanowire-covered fibers, a double-fiber model system was designed (Figure 7.14b). Two fibers, one coated with 300 nm thick gold layer and one without, were entangled to form the core for power generation. The relative brushing between the two fibers was simulated by pulling and releasing the string using an external rotor with a controlled frequency. The gold-coated fiber was connected to the external circuit as the nanogenerator's output cathode. The pulling force ensured a good contact between the two fibers, as shown by an optical microscopy image.

Mechanism of the fiber nanogenerator

The model of two entangled fibers sets the principle of the fiber-based nanogenerator. A coupled piezoelectric and semiconducting properties resulted in a process of charge creation and accumulation and charge release [47, 50]. The

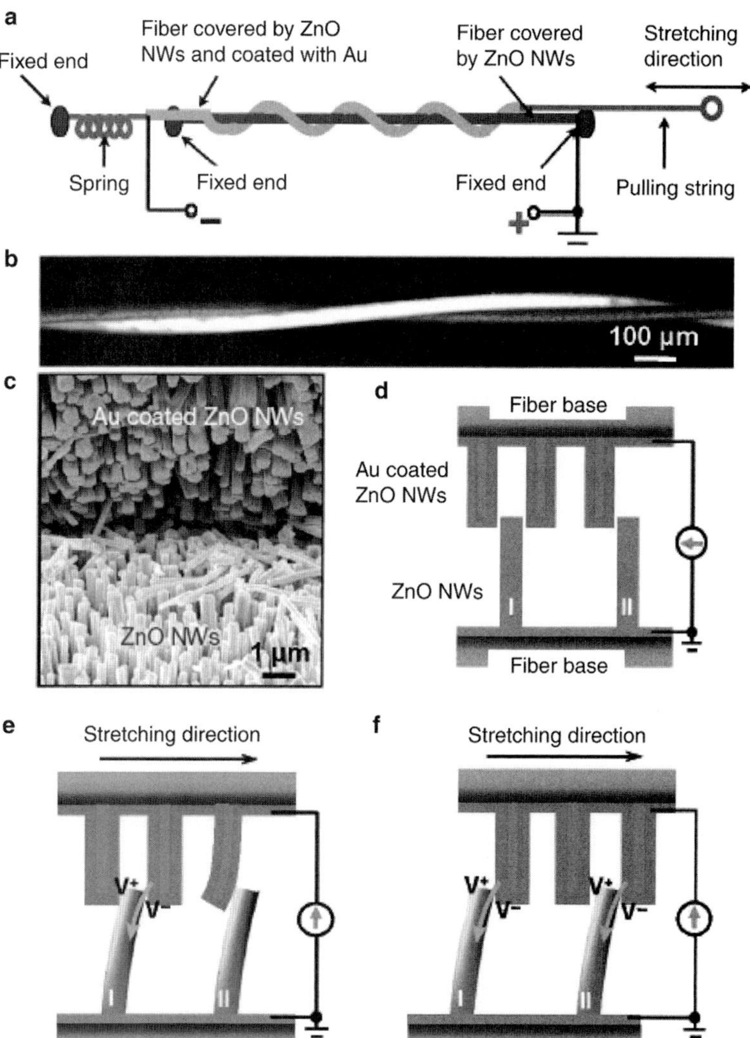

FIGURE 7.14 *(a) Schematic experimental set-up of the fiber-based nanogenerator. (b) Optical micrograph of a pair of entangled fibers. (c) SEM image at the 'teeth-to-teeth' interface of two fibers covered by nanowires with the top one coated with Au. (d) Schematic illustration of the teeth-to-teeth contact between the two fibers covered by nanowires. (e) The piezoelectric potential created across nanowires I and II under the pulling of the top fiber by an external force. (f) When the top fiber is further pulled, the Au-coated nanowires may scrub across the uncoated nanowires.*

gold coating completely covered the ZnO nanowires and formed a continuous layer along the entire fiber. A successful coating was confirmed by its metallic *I–V* characteristic [47, 51]. Once the two fibers were firmly entangled together (Figure 7.14a), some of the gold-coated nanowires penetrated slightly into the spaces between the uncoated nanowires rooted on the other fiber, like a 'teeth-to-teeth' relationship on the two fibers, as shown by the interface image in Figure 7.14(c). By brushing the metal-coated nanowire arrays across the ZnO nanowire arrays, the metal wires act like an array of AFM tips that induce mechanical deformation in the ZnO nanowires and collect the charges.

Figure 7.14(d–f) illustrates the charge generation mechanism of the fiber nanogenerator. Like the deflection of a nanowire by an AFM tip, when the top fiber moves to the right, for example, the gold-coated nanowires bend the uncoated ZnO nanowires to the right. A piezoelectric potential is thus generated across the uncoated nanowire owing to its piezoelectric property, with the stretched surface positive and the compressed surface negative. The side with positive potential has a reverse-biased Schottky contact with the gold, as can be determined from the *I–V* curve of the device, which blocks the flow of current, while the negative-potential side has a forward-biased Schottky contact with the gold that allows the current to flow from the gold to the nanowire. As the density of the nanowires is high, it is very likely that a bent nanowire rooted at the uncoated fiber touches the back side of another gold-coated nanowire after subjecting to bending (such as nanowire I in Figure 7.14e). In this case, the negative-potential surface of the ZnO nanowire contacts the gold layer, so the Schottky barrier at the interface is forward biased, resulting in a current flowing from the gold layer into the ZnO nanowire. Then, when the top fiber keeps moving further towards the right (Figure 7.14f), the gold-coated nanowires scan across the tips of the ZnO nanowires and reach their negatively charged sides (nanowires I and II in Figure 7.14f). Therefore, more current will be released through the forward-biased Schottky barriers (Figure 7.14f). This means that the currents from all of the nanowires will add up constructively regardless of their deflection directions, even in the same cycle of pulling.

Owing to the similar mechanical properties of the top and bottom nanowires, the gold-coated nanowires could also possibly be bent by the nanowires rooted at the uncoated fiber, but this does not affect the mechanism presented in Figure 7.14. For the gold-coated fiber, all of the nanowires are completely covered by a thick gold layer, and they can be considered as an equal-potential electrode connected to the external measurement circuit. Thus, the role played by these ZnO nanowires was only to act as a template for supporting the gold coating, and no piezoelectric charges will be preserved inside the gold-coated nanowires.

Current and voltage output

The output voltage is defined by the characteristic of a single nanowire, while the output current is defined by the overall contribution from all of the active nanowires. The signs of the output voltage/current do not change in response to the deflection configuration of the nanowire owing to the rectifying behavior of the Schottky barrier at the Au–ZnO interface. The same effect is expected if the top fiber is driven to the left. The short-circuit current (I_{sc}) and open-circuit voltage (V_{oc}) were measured to characterize the performance of the fiber nanogenerators. The resulting current signals detected at 80 rpm are shown in Figure 7.15(a). To rule out some systematic artifacts, a 'switching polarity' testing method was applied during the entire measurement as defined in Figure 7.14(a), and 5 pA positive current pulses were detected at each pulling–releasing cycle (blue curve in Figure 7.15a). Negative current pulses with the same amplitude were received (pink curve in Figure 7.15a). The inverted output current signals confirmed that the current had indeed originated from the fiber nanogenerator rather than from the measuring system. The small output current (5 pA) is attributed mainly to the large loss in the fiber due to an extremely large inner resistance. This large inner resistance of the fiber-based nanogenerator is probably due to cracks in the ZnO seed layer directly adjacent to the fiber that are caused by the structural incompatibility and large difference in thermal expansion coefficients of ZnO and microfibers. Open-circuit voltage was also measured by the switching polarity method. Corresponding positive and negative voltage signals were received when the voltage meter was forward and reverse connected to the fiber nanogenerator, respectively (Figure 7.15b). The amplitude of the voltage signal in this case was 1–3 mV.

A relative deflection by a distance as short as one nanowire size is sufficient to generate electricity. This is the fiber-based nanogenerator, with potential for harvesting energy from body movement, muscle stretching, light wind and vibration. It establishes the basis for building a 'power-shirt' to wear in the future.

Current obstacles and future work

Even though nanogenerators have been made on flat polymer film by AFM and microfibers on the basis of two entangled fibers with one coated with Au and one without, the output voltage and current are still too low for any practical applications. A promising way of increasing the magnitude of the output signal would be to increase the number of active nanowires in the current generation process. In the present stage of experiments, less than 5% of the nanowires are estimated to be active in the current generation process [44].

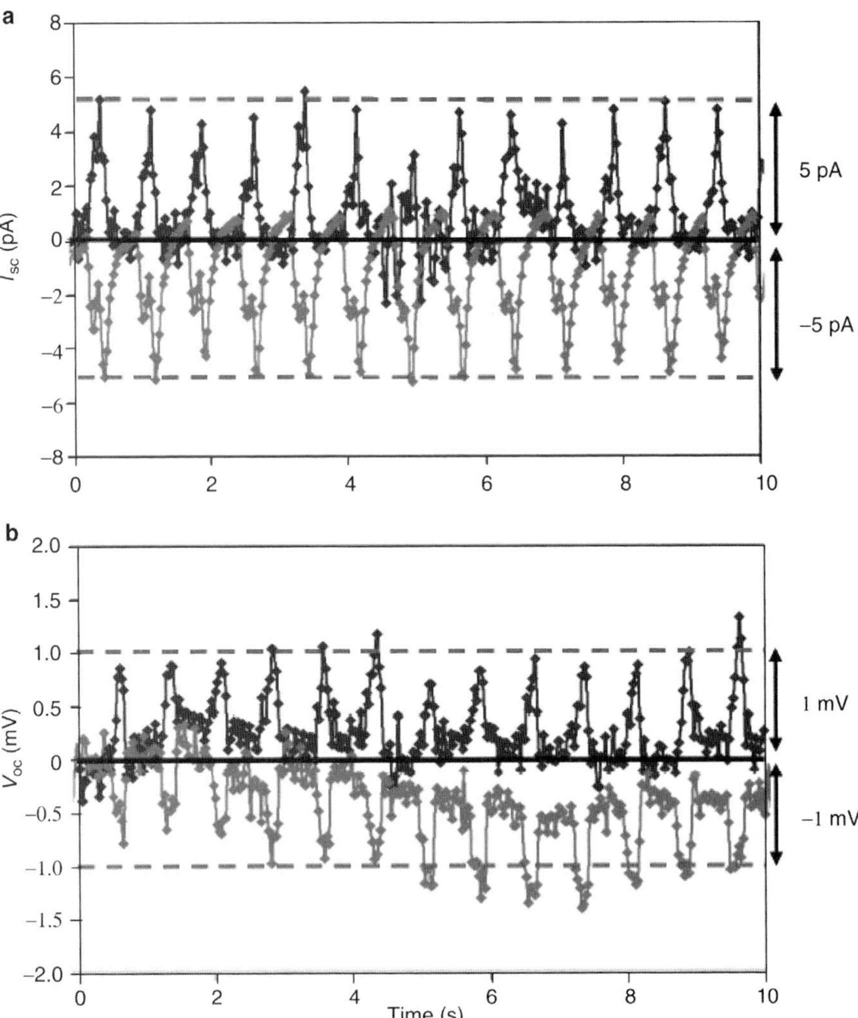

FIGURE 7.15 *Electric output of a double-fiber nanogenerator. (a) The short-circuit output current (I_{sc}) and (b) the open-circuit output voltage (V_{oc}) of a double-fiber nanogenerator measured by applying an external pulling force at a motor speed of 80 rpm. The measurement circuit was forward connected to the nanogenerator (pink curve). The polarity of the connection was reversed (blue curve).*

The rest of them act like parasitic capacitances [52], which only bring down the magnitude of the output voltage. Rational nanogenerator designs require to boost the output signals. Another way of enhancing the output voltage would be to decrease the inner resistance of the ZnO nanowire [52]. If the inner resistance is too significant, as discussed previously, a large portion of the electromotive force of the nanogenerator would be consumed by the inner resistance of the nanogenerator itself. A third approach to improving the performance of the nanogenerator is to fasten the nanowires onto the substrate, making them robust to mechanical deformation [14, 27]. Currently, all of the reported ZnO nanowires on flexible substrates are grown by wet chemical methods. A natural weakness of such methods is that there are a lot of defects inside the nanowire, which impairs the mechanical properties of the nanowires. The nanowires are easy to crack and break from the substrate under deformation. So, if the root of the nanowire could be made stronger on the substrate, the bending angle of the nanowire could probably be increased, and therefore the piezoelectric potential of the nanowire could be improved, and eventually the output voltage of the nanogenerator could be enhanced, because under certain simple assumptions, the piezoelectric potential is proportional to the bending angle of the nanowire, according to theoretical calculations [53].

CONCLUSION

This chapter has presented a brief overview of the chemical growth of ZnO nanowire arrays on flexible substrates. The different strategies have been used to grow ZnO nanowire arrays on flat flexible substrates, such as polyimide and polystyrene, and curved flexible substrates, such as microfibers. By using EBL or photolithography, a patterned growth of ZnO nanowire arrays could also be achieved on both flat substrates and microfibers. Furthermore, based on vertically aligned ZnO nanowire arrays on the flexible substrates, innovative nanotechnology-enabled methods for converting mechanical energy into electrical energy have been demonstrated, with the aim of building self-powered nanosystems. The focus was on the fundamental mechanism of the piezoelectricity and nanogenerators. The core of the nanogenerator is based on coupling the piezoelectric and semiconducting properties of ZnO nanowires and the presence of a Schottky barrier at the metal–semiconductor interface. With the promising approaches already reviewed in this chapter and more to be developed, practical applications of the nanogenerator are imminent.

ACKNOWLEDGMENTS

Research supported by DARPA (Army/AMCOM/REDSTONE AR, W31P4Q-08-1-0009), DARPA STTR with Magnolia Optical Inc., BES DOE (DE-FG02-07ER46394), Air Force Office (FA9550-08-1-0446), KAUST Global Research Partnership, World Premier International Research Center (WPI) Initiative on

Materials Nanoarchitectonics, MEXT, Japan, Emory-Georgia Tech CCNE from NIH, NSF (DMS 0706436, CMMI 0403671). The authors acknowledge contributions from Dr Jinhui Song, Dr Xudong Wang, Dr Yong Qin, Dr Jin Liu and Dr Puxian Gao.

REFERENCES

[1] Wagner RS, Ellis WC. Vapor-liquid-solid mechanism of single crystal growth. Appl Phys Lett 1964;4:89–90.

[2] Westwater J, Gosain DP, Tomiya S, Usui S, Ruda H. Growth of silicon nanowires via gold/silane vapor-liquid-solid reaction. J Vac Sci Technol B 1997;15:554–7.

[3] Morales AM, Lieber CM. A laser ablation method for the synthesis of crystalline semiconductor nanowires. Science 1998;279:208–11.

[4] Huang MH, Wu YY, Feick H, Tran N, Weber E, Yang PD. Catalytic growth of zinc oxide nanowires by vapor transport. Adv Mater 2001;13:113–6.

[5] Park WI, Yi GC, Kim MY, Pennycook SJ. ZnO nanoneedles grown vertically on Si substrates by non-catalytic vapor-phase epitaxy. Adv Mater 2002;14:1841–3.

[6] Wu JJ, Liu SC, Wu CT, Chen KH, Chen LC. Heterostructures on ZnO-Zn coaxial nanocables and ZnO nanotubes. Appl Phys Lett 2002;81:1312–4.

[7] Verges MA, Mifsud A, Serna CJ. Formation of rod-like zinc oxide microcrystals in homogeneous solutions. J Chem Soc Faraday Trans 1990;86:959–63.

[8] Tian ZRR, Voigt JA, Liu J, McKenzie B, McDermott MJ, Rodriguez MA, et al. Complex and oriented ZnO nanostructures. Nat Mater 2003;2:821–6.

[9] Vayssieres L. Growth of arrayed nanorods and nanowires of ZnO from aqueous solutions. Adv Mater 2003;15:464–6.

[10] Chen J, Klaumunzer S, Lux-Steiner MC, Konenkamp R. Vertical nanowire transistors with low leakage current. Appl Phys Lett 2004;85:1401–3.

[11] Goldberger J, Hochbaum AI, Fan R, Yang PD. Silicon vertically integrated nanowire field effect transistors. Nano Lett 2006;6:973–7.

[12] Huang MH, Mao S, Feick H, Yan HQ, Wu YY, Kind H, et al. Room-temperature ultraviolet nanowire nanolasers. Science 2001;292:1897–9.

[13] Park WI, Yi GC. Electroluminescence in n-ZnO nanorod arrays vertically grown on p-GaN. Adv Mater 2004;16:87–90.

[14] Gao PX, Song JH, Liu J, Wang ZL. Nanowire piezoelectric nanogenerator on plastic substrates as flexible power sources for nanodevices. Adv Mater 2007;19:67–72.

[15] Greene LE, Law M, Goldberger J, Kim F, Johnson JC, Zhang YF, et al. Low-temperature wafer-scale production of ZnO nanowire arrays. Angew Chem Int Ed 2003;42:3031–4.

[16] Lin CC, Chen SY, Cheng SY. Nucleation and growth behavior of well-aligned ZnO nanorods on organic substrates in aqueous solution. J Cryst Growth 2005;283:141–6.

[17] Liu TY, Liao HC, Lin CC, Hu SH, Chen SY. Biofunctioal ZnO nanorod arrays grown on flexible substrates. Langmuir 2006;22:5804–9.

[18] Nadarajah A, Word RC, Meiss J, Konenkamp R. Flexible inorganic nanowire light-emitting diode. Nano Lett 2008;8:534–7.

[19] Yang PD, Yan HQ, Mao S, Russo R, Johnson J, Saykally R, et al. Controlled growth of ZnO nanowires and their optical properties. Adv Funct Mater 2002;12: 323–31.

[20] Ng HT, Han J, Yamada T, Nguyen P, Chen YP, Meyyappan M. Single crystal nanowire vertical surround-gate field-effect transistors. Nano Lett 2004;4: 1247–52.

[21] Wang XD, Summers CJ, Wang ZL. Large-scale hexagonal-patterned growth of aligned ZnO nanorods for nanooptoelectronics and nanosensor arrays. Nano Lett 2004;4:423–6.

[22] Fan HJ, Fuhrmann B, Scholz R, Syrowatka F, Dadgar A, Krost A, Zacharias M. Well-ordered ZnO nanowire arrays on GaN substrate fabricated via nanosphere lithography. J Cryst Growth 2006;287:34–8.

[23] Kim YJ, Lee CH, Hong YJ, Yi GC, Kim SS, Cheong H. Controlled selective growth of ZnO nanorod and microrod arrays on Si substrates by a wet chemical method. Appl Phys Lett 2006;89:163128.

[24] Cui JB, Gibson U. Low-temperature fabrication of single-crystal ZnO nanopillar photonic bandgap structures. Nanotechnology 2007;18:155302.

[25] Sounart TL, Liu J, Voigt JA, Hsu JWP, Spoerke ED, Tian Z, Jiang YB. Sequential nucleation and growth of complex nanostructured films. Adv Funct Mater 2006;16:335–44.

[26] Weintraub B, Deng Y, Wang ZL. Position-controlled seedless growth of ZnO nanorod arrays on a polymer substrate via wet chemical synthesis. J Phys Chem C 2007;111:10162–5.

[27] Qin Y, Wang XD, Wang ZL. Microfiber-nanowire hybrid structure for energy scavenging. Nature 2008;451:809–13.

[28] Greyson EC, Babayan Y, Odom TW. Directed growth of ordered arrays of small-diameter ZnO nanowires. Adv Mater 2004;16:1348–52.

[29] Tak Y, Yong K. Controlled growth of well-aligned ZnO nanorod array using a novel solution method. J Phys Chem B 2005;109:19263–9.

[30] Hsu JWP, Tian ZR, Simmons NC, Matzke CM, Voigt JA, Liu J. Directed spatial organization of zinc oxide nanorods. Nano Lett 2005;5:83–6.

[31] Morin SA, Amos FF, Jin S. Biomimetic assembly of zinc oxide nanorods onto flexible polymers. J Am Chem Soc 2007;129:13776–7.

[32] Tada H. Decomposition reaction of hexamine by acid. J Am Chem Soc 1960;82:255–63.

[33] Xu S, Lao C, Weintraub B, Wang ZL. Density-controlled growth of aligned ZnO nanowire arrays by seedless chemical approach on smooth surfaces. J Mater Res 2008;23:2072–7.

[34] Ahuja IS, Yadava CL, Singh R. Tetradentate behavior of hexamethylenetetramine. J Molec Struct 1982;81:289–91.

[35] Govender K, Boyle DS, Kenway PB, O'Brien P. Understanding the factors that govern the deposition and morphology of thin films of ZnO from aqueous solution. J Mater Chem 2004;14:2575–91.

[36] Greene LE, Law M, Tan DH, Montano M, Goldberger J, Somorjai G, Yang PD. General route to vertical ZnO nanowire arrays using textured ZnO seeds. Nano Lett 2005;5:1231–6.

[37] Law M, Greene LE, Johnson JC, Saykally R, Yang PD. Nanowire dye-sensitized solar cells. Nat Mater 2005;4:455–9.

[38] Zhiren Q, Wong KS, Mingmei W, Wenjiao L, Huifang X. Microcavity lasing behavior of oriented hexagonal ZnO nanowhiskers grown by hydrothermal oxidation. Appl Phys Lett 2004;84:2739–41.

[39] Tian ZR, Voigt JA, Liu J, McKenzie B, McDermott MJ. Biomimetic arrays of oriented helical ZnO nanorods and columns. J Am Chem Soc 2002;124: 12954–5.

[40] Wang ZL, Kong XY, Zuo JM. Induced growth of asymmetric nanocantilever arrays on polar surfaces. Phys Rev Lett 2003;91:185502.

[41] Lao CS, Gao PM, Sen Yang R, Zhang Y, Dai Y, Wang ZL. Formation of double-side teethed nanocombs of ZnO and self-catalysis of Zn-terminated polar surface. Chem Phys Lett 2006;417:358–62.

[42] Lee SH, Minegishi T, Park JS, Park SH, Ha J-S, Lee H-J, et al. Ordered arrays of ZnO nanorods grown on periodically polarity-inverted surfaces. Nano Lett 2008;8:2419–22.

[43] Zhao MH, Wang ZL, Mao SX. Piezoelectric characterization of individual zinc oxide nanobelt probed by piezoresponse force microscope. Nano Lett 2004;4:587–90.

[44] Wang XD, Song JH, Liu J, Wang ZL. Direct-current nanogenerator driven by ultrasonic waves. Science 2007;316:102–5.

[45] Wang ZL. Nanopiezotronics. Adv Mater 2007;19:889–92.

[46] Wang ZL. The new field of nanopiezotronics. Mater Today 2007;10(5):20–8.

[47] Wang ZL, Song JH. Piezoelectric nanogenerators based on zinc oxide nanowire arrays. Science 2006;312:242–6.

[48] Xu S, Wei Y, Liu J, Yang R, Wang ZL. Integrated multilayer nanogenerator fabricated using paired nanotip-to-nanowire brushes. Nano Lett 2008;8:4027–32.

[49] Wang ZL. Towards self-powered nanosystems: from nanogenerators to nanopiezotronics. Adv Funct Mater 2008;18:3553–67.

[50] Song JH, Zhou J, Wang ZL. Piezoelectronic and semiconducting coupled power generating process of a single ZnO belt/wire: a technology for harvesting electricity from the environment. Nano Lett 2006;6:1656–62.

[51] Liu J, Fei P, Song JH, Wang XD, Lao CS, Tummala R, Wang ZL. Carrier density and Schottky barrier on the performance of DC nanogenerator. Nano Lett 2008; 8:328–32.

[52] Liu J, Fei P, Zhou J, Tummala R, Wang ZL. Toward high output-power nanogenerator. Appl Phys Lett 2008;92:173105.

[53] Gao Y, Wang ZL. Electrostatic potential in a bent piezoelectric nanowire: the fundamental theory of nanogenerator and nanopiezotronics. Nano Lett 2007; 7:2499–505.

CHAPTER 8

Flexible Energy Storage Devices Using Nanomaterials

Victor L. Pushparaj[1,*], Subbalakshmi Sreekala[1], Omkaram Nalamasu[1,*] and Pulickel M. Ajayan[2]

[1]*Materials Science and Engineering Department, Rensselaer Polytechnic Institute, Troy, New York, USA*

[2]*Department of Mechanical Engineering and Materials Science, Rice University, Houston, Texas, USA*

INTRODUCTION

Renewable energy, and energy storage, has been one of the key issues addressed in this decade. It is also a prime area of interest for start-up companies and giant corporations. Huge attempts are being made to reduce the carbon footprint and, because of their environmental impact, the automobile industry has turned to a new generation of vehicles (electric or hybrid vehicles) [1–3]. The threat to the availability of petroleum has also triggered high costs. In recent years, there has been a shift in focus to inexhaustible energy sources such as wind and solar power [4–7]. Present-day technology can tap renewable energy sources efficiently; however, the storage of energy is a major issue. To address the growing demands requires an efficient energy storage medium and the capability to deliver the power when required.

Energy storage systems are critical components, especially when the harvested renewable energy is to be stored in remote locations. Types of energy storage system include batteries [2, 8], supercapacitors [9–11] and hydrogen storage [12]. In all of the energy storage devices, the materials and technology play a key role. Nanotechnology provides crucial tools to drastically improve the efficiency of energy storage devices based on exploiting the material properties at the nanoscale (e.g. nanoparticles, nanowires and nanotubes) [13, 14]. Nanomaterials have been investigated for the past two decades and the whole scenario was given a boost with the discovery of carbon nanotubes (CNTs) [14]. Nanomaterials have given a new face to the energy sector, especially the

CONTENTS

---

*Present address: Applied Materials Inc., Advanced Technology Group, Santa Clara, California, USA.

way in which the harvested energy is stored efficiently. They play an enormous part in hydrogen storage (nano-magnesium or nano-magnesium alanate with nano-titanium as catalyst) [15], supercapacitors (CNTs) and rechargeable batteries (e.g. using nanostructured anodes like silicon nanowires in lithium-ion batteries) [16–23]. In recent years, there has been a growing demand for flexible electronics, e.g. printed electronics and flexible displays [24], and flexible energy storage devices [25, 26]. In particular, the flexible energy storage device is a nascent technology. In this chapter the possibilities for flexible energy storage devices in different configurations are discussed. Flexible energy storage devices have numerous applications, such as active radio-frequency identification tags, integrated circuit smart cards and portable electronics [27]. The flexibility gives an added advantage as they can be embedded in tiny and flexible electronic devices. For these flexible energy storage devices to be used on a daily basis, they need to exhibit excellent cyclability, and high power and energy densities. The main elements needed in a flexible energy storage device are flexible electrodes [28, 29] and separators [26, 30].

NANOMATERIALS FOR FLEXIBLE ELECTRODES AND SEPARATORS

Sony introduced lithium (Li)-ion batteries in 1991, which led to storage technology advancing in leaps and bounds, especially in mobile computing and cell phones. This new Li-ion battery showed remarkable improvements with its increased capacity: it could store two to three times more energy per unit weight and volume than lead–acid or Ni–Cd batteries, and also had a long life cycle and long shelf-life [2]. This technology does have its disadvantages, such as safety, and the efficiency is limited by the material properties [31, 32]. Hence, there is a constant need to explore new nanomaterials. Importantly, the Li-ion battery, which is also known as the 'rocking chair' or 'swing' battery, has opened up a new arena of flexible energy storage media [33, 34]. Li-ions from the cathode (LiCoO$_2$) intercalate into the crystal structure of the anode (carbon/graphite) during charging, and the ions reverse direction, leaving the anode and re-entering the cathode structure, during discharging. The movement of ions in the anode and cathode hosts should not change or damage the host crystal structure. If it does, this leads to a shorter cycle and life of the battery. This shows that the physical, structural and electrochemical properties of cathode and anode materials are critical to the battery's performance. Hence, nanomaterials play a crucial role in these electrode materials [35]. Nanomaterials such as nanotubes are excellent electrode materials owing to their unique accommodation of the Li-ion insertion/removal, lack of reactivity with the materials, higher electrode/electrolyte contact area (leading to higher charge/discharge rates) and higher electronic conduction.

Carbon nanotubes

CNTs exhibit exceptional electrical conductivity, and are ultralight, low density and highly elastic [28, 36–38]. CNTs exhibit extreme structural flexibility when subjected to large strains and angles without structural failure [38, 39]. This is possible as the vertically aligned CNTs buckle under large strains while still maintaining structural integrity [40, 41]. CNTs exhibit supercompressibility when the CNTs are compressed to 15% strain; the CNTs will recover to nearly their original length after every cycle in a strong cushioning effect. Figure 8.1(a) shows a CNT block (length of CNT ~ 3 mm) before compression and Figure 8.1(b) shows the CNT block after undergoing 100 K cycles of mechanical compression. When the aligned CNTs are repeatedly stressed for a few hundred cycles, the CNTs can recover to nearly their original length. Very few materials are known to exhibit this behavior [42]. Hence, CNTs seem to be an ideal material for flexible energy storage applications that would benefit from a combination of their lightweight, structural and excellent mechanical properties. In addition, the combination of electrical conductivity,

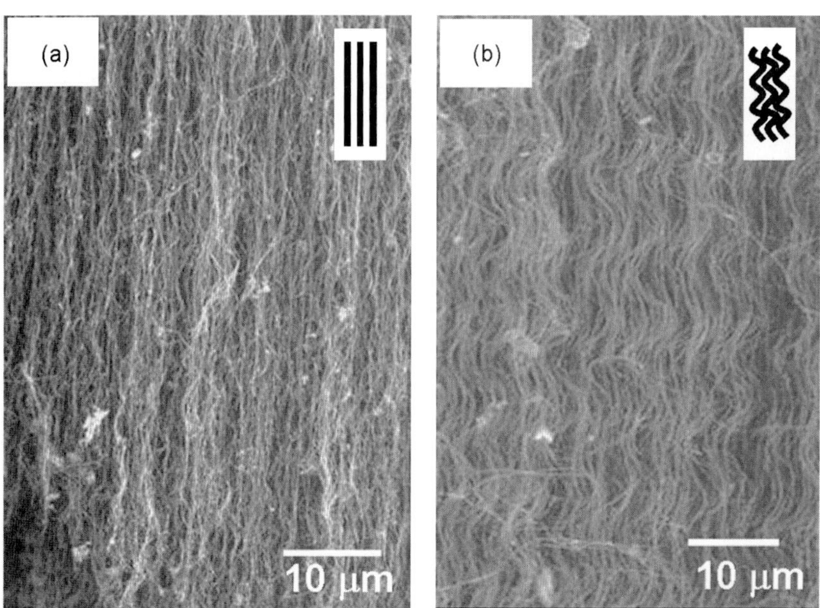

FIGURE 8.1 *SEM image of carbon nanotube (CNT) block grown by thermal chemical vapor deposition. (a) Before compression; (b) after mechanical compression.*

chemical inertness, high conductivity and thermal stability makes them ideal for supercapacitor and battery electrodes [26, 43].

CNTs act as an excellent electrode, but it is difficult to deposit CNTs directly on the current collectors (for a battery or supercapacitor). This is because the growth of CNTs is a high-temperature process and a minimal interaction of the catalyst and substrate is essential in order to grow well-aligned CNTs. Hence, it is easier to grow them on oxide substrates [44]. There are reports on the successful deposition of the CNTs on bulk metal like Inconel® [45] and stainless steel substrates [46]. The growth of aligned multi-walled carbon nanotubes (MWNTs) on a bulk metallic alloy (Inconel 600) was carried out using a vapor-phase catalyst delivery method. Figure 8.2(a) shows aligned MWNTs grown on Inconel substrates/wires. The grown CNTs are well anchored to the substrate and show excellent electrical contact to it [45]. The MWNTs are deposited using a ferrocene–xylene chemical vapor deposition (CVD) system. A xylene–ferrocene solution (1 g of ferrocene in 100 ml of xylene) was injected into a preheated steel bottle (200°C) with a syringe pump at a feeding rate of 0.11 ml/min. A mixture of argon (85%) and hydrogen (15%) was used as a carrier gas. A flow of 100 sccm of this gas mixture was used to

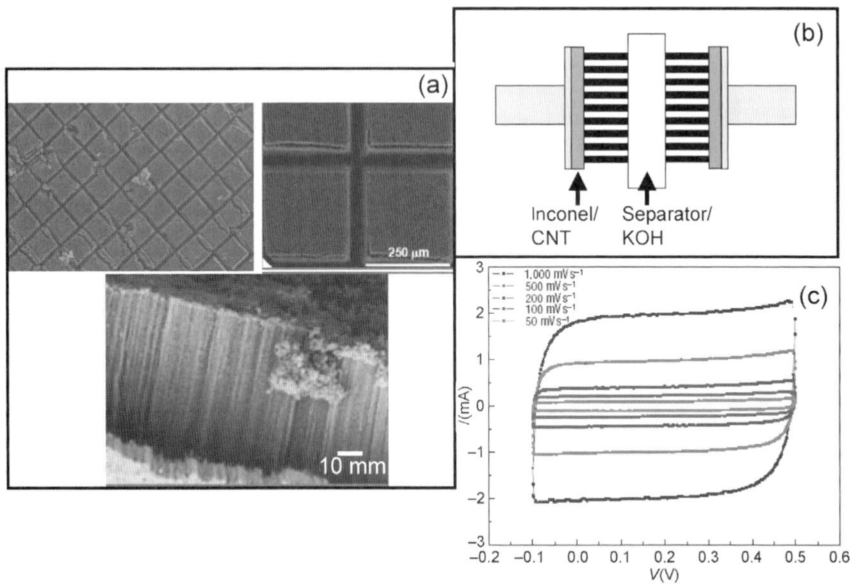

FIGURE 8.2 *(a) SEM image of carbon nanotubes (CNTs) grown on the Inconel substrates/wires. (b) Schematic of electrical double-layer capacitor (EDLC) with CNTs/Inconel electrodes on either ends with KOH electrolyte. (c) Cyclic voltammograms measured for the fabricated EDLC at different scan rates from 50 to 1000 mV/s [45].*

carry the xylene vapors containing the catalyst before entering a tube furnace [45, 47]. The furnace which contained these Inconel substrates was maintained at the desired growth temperatures. A wide temperature window (500–820°C) is available for the CNT growth on the Inconel substrates, with a maximum yield at 770°C with a growth rate of 2.8 µm/min (the growth rate on SiO_2 is higher (~10 µm/min)). The CNT growth was observed to be bulk diffusion of carbon in γ-iron with root growth mechanism [48, 49]. The CNT–Inconel interface exhibits good electrical contact as well as strong mechanical adhesion and can be used directly as an electrode for supercapacitors without any postgrowth processing. The Inconel substrates can be as thin (~100 µm) as possible to give flexibility and mechanical rigidity to the growth of CNTs.

Electrochemical measurements [double-layer capacitor measurement (DLC)] [10, 50] were carried out on the CNTs grown on Inconel (schematic in Figure 8.2b). Figure 8.2(c) shows the cyclic voltammograms of the double-layer capacitor measurement at various scan rates. The rectangular and symmetric shape of the cyclic voltammograms performed at high scan rates of 1000 mV/s exhibited a typical capacitive behavior [45]. It is also observed that there is a low contact resistance between the CNTs and Inconel. Lowering the contact resistance between the nanotubes and current collector electrodes in DLCs has been a major issue in developing CNT-based supercapacitors, and is typically achieved by mixing nanotubes with conductive binders and coating collector electrodes with this composite. This process drastically modifies nanotube electrode properties, leading to adverse effects on the performance of the DLC [10]. In the case of aligned CNTs grown directly on the bulk metallic flexible or rigid Inconel substrate this problem does not arise. The specific capacitance value for the DLCs, calculated from the discharge slope measured at 2 mA, was found to be 18 F/g (comparable to the values obtained for DLCs fabricated using CVD-grown MWNTs) [50] with a power density of 7 kW/kg at 1000 mV/s. This high value of power density obtained on the flexible electrode is well suited for surge-power delivery applications [51]. Hence, the combination of Inconel with CNTs forms an excellent flexible electrode/current collector which can be used in the supercapacitor or anode for Li-ion batteries. The equivalent series resistance (ESR) can thus be reduced, thereby drastically improving the energy and power density. One of the major disadvantages in having this electrode configuration is that CNTs might peel off or fall off from the Inconel sheets when subjected to high shear stress, because the adhesion between the CNTs and Inconel is not significant [45]. Hence, the correct choice of polymer and embedding CNTs in it not only avoids the loss of the CNTs, but also gives mechanical robustness and flexibility.

Nanocomposites

Integrated CNT–polymer composite [26, 52–54] is one of the best integrated structures for mechanically flexible energy storage devices. These nanocomposites enable various functional devices in a wide range of innovative products such as smart cards, displays and implantable medical devices [27, 54]. In recent years, supercapacitors and batteries have been demonstrated based on CNT–polymer composites. One interesting form of thin flexible energy storage device is to build a variety of storage systems based on the basic building block, i.e. an integrated nanocomposite sheet consisting of electrode, spacer and electrolyte. A nanocomposite which consists of cellulose, room-temperature ionic liquid (RTIL) and CNTs fits the characteristics of a spacer, electrolyte and electrode. These provide inherent flexibility as well as porosity to the system.

Cellulose, the main constituent of paper and an inexpensive insulating separator structure with excellent biocompatibility, can be made with adjustable porosity. As mentioned earlier, CNTs exhibit extreme flexibility as a mechanical structure and have been used as electrodes in electrochemical devices [18, 20–22, 55–57]. The major challenge in fabricating CNT-integrated cellulose composites is that cellulose is insoluble in most of the common solvents. This issue is solved here by using an RTIL, 1-butyl,3-methylimidazolium chloride ([bmIm][Cl]) [58], which dissolves up to 25% (w/w) of unmodified cellulose by using microwave irradiation [59]. The ionic nature of RTIL [60] permits it to be used as an electrolyte in supercapacitors [61], allowing the assembly of all three components via a simple scalable process.

The nanocomposite fabrication proceeded as follows: a 5–20 nm thick aluminum layer was deposited on an Si or Si/SiO$_2$ wafer as a buffer layer, on top of which a 1–3 nm thick iron catalyst layer (which forms nanosized particles for catalytic growth of CNTs) was deposited. The predeposited substrate was loaded into the CVD chamber. Aligned CNT films were prepared by thermal decomposition of ethylene (with a flow rate of 50–150 sccm) on the catalyst film at 750–800°C. An Ar/H$_2$ gas mixture (15% H$_2$) with a flow rate of 1300 sccm was used as the buffer gas during the whole CVD run. The length of the MWNTs grown on the Si substrate was 200 μm. Figure 8.3(a) shows the as-grown aligned MWNTs by thermal CVD and Figure 8.3(b) shows the TEM image of MWNTs. Cellulose (30 mg) was dissolved in RTIL (1.0 g) by preheating the RTIL to 70 °C. The contents were then vortex-mixed and microwaved for 4–5 s, to afford a 3% (w/w) cellulose in RTIL ([bmIm][Cl]). The resulting solution was drop-coated on the preheated (∼70°C) MWNT arrays and kept for 5 min for the infiltration. After infiltration, the substrate was kept in dry ice until it solidified (∼15 min). The formed nanocomposite film was washed

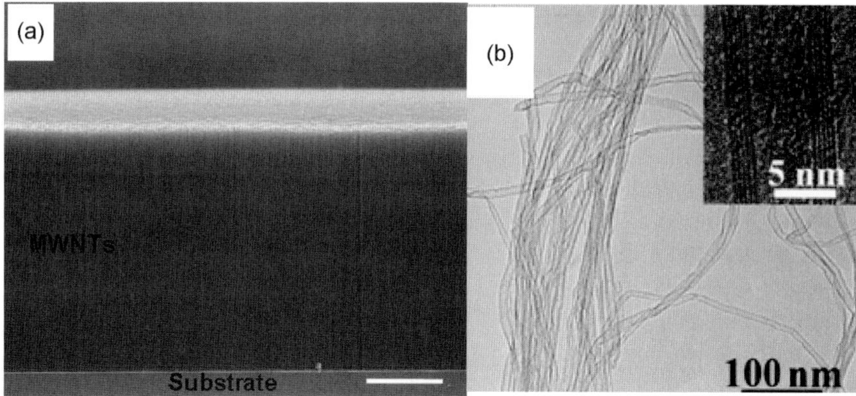

FIGURE 8.3 *(a) SEM image of vertically aligned carbon nanotubes (CNTs) on Si substrate. MWNTs: multiwalled carbon nanotubes. Scale bar: 100 µm. (b) TEM image of CNTs.*

with ethanol to extract the entire RTIL and the nanocomposite film was then dried in a vacuum. The nanocomposite film on the Si substrate was peeled off and directly used as electrodes for batteries and supercapacitors. The peeled-off film comprises the MWNTs, cellulose and RTIL. This is the basic building block for the energy storage device.

The supercapacitor and battery devices were assembled as shown in the schematic of Figure 8.4. In the case of the battery, either a thin film of Li was deposited by thermal evaporation or the thin Li metal foil was used as another electrode. In the ideal scenario, it is best to have $LiCoO_2$ thin films for the

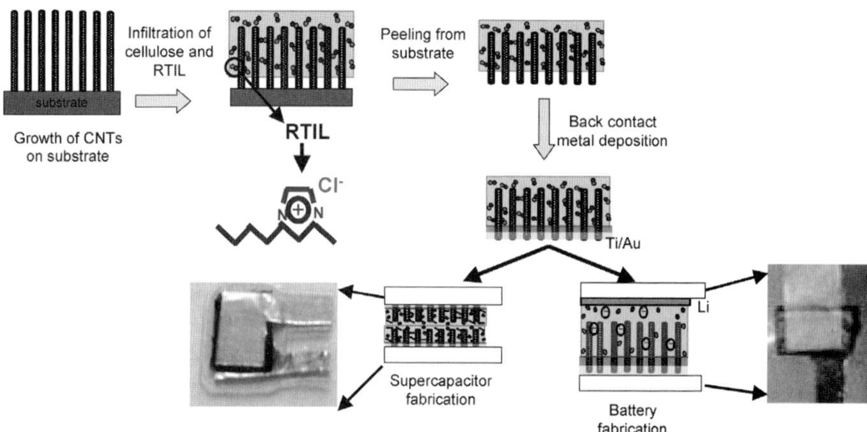

FIGURE 8.4 *Flow diagram to fabricate flexible energy storage devices. The fabricated nanocomposite is the basic building block for the energy storage devices. CNTs: carbon nanotubes; RTIL: room-temperature ionic liquid.*

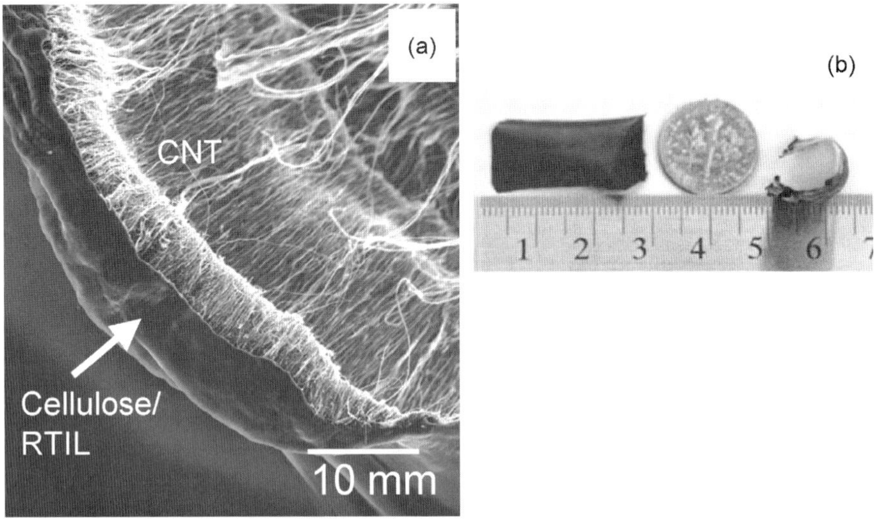

FIGURE 8.5 *(a) SEM image of the nanocomposite sheet. CNT: carbon nanotube; RTIL: room-temperature ionic liquid. (b) Photograph showing the nanocomposite sheet and a curled nanocomposite sheet, indicating flexibility.*

cathode. The excellent mechanical flexibility of the nanocomposite paper (CNT cellulose–RTIL) (Figure 8.5a) is shown in Figure 8.5(b) [26]. The nanocomposite paper can be rolled up, twisted or bent to any curvature and is completely recoverable. The nanocomposite paper, which can be typically a few tens of micrometers thick, contains MWNTs as the working electrode and the cellulose surrounding individual MWNTs, as well as the extra layer as the spacer and the RTIL in cellulose as the self-sustaining electrolyte. The thin, lightweight ($\sim$15 mg/cm^2) design of the device results from avoiding the use of a separate electrolyte and spacer, generally used in conventional supercapacitors; therefore it is also extremely flexible and relatively environmentally benign. Two of the nanocomposite units bonded back-to-back make a single supercapacitor device. The specific capacitance calculated from the electrochemical measurements on this integrated supercapacitor with RTIL (nonaqueous) and KOH (aqueous) electrolytes was found to be 22 F/g and 36 F/g, respectively, for the flexible separator and flexible electrodes. The operating voltage in the case of the RTIL electrolyte was 2.3 V, while for the KOH it was 0.95 V. The specific capacitance of the device in the KOH solution is higher compared to RTIL, because the dielectric constant and ionic mobility of the latter are lower. A power density of 1.5 kW/kg (energy density 13 Wh/kg) was obtained at room temperature for the nanocomposite (RTIL) supercapacitor, which was within the reported ranges (0.01–10 kW/kg) of commercial supercapacitors and comparable to reported flexible devices [26]. An important point regarding these flexible nanocomposites is that not only are they flexible but also the operating temperature regime ranges from 195 to 450 K in the case of the RTIL electrolyte, which is unique, and makes it useful as a flexible storage device in extreme conditions.

Using this basic building nanocomposite block, the Li-based battery is fabricated. The fabricated flexible battery consists of RTIL-free nanocomposite films and a thin Li–metal layer. The excess cellulose layer in the nanocomposite (cathode) acts as the spacer. Drops of aqueous electrolyte [1 M $LiPF_6$ in ethylene carbonate and dimethyl carbonate (1:1 v/v)] are used for the battery operation. No stand-alone spacer was used since the excess cellulose layer in the nanocomposite served as the spacer. It was observed to have a large irreversible capacity of 430 mAh/g and a reversible capacity of ~110 mAh/g. This device operates under full mechanical flexibility.

So far we have seen that the cellulose, a non-conducting polymer, has been used for the CNT composite, and replacing it with a conducting polymer to form CNT composites opens up numerous possible applications [62–66]. In an interesting work on CNT–organic conducting polymer, it was found that the polymer keeps intact the vertically aligned CNTs (as in the case for CNT–cellulose), where 90% of the CNT [67, 68] tube length is exposed and the remainder is embedded in the polymer. The organic conducting polymer used here, poly(3,4-ethylenedioxythiophene) (PEDOT), was deposited on vertically aligned CNTs by chemical vapor-phase polymerization. Another coating of polyvinylidene fluoride (PVDF) was deposited on PEDOT to give mechanical stability for the composite to form a flexible free-standing film. The free-standing film electrode had an electronic conductivity over 200 S/cm, which is higher than that measured for the same membrane without a PEDOT layer. The organic conducting polymer–CNT is an excellent flexible electrode (anode) in Li-ion batteries. The electrolyte used was 1 M $LiPF_6$ in a 50:50 (v/v) mixture of ethylene carbonate and dimethyl carbonate. The assembled cell was cycled between 0 and 2 V at a constant current density of 0.1 mA/cm^2 at room temperature. The first cycle for this nanostructured composite electrode exhibited a high irreversible capacity, which can be attributed to the formation of a solid electrolyte interface layer on the surface of the electrodes. Even after 50 cycles a highly stable discharge capacity (265 mAh/g) was observed, which is higher than that observed for single-walled carbon nanotube (SWNT) paper (~173 mAh/g) under identical working conditions [20, 69]. This can mostly be attributed to the higher surface area of electrodes available owing to the alignment of CNTs retained by the PEDOT polymer. No degradation was observed in the electrodes even after 50 cycles. These nanostructure electrodes provide mechanical stability and flexibility owing to the presence of polymer and CNTs. It was also observed that the capacity of this flexible nanocomposite was 50% higher than the free-standing SWNT paper.

Nanoparticles and non-carbon-based nanotubes

Nanoparticles have the advantage of quickly absorbing and storing vast amounts of Li-ions without causing deterioration to the electrode. Traditionally most of these nanomaterials are rigid, but using either an active or inactive polymer these nanotubes (metallic or intermetallic) can be embedded in a matrix that will provide flexibility to the device combined with high performance. As mentioned in the nanocomposite section based on the CNT/PEDOT/PVDF device, the PEDOT is eliminated, the Pt nanoparticles were used on the aligned CNTs and then PVDF deposition was carried out. The electrochemical capacity for this configuration is 133 F/g with a electrochemical surface area of 143 m^2/g for Pt nanoparticles [29, 62, 70].

In the case of Li-ion battery technology, the MWNTs exhibited a reversible capacity in the range of 100–400 mAh/g based on the processing conditions [20, 71, 72]. MWNTs prepared by CVD demonstrated a reversible lithium storage capacity of 340 mAh/g with good cyclability at moderate current density [73]. But when the CNTs were synthesized by catalytic decomposition of acetylene or ethylene at 70°C over iron nanoparticles [74], the charge capacity of CNT was 640 mAh/g compared with well-graphitized CNT (282 mAh/g), but had better rate capability and cycle life. It has been observed that nanocomposites of CNTs and Sn_2Sb alloys as electrodes exhibited higher specific capacity than CNTs and improved the cyclability relative to the original Sn–Sb alloy particles.

Other nanomaterials that allow effective Li interaction are WS_2 nanotubes synthesized at high temperature [75] and multiwalled TiS_2 nanotubes [76] synthesized at low temperature. The anatase TiO_2 nanotubes exhibited high initial capacity (170 mAh/g), excellent high-rate discharge capability (640 mAh/g) and good reversibility (118 mAh/g after 100 cycles) for electrochemical Li insertion and extraction.

The Sn-based anode battery, which consists of Sn, Co and C nanoparticles, was commercialized by Sony (2005). This resulted in a 50% higher per unit volume ratio of Li-ion density compared to conventional graphite anodes, resulting in a dramatic increase in overall battery capacity, by 30% [77]. Another breakthrough in battery technology is Toshiba Corporation's new battery, which can recharge 80% of battery energy capacity in only 1 min, approximately 60 times faster than typical Li-ion batteries. Hybrid nanotubes have also been used in Li rechargeable batteries and in supercapacitors [75, 76].

Hybrid nanostructures

Engineering of nanomaterials has been used in making multisegmented nanotubes and/or nanowires. This tailoring helps in reducing the electrode interface resistance, increasing electrical conductivity and increasing the power/energy density. There are a variety of methods available to grow nanotubes, nanoparticles and nanowires. Combining any of these to form hybrid structures can result in electrodes for energy storage devices (supercapacitors and batteries). One of the most common methods available is a template method, which is a powerful tool for fabricating one-dimensional nanomaterials [18, 78]. The template method opens up the possibility of spatial arrangement of nanostructures allowing complex structures to be fabricated. Since it has been a challenge to grow CNTs on metal substrates, this unique method of fabricating multisegment nanowires/nanotubes opens up the possibility of eliminating the necessity to grow CNTs on metal substrates for energy storage device applications.

An interesting hybrid structure of CNTs is two-segmented (metal–CNT) and three-segmented (CNT–metal–CNT) hybrid structures which can be grown using the template method (Figure 8.6a) [79]. This is accomplished

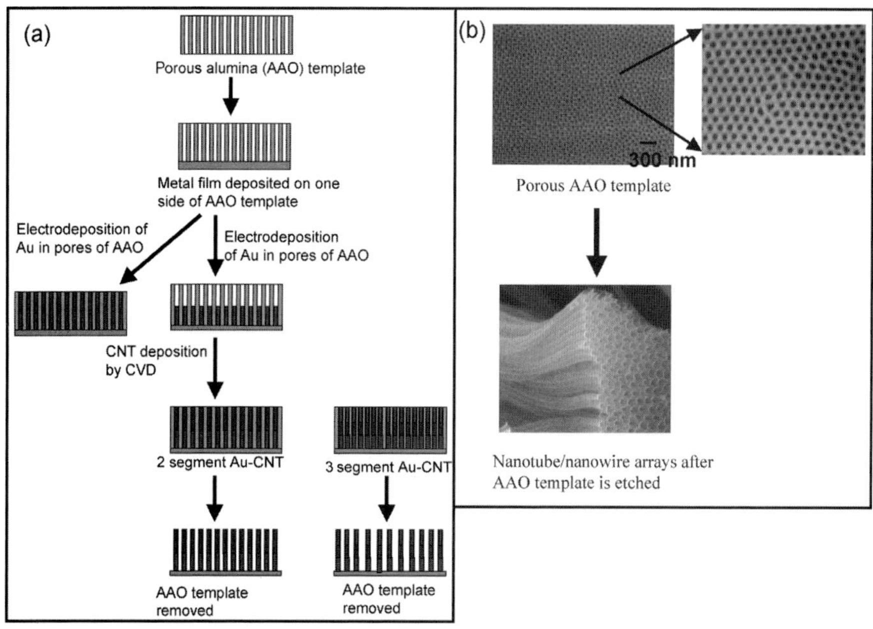

FIGURE 8.6 *(a) Gold–carbon nanotube (CNT) deposited in the pores of an anodized alumina (AAO) template. (b) SEM image of porous AAO template and etching out AAO template leaving the aligned CNTs/Au structures.*

by a combination of electrochemical deposition (to deposit the metal of interest) and CVD techniques (for growth of CNTs). The metal–CNT hybrid nanostructures give good metal contact for each individual CNT. The porous anodized alumina template (AAO) was prepared by a modified two-step anodizing process with pores of 200 nm and length of 50 µm [80–82]. Au nanowires were deposited by electrodeposition in the AAO templates, then the remaining area of the pores in AAO template was deposited with CNT by pyrolysis of acetylene at 650°C for 1–2 h with a flow of a gas mixture containing Ar (85%) and C_2H_2 (15%) at a rate of 35 ml/min. The flowchart of the preparation of the hybrid CNT–Au nanostructures is shown in Figure 8.6(a). A metal thin film was deposited on the side of the Au nanowire to prevent the nanostructure collapsing on removal of the AAO template. Figure 8.6(b) shows the SEM image of the AAO template and the nanotubes/nanowires after removal of the template. The grown MWNTs fused well with the gold nanowire, forming a well-adhered interface. The inner diameter of the CNTs was 170 nm with a wall thickness of 15 nm, as shown in the TEM image. The CNT/Au nanostructures were uniform and aligned after the removal of the AAO template owing to the deposited thin films, as mentioned before.

The hybrid electrode structures are quite flexible and supercapacitive measurements were taken. The capacitance-voltage behavior was found to be rectangular in shape, indicating excellent capacitance properties when the electrodes on both CNT–Au nanostructure and CNT-alone electrode structures were measured for their electrochemical properties. The equivalent series resistance (ESR) for the CNT–Au hybrid nanostructure is low compared to the CNT electrode alone. The ESR generally arises from the resistance of the electrode, the electrolyte, and contact resistance between the electrodes and current collectors. The ESRs were 480 mΩ and 3.4 Ω for the CNT–Au hybrid nanostructure and CNT electrodes, respectively. Low ESR results are due to the extremely small contact resistance, leading to the efficient power density. The calculated specific capacitance was 72 F/g and 38 F/g, respectively, for the CNT–Au hybrid nanostructure and CNT electrodes. Both these structures exhibit good stability over 1000 charge discharge tests. The power density of 48 kW/kg was obtained at room temperature for a supercapacitor device with a CNT–Au hybrid nanostructure, while for the CNTs the reported values were 18 kW/kg [83, 84]. This novel flexible device can be used for high-power applications.

Intermetallic alloys were also explored in the case of Li-ion batteries, which essentially demonstrate higher energy density but are limited by cyclability issues. For instance, nanocomposites consisting of intermetallic alloys (MM′) with an inactive element (M) and an active element (M′) can be used as anode materials. Hence, M′ can alloy with Li to form the compound Li_xM' to result in higher energy density and also lead to volume expansion due to alloying of M′ with Li. The M matrix can act as a buffer to the volume expansion. The expansion varies from 100 to 600% on insertion of Li [85–88]. There are many active elements which can alloy with Li, such as B, Mg, Si, As, Al, Sn, Sb and Pb, and many inactive elements, such as Fe, Co, Ni and Cu. Hence, the intermetallic (MM′) alloys with various combinations of inactive element and active element are chemically versatile, e.g. FeSn, $FeSn_2$, Cu_6Sn_5 and SnSb for Li storage. There are reports on the initial discharge capacities of approximately 800 mAh/g and 610 mAh/g for nanocrystalline $FeSn_2$ and Cu_6Sn_5 alloys respectively, with a limited cyclability. Therefore, there is still room for improvement based on tailoring the composition of alloys with relevant nanostructures.

NANOMATERIALS FOR CATHODES

Cathode nanomaterials have been extensively investigated in recent years. One of the prominently explored materials is V_2O_5 nanowires prepared by template synthesis, which shows an equivalent Li storage capacity to thin-film electrodes at a low discharge rate (C/20). However, at 200°C the nanostructure electrode shows three times the capacity of thin-film electrodes, and above 500°C the nanostructured electrode delivers four times the capacity of thin-film electrodes. Undoped and doped $Li_{1-x}M_xFePO_4$ (with M = Mg, Al, Ti, Nb or W) [89] synthesized by solid-state reaction and doping by metals increases the electronic conductivity of $LiFePO_4$ by a factor of $\sim 10^8$. $LiFePO_4$ nanoparticles (cathode) embedded in amorphous carbon prepared by carbon aerogel synthesis [90] or by a spray solution [91, 92] demonstrated high capacity (where the carbon significantly enhances the electronic conductivity of $LiFePO_4$) and stable cyclability. This nanocomposite delivered nearly 100% of its theoretical discharge capacity at a high discharge rate of 3 C, and 36% of its theoretical capacity at an enormous discharge rate of 65 C. This indicates that nanocrystalline $LiFePO_4$ cathode active material is highly suited for electric vehicles, hybrid electric vehicles and stationary storage batteries. $LiMnO_2$ nanoparticles have a profound and beneficial effect on the cyclability of materials [93]. $LiCoO_2$ nanoparticles (5 nm) were found to exhibit a higher rate capability than the usual 5 μm $LiCoO_2$ [94]. To utilize the rate power and higher capacity, these nanomaterials were embedded in the active or passive matrix to make it more flexible and mechanically robust. Among the polymers, polypyrole, polythiophene and polyaniline show high electroactivity with good reversibility and chemical stability. Hence, by including a polymer cathode, the battery may exhibit higher energy density and mechanical flexibility [95, 96].

In the case of conducting polymer films [e.g. polypyrole (PPy)] with liquid electrolytes (based on electroactive polymers), the specific energy values range from 80 to 390 Wh/kg. The specific energy of the $Li/LiClO_4$–PC/PPy cell was 151 Wh/kg with excess electrolyte containing a PPy film about 1 μm thick. For Li/polyaniline cells with $LiCoO_4$–PC electrolytes [97–99] the specific energy ranges from 87 to 540 Wh/kg based on the weight of electroactive polymer only. The capacity was 30% higher when using 2 M $LiClO_4$ in ethylene carbonate instead of propylene carbonate. For electrolyte salts such as $LiPF_6$ or $LiBF_4$ the capacity was significantly lower than for the cell using $LiClO_4$.

CONCLUSION

Traditional energy storage devices are not fully compatible with flexible electronics devices such as smart cards and electronic paper. Hence, advances in flexible energy storage devices are important to meet this demand. The research and development of new and innovative nanomaterials is progressing well to address the requirements of the flexible electronics industry. The flexible energy storage device is still in its infancy and hence there is still plenty of room available in the materials exploratory domain; for instance, making a flexible, mechanically robust device of metal nanowires/nanoparticles (e.g. Ag, Cu or metal alloys) for the flexible electronics market. Likewise, Si nanowires has been well received in recent years owing to their higher capacity, but it is still below the traditional rigid batteries. Efforts are underway to utilize these extraordinary materials as flexible energy storage devices.

REFERENCES

[1] Rand DAJ, Woods R, Dell RM. Batteries for electric vehicles. Taunton: Research Studies Press; 1998.
[2] Dell RM, Rand DAJ. Understanding batteries. Cambridge: Royal Society of Chemistry; 2001.
[3] Electric and Hybrid Vehicle Technology International. Annual Review 1995–1998. Dorking: UK & International Press.
[4] Della RM, Rand DAJ. Energy storage – a key technology for global energy sustainability. J Power Sources 2001;100:2–17.
[5] Key world energy statistics. Paris: International Energy Agency Press; 2000.
[6] World energy outlook 2000. Paris: International Energy Agency Press; 2000.
[7] New and renewable energy: prospects in the UK for the 21st century. UK Department of Trade and Industry Report; March 1999.
[8] Long J, Dunn B, Rolison D, White H. Three-dimensional battery architectures. Chem Rev 2004;104:4463–92 (references therein).
[9] Conway BE. Electrochemical supercapacitors. Norwell, MA: Kluwer; 1999.

[10] Burke A. Ultracapacitors: why, how, and where is the technology. J Power Sources 2000;91:37–50.

[11] Conway BE. Transition from supercapacitor to battery behavior in electrochemical energy storage. J Electrochem Soc 1991;138:1539–48.

[12] Conte M, Prosini PP, Passerini S. Overview of energy/hydrogen storage: state-of-the-art of the technologies and prospects for nanomaterials. Mater Sci Eng B 2004;108:2–8.

[13] Bruce PG, Scrosati B, Tarascon JM. Nanomaterials for rechargeable lithium batteries. Angew Chem Int Ed 2008;47:2930–46.

[14] Iijima S. Helical microtubules of graphitic carbon. Nature 1991;354:56–8.

[15] http://www.wtec.org/loyola/nano/US.Review/04_06.htm.

[16] Niu CM, Sichel EK, Hoch R, Moy D, Tennent H. High power electrochemical capacitors based on carbon nanotube electrodes. Appl Phys Lett 1997;70: 1480–2.

[17] Wu GT, Weng CS, Zhang XB, Yang HS, Qi ZF, He PM, Li WZ. Structure and lithium insertion properties of carbon nanotubes. J Electrochem Soc 1999; 146:1696–701.

[18] Che GL, Lakshmi BB, Fisher ER, Martin CR. Carbon nanotube membranes for electrochemical energy storage and production. Nature 1998;393:346–9.

[19] Endo M, Kim YA, Hayashi T, Nishimura K, Matusita K, Miyashita K, Dresselhaus MS. Vapor-grown carbon fibers (VGCFs)-basic properties and their battery applications. Carbon 2001;39:1287–97.

[20] Frackowiak E, Gautier S, Gaucher H, Bonnamy S, Beguin F. Electrochemical storage of lithium multiwalled carbon nanotubes. Carbon 1999;37:61–9.

[21] Frackowiak E, Metenier K, Bertagna V, Beguin F. Supercapacitor electrodes from multiwalled carbon nanotubes. Appl Phys Lett 2000;77:2421–3.

[22] Frackowiak E. Carbon materials for supercapacitor application. Phys Chem Chem Phys 2007;9:1774–85.

[23] Chan CK, Peng H, Liu G, McIlwrath K, Zhang XF, Huggins RA, Cui Y. High-performance lithium battery anodes using silicon nanowires. Nat Nanotechnol 2008;3:31–5.

[24] Street RA, Wong WS, Ready SE, Chabinyc ML, Arias AC, Limb S, et al. Jet printing flexible displays. Mater Today 2006;9(4):32–7 (references therein).

[25] Saunier J, Alloin F, Sanchez JY, Caillon G. Thin and flexible lithium-ion batteries: investigation of polymer electrolytes. J Power Sources 2003;119:454–9 (references therein).

[26] Pushparaj VL, Shaijumon MM, Kumar A, Murugesan S, Ci L, Vajtai R, et al. Flexible energy storage devices based on nanocomposite paper. Proc Natl Acad Sci USA 2007;104:13574–7.

[27] Wyld DC. Radio frequency identification technology. In: Garson GD, Khosrow-Pour M, editors. Handbook of research on public information technology. Hershey, PA: Information Science Reference (an imprint of IGI Global); 2007. p. 425.

[28] Cao A, Dickrell PL, Sawyer WG, Ghasemi-Nejhad MN, Ajayan PM. Super-compressible foamlike carbon nanotube films. Science 2005;310:1307–10.

[29] Chen J, Minett AI, Liu Y, Lynam C, Sherrell P, Wang C, Wallace GG. Direct growth of flexible carbon nanotube electrodes. Adv Mater 2008;20:566–70.

[30] Arora P, Zhang Z. Battery separators. Chem Rev 2004;104:4419–62.

[31] Linden D, Reddy TB, editors. Handbook of batteries. 3rd ed. New York: McGraw Hill; 2002.

[32] Sadoway DR, Mayes AM. Portable power: advanced rechargeable lithium batteries. MRS Bull 2002;27:590–2.

[33] Gao B, Kleinhammes A, Tang XP, Bower C, Fleming L, Wu Y, Zhou O. Electrochemical intercalation of single-walled carbon nanotube with lithium. Chem Phys Lett 1999;307:153–7.

[34] Tans SJ, Verschueren ARM, Dekker C. Room-temperature transistor based on a single carbon nanotube. Nature 1998;393:49–52.

[35] Arico AS, Bruce P, Scrosati B, Tarascon JM, van Schalkwijk W. Nanostructured materials for advanced energy conversion and storage devices. Nat Mater 2005;4:367–7.

[36] Dresselhaus MS, Dresselhaus G, Eklund PC. Science of fullerenes and carbon nanotubes. San Diego, CA: Academic Press; 1996.

[37] Baughman RH, Zakhidov AA, de Heer WA. Carbon nanotubes – the route toward applications. Science 2002;297:787–92.

[38] Qian D, Wagner GJ, Liu WK, Yu MF, Ruoff RS. Mechanics of carbon nanotubes. Appl Mech Rev 2002;55:495–533.

[39] Iijima S, Brabec C, Maiti A, Bernholc J. Structural flexibility of carbon nanotubes. J Chem Phys 1996;104:2089–92.

[40] Yakobson BI, Brabec CJ, Bernholc J. Nanomechanics of carbon tubes: instabilities beyond linear response. Phys Rev Lett 1996;76:2511–4.

[41] Lourie O, Cox DM, Wagner HD. Buckling and collapse of embedded carbon nanotubes. Phys Rev Lett 1998;81:1638–41.

[42] Suhr J, Victor P, Ci L, Sreekala S, Zhang Z, Nalamasu O, Ajayan PM. Fatigue resistance of aligned carbon nanotube arrays under cyclic compression. Nat Nanotechnol 2007;2:417–21.

[43] Pushparaj VL, Ci L, Sreekala S, Kumar A, Kesapragada S, Gall D, Nalamsu O, Pulickel AM. Effects of compressive strains on electrical conductivities of a macroscale carbon nanotube block. Appl Phys Lett 2007;91:153116.

[44] Wei BQ, Vajtai R, Jung Y, Ward J, Zhang R, Ramanath G, Ajayan PM. Organized assembly of carbon nanotubes-cunning refinements help to customize the architecture of nanotube structures. Nature 2002;416:495–6.

[45] Talapatra S, Kar S, Pal SK, Vajtai R, Ci L, Victor P, et al. Direct growth of aligned carbon nanotubes on bulk metals. Nat Nanotechnol 2006;1:112–6.

[46] Masarapu C, Wei B. Direct growth of aligned multiwalled carbon nanotubes on treated stainless steel substrates. Langmuir 2007;23:9046–9.

[47] Puetzky AA, Geohegan DB, Jesse S, Ivanov IN, Eres G. In situ measurements and modeling of carbon nanotube array growth kinetics during chemical vapor deposition. Appl Phys A 2005;81:223–40.

[48] Singh C, Shaffer MSP, Windle AH. Production of controlled architectures of aligned carbon nanotubes by an injection chemical vapor deposition method. Carbon 2003;41:359–68.

[49] Pal SK, Talapatra S, Kar S, Ci L, Vajtai R, Borca-Tasciuc T, et al. Time and temperature dependence of multi-walled carbon nanotube growth on Inconel 600. Nanotechnology 2008;19:045610.

[50] Du C, Yeh J, Pan N. High power density supercapacitors using locally aligned carbon nanotube electrodes. Nanotechnology 2005;16:350–3.

[51] Shaijumon MM, Ou FS, Ci L, Ajayan PM. Synthesis of hybrid nanowire array and their application as high power supercapacitor electrodes. Chem Commun 2008;44:2373–5.

[52] Jung J, Kar S, Talapatra S, Soldano C, Vishwanathan G, Li X, et al. Aligned nanotube-polymer hybrid architectures for diverse flexible electronic applications. Nano Lett 2006;6:413–8.

[53] Dai H, Kong J, Zhou C, Franklin N, Tombler T, Cassell A, et al. Controlled chemical routes to nanotube architectures, physics, and devices. J Phys Chem B 1999;109:11246–55.

[54] Meadows PM, Woods CM, Chen J, Tsukamoto H. US Patent 7295878; 2007.

[55] Niu CM, Sichel EK, Hoch R, Moy D, Tennent H. High power electrochemical capacitors based on carbon nanotube electrodes. Appl Phys Lett 1997;70:1480–2.

[56] Wu GT, Weng CS, Zhang XB, Yang HS, Qi ZF, He PM, Li WZ. Structure and lithium insertion properties of carbon nanotubes. J Electrochem Soc 1999; 146:1696–701.

[57] Endo M, Kim YA, Hayashi T, Nishimura K, Matusita T, Miyashita K, Dresselhaus MS. Vapor-grown carbon fibers (VGCFs) – basic properties and their battery applications. Carbon 2001;39:1287–97.

[58] Welton T. Room-temperature ionic liquids: solvents for synthesis and catalysis. Chem Rev 1999;99:2071–83.

[59] Swatloski RP, Spear SK, Holbrey JD, Rogers RD. Dissolution of cellulose with ionic liquids. J Am Chem Soc 2002;124:4974–5.

[60] Howlett PC, MacFarlane DR, Hollenkamp AF. High lithium metal cycling efficiency in a room-temperature ionic liquid. Electrochem Solid State Lett 2004;7: A97–101.

[61] Kim YJ, Matsuzawa Y, Ozaki S, Park KC, Kim C, Endo M, et al. High energy-density capacitor based on ammonium salt type ionic liquids and their mixing effect by propylene carbonate. J Electrochem Soc 2005;152:A710–5.

[62] Chen J, Liu Y, Minett AI, Lynam C, Wang J, Wallace GG. Flexible, aligned carbon nanotube/conducting polymer electrodes for a lithium-ion battery. Chem Mater 2007;19:3595–7.

[63] Migahed MD, Fahmy T, Ishra M, Barakat A. Preparation, characterization, and electrical conductivity of polypyrrole composite films. Polym Test 2004;23: 361–5.

[64] Zhang X, Zhang J, Liu Z. Conducting polymer/carbon nanotube composite films made by in situ electropolymerization using an ionic surfactant as the supporting electrolyte. Carbon 2005;43:2186–91.

[65] Song HK, Palmore GTR. Redox-active polypyrrole: toward polymer-based batteries. Adv Mater 2006;18:1764.

[66] Chen J, Liu Y, Minett AI, Lynam C, Wang J, Wallace GG. Flexible, aligned carbon nanotube/conducting polymer electrodes for a lithium-ion battery. Chem Mater 2007;19:3595–7.

[67] Yang Y, Huang S, He H, Mau AWH, Dai L. Patterned growth of well-aligned carbon nanotubes: a photolithographic approach. J Am Chem Soc 1999;121: 10832–3.

[68] Dai L, Patil A, Gong X, Guo Z, Liu L, Liu Y, Zhu D. Aligned nanotubes. Chem Phys Chem 2003;4:1150–69.

[69] Ng SH, Wang J, Guo ZP, Chen J, Wang GX, Liu HK. Single wall carbon nanotube paper as anode for lithium-ion battery. Electrochim Acta 2005;51:23–8.

[70] Liu Y, Chen J, Zhang W, Ma Z, Swiegers GF, Too CO, Wallace GG. Nano-Pt modified aligned carbon nanotube arrays are efficient, robust, high surface area electrocatalysts. Chem Mater 2008;20:2603–5.

[71] Liu HK, Wang GX, Guo Z, Wang J, Konstantinov K. Nanomaterials for lithium-ion rechargeable batteries. J Nanosci Nanotechnol 2006;6:1–15.

[72] Yin J, Wada M, Kitano Y, Tanase S, Kajita O, Sakai T. Nanostructured Ag-Fe-Sn/carbon nanotubes composites as anode materials for advanced lithium-ion batteries. J Electrochem Soc 2005;152:A1341–6.

[73] Wang GX, Ahn JH, Yao J, Lindsay M, Liu HK, Dou SX. Preparation and characterization of carbon nanotubes for energy storage. J Power Sources 2003;119: 16–23.

[74] Wu GT, Wang CS, Zhang XB, Yang HS, Qi ZF, He PM, Li WZ. Structure and lithium insertion properties of carbon nanotubes. J Electrochem Soc 1999; 146:1696–701.

[75] Wang GX, Bewlay S, Yao J, Liu HK, Dou SX. Tungsten disulfide nanotubes for lithium storage. Electrochem Solid State Lett 2004;7:A321–3.

[76] Chen J, Tao Z-L, Li S-L. Lithium intercalation in open-ended TiS_2 nanotubes. Angew Chem Int Ed 2003;42:2147–51.

[77] http://www.physorg.com/news3061.html.

[78] Martin CR. Nanomaterials – a membrane-based synthetic approach. Science 1994;266:1961–6.

[79] Ou FS, Shaijumon MM, Ci L, Benicewicz D, Vajtai R, Ajayan PM. Multisegmented one-dimensional hybrid structures of carbon nanotubes and metal nanowires. Appl Phys Lett 2006;89:243122.

[80] Masuda H, Fukuda K. Ordered metal nanohole arrays made by a 2-step replication of honeycomb structures of anodic alumina. Science 1995;268:1466–8.

[81] Meng G, Jung Y, Cao A, Vajtai R, Ajayan PM. Controlled fabrication of hierarchically branched nanopores, nanotubes, and nanowires. Proc Natl Acad Sci USA 2005;102:7074–8.

[82] Shaijumon MM, Ou FS, Ci L, Ajayan PM. Synthesis of hybrid nanowire arrays and their application as high power supercapacitor electrodes. Chem Commun 2008;44:2373–5.

[83] Futaba DN, Hata K, Yamada T, Hiraoka T, Hayamizu Y, Kakudate Y, et al. Shaped-engineerable and highly densely packed single-walled carbon nanotubes and their application as super-capacitor electrodes. Nat Mater 2006;5:987–94.

[84] Du C, Yeh J, Pan N. High power density supercapacitors using locally aligned carbon nanotube electrodes. Nanotechnology 2005;16:350–3.

[85] Winter M, Besenhard JO. Electrochemical lithiation of tin and tin-based intermetallics and composites. Electrochim Acta 1999;45:31–50.

[86] Besenhard JO, Hess M, Komenda P. Dimensionally stable Li-alloy electrodes for secondary batteries. Solid State Ionics 1990;40/41:525–9.

[87] Shodai T, Okada S, Tobishima S, Yamaki J. Anode performance of a new layered nitride Li_3-xCoxN (x = 0.2–0.6). J Power Sources 1997;68:515–8.

[88] Yang J, Wachtler M, Winter M, Besenhard JO. Sub-microcrystalline Sn and Sn-SnSb powders as lithium storage materials for lithium-ion batteries. Electrochem Solid State Lett 1999;2:161–3.

[89] Chung SY, Bloking JT, Chiang YM. Electronically conductive phosphor-olivines as lithium storage electrodes. Nat Mater 2002;1:123–8.

[90] Wang GX, Yang L, Bewlay SL, Chen Y, Liu HK, Ahn JH. Electrochemical properties of carbon coated LiFePO$_4$ cathode materials. J Power Sources 2005;146:521–4.

[91] Bewlay S, Konstantinov K, Wang GX, Dou SX, Liu HK. Conductivity improvements to spray-produced LiFePO$_4$ by addition of a carbon source. Mater Lett 2004;58:1788–91.

[92] Konstantinov K, Bewlay S, Wang GX, Lindsay M, Wang JZ, Liu HK, et al. New approach for synthesis of carbon-mixed LiFePO$_4$ cathode materials. Electrochim Acta 2004;50:421–6.

[93] Paterson AJ, Armstrong AR, Bruce PG. Stoichiometric LiMnO$_2$ with a layered structure-charge/discharge capacity and the influence of grinding. J Electrochem Soc 2004;151:A1552–32.

[94] Kawamura T, Makidera M, Okada S, Koga K, Miura N, Yamaki JI. Effect of nano-size LiCoO$_2$ cathode powders on Li-ion cells. J Power Sources 2005;146:27–32.

[95] Novak P, Muller K, Santhanam KSV, Haas O. Electrochemically active polymers for rechargeable batteries. Chem Rev 1997;97:207–81.

[96] Arbizzani C, Mastragostino M, Scrosati B. *Handbook of organic conductive moleculars and polymers*, Nalwa H. S. (ed.). Wiley, Chichester, vol. 4595.

[97] Tanaka K, Shichiri T, Yamabe T. Influence of polymerization temperature on the characteristics of polythiophene films. Synth Met 1986;16:207–14.

[98] MacDiarmid AG, Yang LS, Huang WS, Humphrey BD. Polyaniline – electrochemistry and application to rechargeable batteries. Synth Met 1987;18:393–8.

[99] Desilvestro J, Scheifele W, Haas O. In situ determination of gravimetric and volumetric charge-densities of battery electrodes-polyaniline in aqueous and non-aqueous electrolytes. J Electrochem Soc 1992;139:2727–36.

CHAPTER 9

Flexible Chemical Sensors

H. Hau Wang

Materials Science Division, Argonne National Laboratory, Argonne, Illinois, USA

INTRODUCTION

A sensor is a device that measures a physical quantity and converts it into a signal which can be read by an observer or by an instrument. There are many different types of sensors; based on their detection targets as well as detection mechanisms they can be categorized as mechanical, electromagnetic, thermal, optical, acoustic, radiation and chemical sensors, etc. [1]. There are clearly overlapping areas such as using mechanical or optical effects to monitor chemical species. The trend is moving toward sensors with better sensitivity and selectivity and faster response time, which are easier to build, more portable, remotely capable and ideally less expensive. More advanced sensor design also tends to be multifunctional, i.e. in addition to its main sensing function, it keeps track of many environmental factors such as temperature, humidity, time, location and event history. Better, faster and more compact integrated electronics, as well as rapid developments in nanoscience and engineering, are bringing these goals closer to reality. Many new materials as well as known materials that can be made on the nanoscale, such as carbon nanotubes (CNTs), graphene, various semiconductive nanowires and nanotubes, are now becoming available for building novel sensors. The scope of this chapter will be limited to chemical sensors [2–4] in general and focused on recent advances related to flexible chemical sensors.

Why is there a need for flexible chemical sensors? A somewhat simplistic analogy is between a desktop and a laptop computer. Many similar tasks can be carried out on both computer platforms. The laptop may not be as heavy duty but it is portable and one can use it nearly anywhere at anytime. In fact, in late 2008 laptop sales surpassed those of desktop computers. While stationary sensing devices and equipment will never be replaced, mobility and timing are certainly important factors in sensing applications. The goals for a flexible chemical sensor are portability, high sensitivity, low false alarm rate and fast

CONTENTS

response. After initial positive identification, detailed analysis can be carried out in a laboratory environment. At present, there are a few niche areas that demand flexible chemical sensors for compact electronic device design; for example, leak inspection for underground piping, flammable and/or toxic gas detection in waste storage areas and confined spaces, security checkpoints for improvised explosive devices, sensing functions in wearable smart garments, and sensing and health monitoring in biomedical devices and tools. For any field application, small form factor, light weight and portability, with uncompromised sensitivity and selectivity, are essential. In general, for green and better technology, current research is investigating compactness, portability, reduced energy consumption, higher efficiency, etc. To build up such capabilities, miniaturization with integrated and compact design, utilization of nanoscale materials, sensing and electronic components that allow mechanical bendability, etc., are all at various stages of development. In this chapter, the available sensing elements, types of flexible chemical sensors, related signal transmission, selectivity and integration of these nanoscale materials toward compact devices will be briefly reviewed.

MATERIALS FOR FLEXIBLE CHEMICAL SENSORS

Carbon nanotubes

The discovery of CNTs by Iijima in 1991 opened up a brand new era in materials science and nanotechnology [5]. The intrinsic electronic properties of single-walled carbon nanotubes (SWNTs) are such that they may be metallic or semiconductive depending on their diameter and the graphitic ring arrangement around the tube wall [6]. Furthermore, CNTs show exceptionally good thermal and mechanical properties. It is expected that CNTs could solve the thermal dissipation problem of nanodevices due to their high thermal conductivity. CNTs can transport a significant amount of electric current without the doping problem encountered in silicon (Si)-based field-effect transistors (FETs) because of their graphitic nature and because the covalent bonds among carbon atoms are much stronger. CNTs are among the most promising materials anticipated to impact future nanotechnology owing to their unique structural and electronic properties, which have generated great interest for application in a broad range of potential nanodevices. Around 2000, individual SWNTs were recognized as being extremely sensitive chemical sensors toward NO_2, NH_3 and O_2 (Table 9.1) [7, 8]. The highest mobility at room temperature reaches 10^4 $cm^2/V \cdot s$ [9]. The combined mechanical strength, semiconductive electrical properties with very high mobility, and extremely high chemical sensitivity of SWNTs have led to the study of their use in flexible chemical sensors.

Table 9.1 Recent chemical sensors with measurement type and detection limit

Target analyte	Sensing material	Measurement	Detection limit	Refs
Moisture, H_2O	Poly(cellulose-acetate-butyrate)	Capacitance	0–100% RH	[54]
	PANI	Chemiresistor	25 ppm	[85]
	PPy/TiO_2/electrolyte	Impedance	30–90% RH	[86]
	PAM single fiber	Optical	35–88% RH	[60]
H_2	Pd/SWNTs	Chemiresistor	30 ppm–2%	[46, 49, 70]
	Pd/SWNT	Chemiresistor	100 ppm	[87]
	Pd thin film/surfactant	Chemiresistor	25 ppm	[48]
	TiO_2 nanotubes	Chemiresistor	1000 ppm	[88]
O_2	SWNTs film	Chemiresistor	20%	[8]
CO_2	Trapped in seawater under pressure	Raman	18 mmol/kg	[89]
	CNTs/MEMS	Resonant transduction	0–15 vol%	[90]
CO	$PANI/FeCl_3-AlCl_3$	Chemiresistor	10 ppm	[91]
	$PANI/In_2O_3$	SAW	<60 ppm	[92]
	SWNTs-COOH	CFET	1 ppm	[93]
	$SnO_{1.8}$:Ag 5% (400°C)	Chemiresistor	100 ppm	[94]
	SnO_2 nanowires (150°C)	Chemiresistor	10 ppm	[95]
	Single SnO_2 nanowire (260°C)	Chemiresistor	5 ppm	[96]
NO	SWNTs/PEI	CFET	5 ppb	[97]
NO_2	PANI/PSSA	Chemiresistor	<20 ppm	[98]
	$PANI/In_2O_3$	SAW	510 ppb	[92]
	PANI/PS single nanofiber	Optical	100 ppb	[60]
	PEI/SWNTs	CFET	100 ppt	[99]
	PSSMA/PAH/MWNTs	Chemiresistor	1 ppm	[100]
	SWNTs	CFET	44 ppb	[101]
	SWNTs	CFET	2 ppm	[7]
	Graphene	Chemiresistor	5 ppm	[29]
	In_2O_3 nanowires (RT)	CFET	5 ppb	[37]
	SnO_2 nanowires (100°C)	Chemiresistor	300 ppb	[95]
	SnO_2 nanobelts (RT)	UV irradiation	3 ppm	[35]
NH_3	PANI/SWNT	Chemiresistor	50 ppb	[102]
	PANI nanofibers ~100 nm	Chemiresistor	100 ppm	[59]
	BTB doped/PMMA fiber	Optical	3 ppm	[60]
	Nafion/SWNTs	CFET	100 ppm	[99]
	SWNTs	CFET	0.1%	[7]
	SWNTs + PABS	Chemiresistor	5 ppm	[103]
	Graphene	Chemiresistor	5 ppm	[29]
	Graphene	CFET	100 ppm	[104]
	PANI/reduced graphene oxide	Chemiresistor	100 ppm	[28]
	In_2O_3 nanowires (RT)	CFET	200 ppm	[105]

Table 9.1 *(continued)*

Target analyte	Sensing material	Measurement	Detection limit	Refs
N_2H_4	PANI nanofibers ~100 nm	Chemiresistor	3 ppm	[59]
HCl	PANI nanofibers ~100 nm	Chemiresistor	100 ppm	[59]
H_2S	PANI/$ZnCl_2$, $CuCl_2$,	Chemiresistor	<10 ppm	[106]
	Biotemplated ZnO	Chemiresistor	50 ppm	[107]
	SWNTs/Pd	CFET	50 ppm	[50]
	α-Fe_2O_3	Chemiluminescence	10 ppm 109°C	[108]
SO_2	MWNTs	Ac impedance	10 ppm	[109]
CH_4	SWNTs/Pd/UV	Chemiresistor	6 ppm	[110]
	$SnO_{1.8}$:Ag 5% (400°C)	Chemiresistor	1000 ppm	[94]
Methanol	PANI/Pd	Chemiresistor	1 ppm	[111]
	PANI	Chemiresistor	25 ppm	[85]
	PANI nanofibers ~100 nm	Chemiresistor	2%	[59]
Acetone	SWNTs + PQT	CFET	10 ppb	[112]
Cl_2	In_2O_3 thin film (250°C)	Chemiresistor	5 ppm	[113]
	WO_3 thin film (150°C)	Chemiresistor	0.05 ppm	[114]
Chloroform	PANI nanofibers ~100 nm	Chemiresistor	3%	[59]
Nitroaromatics	DDFTTF/PVP-HDA	CFET	100 ppb	[52, 53]
	SWNTs	CFET	262 ppb	[101]
	Graphene	Chemiresistor	52 ppb	[29]
	Reduced graphene oxide	Chemiresistor	0.5 ppb	[30]
	PV	Lasing intensity	5–100 ppb	[115]
	SWNTs/Cu n.p.	Electrochemical	5 ppb	[116]
Nerve agents	SWNTs/PET	Chemiresistor	10 ppm	[117]
DIMP	Reduced graphene oxide	Chemiresistor	5 ppb	[30]
DMMP				
OH^- pH	SWNTs	CFET	pH 1–11	[118]

PANI: polyaniline; PAM: polyacrylamide; PS: polystyrene; PSSA: polystyrenesulfonic acid; PEI: poly(ethyleneimine); PPy: poly(pyrrole); PSSMA: poly(4-styrenesulfonic acid-*co*-maleic acid); PAH: poly(allylamine hydrochloride); DDFTTF: 5,5′-bis-(7-dodecyl-9H-fluoren-2-yl)-2,2′-bithiophene; PVP: poly(4-vinylphenol); HDA: 4,4′-(hexafluoroisopropylidene)diphthalic anhydride; PV: derivatized phenylenevinylene; DIMP: diisopropyl methylphosphonate (Sarin); DMMP: dimethyl methylphosphonate (Soman); PQT: poly(3,3′-dialkyl-quarterthiophone); PABS: poly(*m*-aminobenzene sulfonic acid); BTB: bromothymol blue; PMMA: poly(methylmethacrylate); PET: polyethylene terephthalate; CNT: carbon nanotube; SWNT: single-walled carbon nanotube; MWNT: multiwalled carbon nanotube; MEMS: microelectromechanical system; RT: room temperature; RH: relative humidity; CFET: chemical field effect transistors.

Extensive efforts have been made to control the growth and properties of CNTs since their discovery in 1991 [5]. Large quantities of CNTs can now be produced by arc discharge, laser ablation, chemical vapor deposition (CVD) and flame synthesis [10]. Single-, double- and multiwalled carbon nanotubes are commercially available in small quantities. However, the applications of CNTs have been limited because of problems with catalyst residues, difficulties with the alignment and diameter uniformity of the nanotubes, mixed metallic and semiconductive properties, etc. Typical commercial SWNTs contain impurities such as graphitic carbon, amorphous carbon, fullerenes, polyaromatic hydrocarbon, transition metal catalysts and catalyst supports (silica and alumina). These impurities, especially the metal catalysts, may have detrimental effects on the SWNTs' sensing capability. Depending on the CNT preparation process, different strategies have been reported to remove these impurities. In general, fullerenes and polyaromatic hydrocarbons are removed through CS_2 extraction [11]. For multiwalled carbon nanotubes (MWNTs), carbon nanoparticles are burnt out at 500°C in dry air [11]. SWNTs may be suspended in aqueous solution with the use of surfactants, benzalkonium chloride [11] or sodium dodecyl sulfate (SDS) [12]. Alternatively, SWNTs may also be suspended in organic solvents such as 1,2-dichlorobenzene [13]. Metal catalysts and catalyst supports can be removed through chemical etching by treating with various acids [11,14]. Mechanical treatment such as sonication in solution with ZrO_2 or $CaCO_3$ nanoparticles can release encapsulated metal catalysts from graphene sheets. These released magnetic nanoparticles containing Co and Ni may be separated with the use of strong permanent magnets [14]. Suspended SWNTs are further purified through centrifugation to separate large aggregates. The well-dispersed SWNTs can then be drop- or spin-cast over a substrate and made ready for sensing applications.

The chemically purified SWNTs consist of a range of nanotubes with different diameters. The electrical properties of these SWNTs are one-third metallic and two-thirds semiconductive [6]. For chemical sensor applications, a dilute SWNT solution is cast over a substrate to prepare a percolation network. For a sparsely populated network, since metallic SWNTs constitute only one-third of the total population, the percolation network is typically semiconductive, which is suitable for sensing. When metallic contact is desired, such as to form a contact electrode, a simple design strategy is to prepare a densely populated SWNT network. For nanoscale sensors and electronic devices where the highest sensitivity, selectivity and mobility are sought, it is desirable to have monodispersed SWNTs. Toward that goal, there are active ongoing research efforts for the separation of SWNTs. The as-prepared SWNTs have different lengths, diameters and electronic properties, i.e. metallic or semiconductive. The general approach for separation typically consists of chemical functionalization and chromatography separation. The functionalization may be surfactant encapsulation (SDS or sodium cholate), single-stranded DNA wrapping and covalent chemical modification. Various chromatographic techniques (ion exchange, gel permeation and size exclusion) as well as ultracentrifugation have led to significant progress [15, 16]. Separation of metallic from semiconductive CNTs approaching 99% has been demonstrated [15]. Scaling up beyond the milligram scale is possible [16].

Graphene

Graphene is a single layer of graphite. A few-layer-graphene has been shown to be highly crystalline and semimetallic in its electronic properties. The atomically flat thin film has been found to be ambipolar with a tiny overlap between valence and conduction bands [17]. Both electron and hole carriers can be induced by applying gate voltage to show mobility as high as 10^4 cm^2/V·s at ambient temperature. When a two-dimensional graphene sheet is reduced to a one-dimensional graphene nanoribbon (GNR), as graphene width decreases both quantum confinement and edge effects become pronounced. In contrast to earlier studies, which indicated that the electronic properties of GNRs were similar to those of the CNTs, i.e. alternating metallic and semiconductive properties [18, 19], recent ab initio calculations revealed that there were energy gaps in all GNRs [20]. The calculations were carried out with a 12–14 fused aromatic ring system and the widths of the GNRs were ~3 nm. Experimental work observed that the energy gap scaled inversely with the ribbon width (~15–70 nm) according to:

$$E_{\text{gap}} = \alpha/(W - W^\star) \tag{9.1}$$

where α is a constant, and W and $W^\star$ are the GNR width and inactive GNR width, respectively [21].

The very high mobility, gate voltage-controlled electron or hole carriers, and tuneable bandgap suggest that GNRs may be ideal for nanoscale electronics and have triggered immense interest in the large-scale synthesis of GNRs. GNRs with energy gaps are beneficial for nanodevices compared with SWNTs where one-third of the nanotubes are expected to be metallic. This intrinsic semiconductive nature of GNRs may significantly simplify the fabrication of high-performance fast semiconductor nanodevices. GNR synthesis on a relatively large scale has been realized recently, using a procedure similar to the purification and separation of SWNTs. The process started by exfoliating expandable graphite made by intercalating ~350 μm-scale graphite flakes with sulfuric acid and nitric acid at 1000°C in forming gas for 1 min,

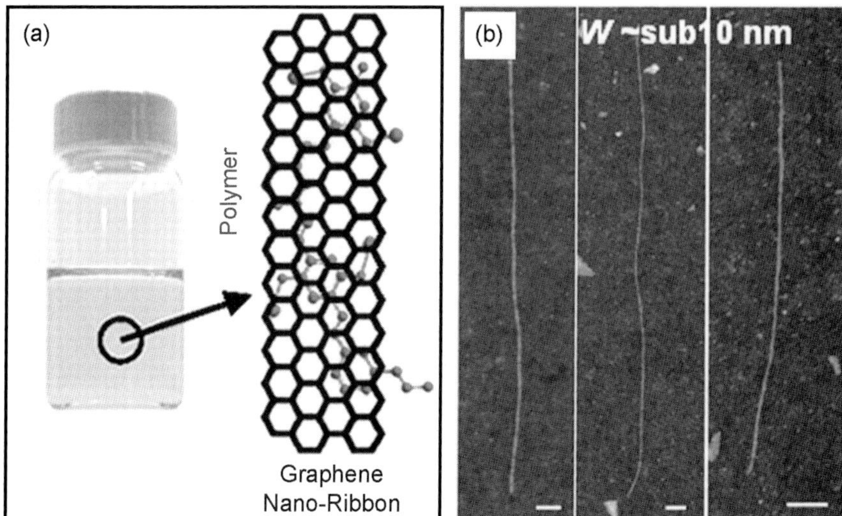

FIGURE 9.1 *(a) Graphene nanoribbons (GNRs) in organic solution with polymer and (b) AFM images of selected GNRs with width less than 10 nm. All scale bars represent 100 nm. (Reprinted with permission from [22], © 2008 AAAS.)*

sonicating the resulting exfoliated material in a 1,2-dichloroethane (DCE) solution of poly(m-phenylenevinylene-co-2,5-dioctoxy-p-phenylenevinylene) (PmPV) to disperse the GNRs, and centrifuging the suspension to retain the GNRs in the supernatant. The supernatant consisted of single and few-layer-graphene ribbons and sheets (Figure 9.1) [22]. The transport measurement indicated that all sub-10 nm GNRs were semiconductors and afforded GNR FET with an on/off ratio of $\sim 10^7$ at room temperature. Very recently, two articles reported longitudinal unzipping of CNTs, producing narrow and uniform GNRs. The first one suspended MWNTs in concentrated sulfuric acid followed by classic potassium permanganate ($KMnO_4$) oxidation that attacked $C\!=\!C$ double bonds to afford GNRs. A chemical reduction step is required to restore electrical conductivity [23]. The second one also started with MWNTs, but spread over Si wafer and coated with poly(methylmethacrylate) (PMMA). The MWNT-embedded PMMA was peeled off and exposed to Ar plasma that etched away part of the MWNTs layer by layer. GNRs with sub-10 nm width have been prepared and suggest that large-scale well-aligned semiconductor nanodevices for practical applications may be possible [24].

The high mobility, gate voltage-controlled electron or hole carriers, and tuneable bandgap certainly invite chemical sensing applications. With large-scale synthesis of graphene from graphene oxide becoming available [25–28], graphene sheets have been tested for the detection of NO_2, NH_3, nitroaromatics, etc. (Table 9.1) [29, 30]. The reported detection limits are within the ppm or even ppb range. With additional doping or defect control, graphene is expected to perform even better [31]. Highly conducting and transparent graphene sheets that are transferable, flexible, stretchable and foldable have recently been demonstrated [32]. The future of graphene- or GNR-based high-performance flexible chemical sensors is very promising.

Nanoscale materials: nanowires, nanotubes and nanoparticles

Ceramic metal oxides have long been used for sensing gases, such as combustible gases, toxic gases and volatile organic compounds (VOCs). Typical operation requires an elevated temperature, around 400–600°C, and under these conditions various gases can be oxidized over surfaces of metal oxides and/or metal oxides with catalysts. The sensor resistance changes are tracked and calibrated against known concentrations of various gases to correlate with quantitative sensing measurements. With the recent development of high-quality ceramic nanowires and nanotubes, many interesting phenomena are observed at room temperature that may bring significant changes to new sensor development. For example, vapor-phase catalyzed epitaxially grown ZnO nanorods (20–150 nm diameter) were reported to show lasing action at 385 nm under 325 nm optical pumping at room temperature [33]. The conductivity of ZnO nanowires is extremely sensitive to ultraviolet (UV) irradiation. With low-power UV lamp irradiation at 365 nm, the conductivity increases by four orders of magnitude at ambient temperature and the ZnO nanowires can be reversibly switched between low (dark) and high (under UV) conductivity states (Figure 9.2) [34]. Crystalline SnO_2 nanobelts (80–120 nm wide and 10–30 nm thick) behave similarly at room temperature. Upon UV irradiation (254 and 365 nm) in air or under a 100 ppm NO_2 environment, the conductivity of a single SnO_2 nanobelt increases by about three orders of magnitude. It is observed that O_2 and NO_2 gases can adsorb onto SnO_2 nanobelts. These metal oxide nanowires are n-carriers in nature. Under UV irradiation, the photogenerated holes will

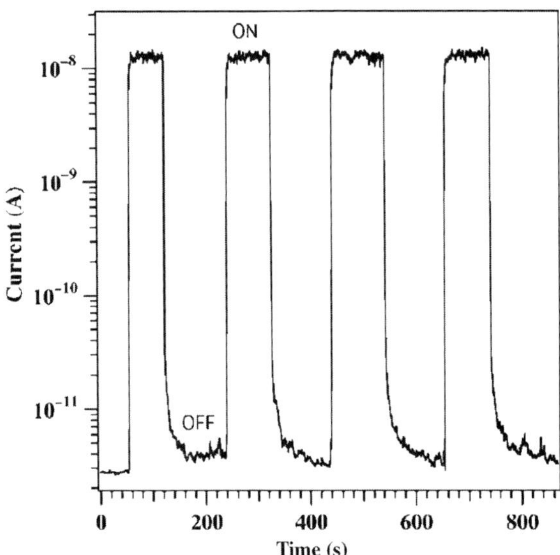

FIGURE 9.2 *A ZnO nanowire switched between high and low conductivity states with an ultraviolet lamp ON and OFF, respectively. (Reprinted with permission from [34], © Wiley-VCH Verlag GmbH, D-69469 Weinheim, 2002.)*

recombine with surface-trapped anions such as O_2^- and NO_2^- to release the adsorbed gases and increase the current flow (Figure 9.3) [35]. The reported NO_2 detection limit using SnO_2 nanobelts reaches 3 ppm at ambient temperature.

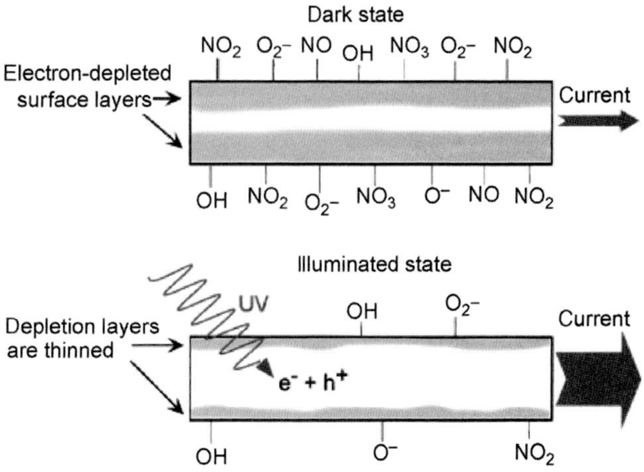

Metal oxide nanowires continue to attract sensor research interest owing to their high intrinsic surface-to-volume ratios. Another driving force is that certain pollutants such as NO_2 can cause respiratory illnesses and the Environmental Protection Agency highway emission limits are continuously being reduced (federal, US: 53 ppb; CA: 30 ppb). In order to improve the sensitivity, high-quality nanowires and nanotubes, with uniform diameters less than 30 nm are desired. Uniform diameter control of the nanowires is not a simple matter. Conventional vapor-phase growth cannot control the diameter. With the development of uniform Au nanoclusters together with a vapor–liquid–solid (VLS) growth approach, where the nanowire diameter can be controlled by the catalyst nanoparticle size, diameter-controlled high-quality crystalline In_2O_3 nanowires ($\sim$10, 20 and 30 nm) are becoming available [36]. Chemical field-effect transistors (CFETs), made with either single or multiple crystalline In_2O_3 nanowires (10 nm diameter), lead to very sensitive NO_2 sensors (Figure 9.4). A detection limit as low as 5 ppb at room temperature has been reported (Table 9.1) [37].

There are a few common features in these successful results with the use of ZnO, SnO_2 and In_2O_3 metal oxide nanowires. The use of nearly single-crystalline (high-quality) nanowires appears to be essential. A small diameter ($\sim$10 nm), with a high surface-to-volume ratio, is also important for reaching high

FIGURE 9.3 *Schematic drawing showing that n-carrier nanowires in the presence of air or NO_2 gas under dark or illuminated states carry low or high current, respectively. With ultraviolet (UV) irradiation, the photogenerated holes will combine with the surface trapped anions such as O_2^- and NO_2^- to release the adsorbed O_2 and NO_2. (Reprinted with permission from [35], © Wiley-VCH Verlag GmbH, 69451 Weinheim, Germany 2002.)*

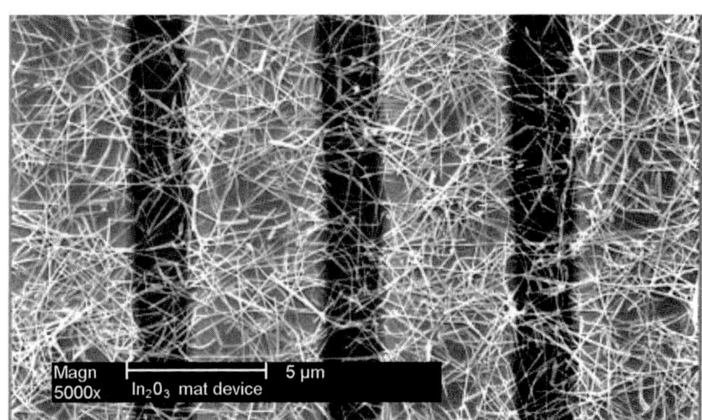

FIGURE 9.4 *SEM image of an In_2O_3 multinanowire field-effect transistor device over Si/SiO_2 substrate with Ti/Au contact electrodes. (Reprinted with permission from [37], © 2004 American Chemical Society.)*

sensitivity at room temperature. The use of metal oxides for gas sensing has been extensively reviewed [38–42]. Although the majority of the work uses polycrystalline metal oxides and requires temperature higher than 300°C, promising results based on high-quality single-crystalline metal oxides have been achieved at ambient temperature (see above). It is foreseeable that high-quality metal oxide nanowires, nanotubes and possibly nanoparticles may be attached to flexible substrates and developed into flexible chemical sensors. A Si nanowire array, prepared separately, has been transferred onto a polyethylene terephthalate (PET) substrate with use of a poly(dimethylsiloxane) (PDMS) contact printing technique. The flexible Si nanowire FET sensor can detect NO_2 down to 20 ppb [43]. Both Si and Ge nanowires have been transferred onto hard (Si/SiO_2) and soft (Kapton, polyimide) substrates on a wafer scale through contact printing with sliding and lubrication. The process prepares nanowire arrays with high density ($\sim$8 nanowires/µm) and 95% directional alignment [44]. Therefore, it is likely that flexible chemical sensors based on crystalline metal oxide nanowires will be developed in the future.

TYPES OF FLEXIBLE CHEMICAL SENSORS

Chemiresistive sensors

Among different detection techniques, chemiresistive detection is one of the most straightforward methods. The sensing mechanism depends on chemical species A; upon exposure to a second species B, a chemical reaction triggers a resistance change in the sample. The resistance change may be the result of a chemical reaction, such as a donor species A losing an electron to the acceptor B to form a redox adduct A^+B^-. In the extreme case, the resulting resistance may be several orders of magnitude less than the original resistance owing to a complete charge transfer type of chemical reaction. The chemiresistive detection is more commonly used to measure a reversible chemisorption reaction such as hydrogen or oxygen molecules being adsorbed onto a metal/catalyst surface and subsequently dissociated into atoms. The heat evolved in this type of reaction is in the range of chemical reactions and values of 10–100 kcal/mol have been observed [45]. The chemiresistive measurements can also be applied to detect a weaker interaction such as organic solvent molecules physically adsorbed onto an SWNT. The heat change will be in the van der Waals' adsorption region and is usually less than 5 kcal/mol [45]. The resistance change can be monitored with a multimeter through two-terminal contact electrodes (Figure 9.5) [46]. The two-terminal design is simple and easier to build and is suitable for many chemical reactions; however, it is less sensitive and provides limited probe information.

A special kind of chemiresistive sensor is worth noting here. When gas molecules interact with an active nanoparticle, the gas molecules will infiltrate the crystal lattice and fill the interstitial sites. The gas molecules may also dissociate into atoms. The resulting volume of the crystal lattice is expected to expand. A novel hydrogen sensor was reported based on such a phenomenon [47]. Palladium nanowires of 55–150 nm diameter with coarse grains were first electrochemically deposited along the step edges of a crystalline graphite surface (Figure 9.6). A cyanoacrylate film was cast over these Pd nanowires and transferred over to a glass substrate. A two-terminal chemiresistive Pd nanowire sensor similar to that in Figure 9.5 was built. The loosely connected Pd coarse grains showed high resistance over 10 MΩ. Upon exposure to hydrogen, at concentrations above 2%, the volume expansion induced better contacts among Pd grains and the resistance decreased significantly. This break junction sensing behavior was also directly observed with use of an atomic force microscope (AFM) before and after hydrogen intercalation. This novel hydrogen sensor was reported to have a fast response time, in the tens of milliseconds. The drawback is that owing to the physical expansion working mechanism, a significantly large hydrogen concentration (2%) is required to trigger the sensing response. The limitation on high detection concentration is largely resolved with the use of a sparsely decorated Pd thin film over glass (Figure 9.7). A water-repellent surfactant was first applied over the glass surface, followed by Pd evaporation to facilitate the movement of Pd nanoparticles upon hydrogen intercalation. With a two-dimensional thin-film coarse grain approach, the hydrogen sensing can be extended to 25 ppm [48].

For a flexible sensor design, the sensing mechanism depending on break junctions over a rigid substrate is not suitable. A percolation network of

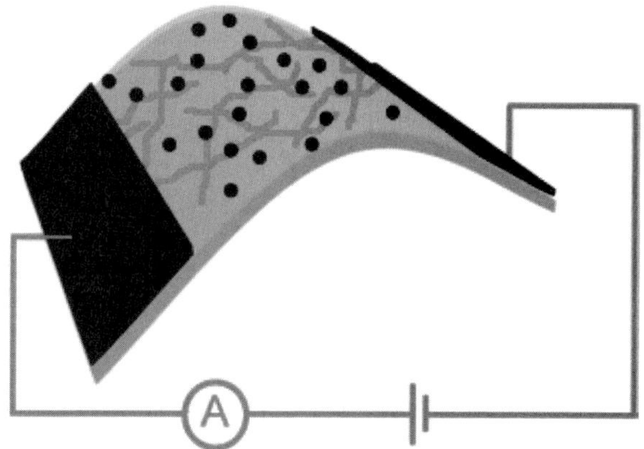

FIGURE 9.5 *Schematic drawing showing a two-terminal chemiresistive set-up.*
(Reprinted with permission from [46], © 2007 Wiley-VCH Verlag GmbH & Co. KGaA, Weinheim.)

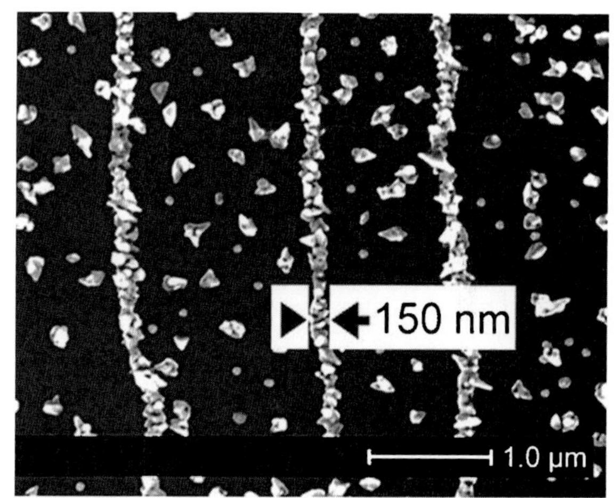

FIGURE 9.6 *SEM of Pd mesowires electrodeposited over a graphite surface.*
(Reprinted with permission from [47], © 2001 AAAS.)

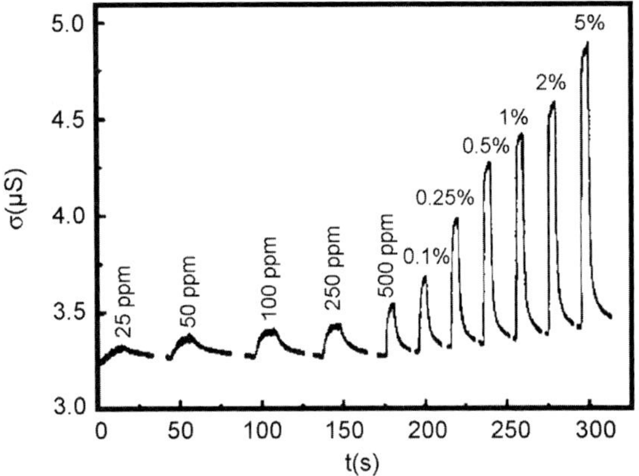

FIGURE 9.7 *Conductance of a 3.3 nm Pd thin film over surfactant-treated glass showing responses to H_2 to 25 ppm. (Reprinted with permission from [48], © 2005 American Institute of Physics.)*

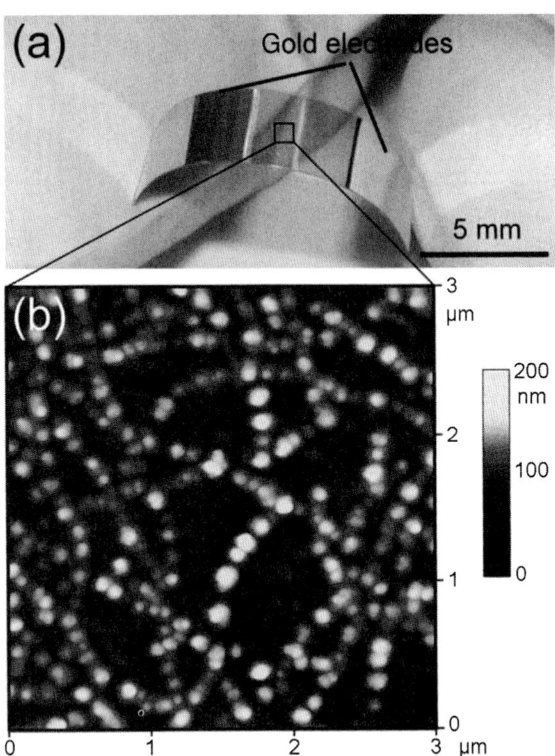

FIGURE 9.8 *(a) SWNTs/Pd sensor over flexible PET. (b) AFM image of Pd nanoparticles over SWNTs. (Reprinted with permission from [49], © 2007 American Institute of Physics.)*

SWNTs was first transfer-printed with a PDMS stamping technique onto an epoxy resin SU-8-coated flexible PET substrate (Figure 9.8). The epoxy layer enhances the bonding between CNTs and the PET substrate. The density of the SWNTs should be high enough to establish a percolation network that allows signal transmission, but low enough to avoid overall metallic conductivity. The as-synthesized SWNTs, typically consisting of one-third metallic and two-thirds semiconductive tubes, naturally fulfill the requirement. Decoration of metal nanoparticles, in these cases Pd, for high hydrogen selectivity, can be accomplished by electrodeposition, electron-beam or thermal evaporation [49, 50]. The resulting SWNTs/Pd H_2 sensors showed high sensitivity and good response time, with a low detection limit around 30 ppm (Table 9.1). The SWNT percolation network approach demonstrated good device robustness. After bending 1000 times, the sensitivity retained 87% of its original value (Figure 9.9).

Chemical field-effect transistors

The sensing element design for a three-terminal CFET is very similar to a two-terminal set-up. The sensing element, conductive polymer or CNTs, etc., is cast over a thin insulating substrate. Two detection electrodes, source and drain, are built on top of the sensing layer. One additional back gate is capacitively coupled to the two-terminal sensing electrodes across a thin dielectric layer such as SiO_2 (Figure 9.10) [51]. The conduction between source and drain can be modulated by the gate potential. The three-terminal design allows more detailed sensor characterization. Besides being able to amplify the sensing current and therefore detect weaker chemical interactions, through the biased potential, the nature of the charge carriers, either p- or n-type, can be determined. The carrier mobility can be obtained from the slope of the conductance versus biased potential (V_g) plot through the following correlation:

$$g_{ds} = \frac{\partial I_{ds}}{\partial V_{ds}} \approx \frac{W}{L} \mu C_i (V_g - V_t) \qquad (9.2)$$

where g_{ds}, I_{ds} and V_{ds} are the conductance (Ω^{-1}), current (A) and voltage (V) between source and drain, W and L are the channel width and length, μ $(cm^2/V \cdot s)$ is the mobility, C_i (F/cm^2) is the dielectric capacitance per unit area, V_g (V) is the gate potential, and V_t (V) is the threshold potential for the device to be on the ON stage [51]. For a flexible chemical FET design, while new materials are continuously being developed, the rigid Si substrate is typically replaced by a conductive indium–tin oxide (ITO) film over a flexible substrate such as PET [52]. Thin dielectric materials also play an important role in fabrication of the flexible FET device. An ultrathin cross-linkable polymer dielectric layer based on poly(4-vinylphenol) (PVP) and cross-linker 4,4′-(hexafluoroisopropylidene)diphthalic an-hydride (HDA) has been developed as a new gate dielectric that allows a low operating voltage of less than 1 V over 10^4 electrical cycles [53]. The active sensing materials may be conductive polymers, SWNTs or composites, etc. The source and drain electrodes may be gold or a conductive polymer such as poly(3,4-ethylenedioxythiophene)-poly(styrenesulfonate) (PEDOT-PSS). The flexible substrate design is not limited to gas sensing but also allows aqueous analytes [52].

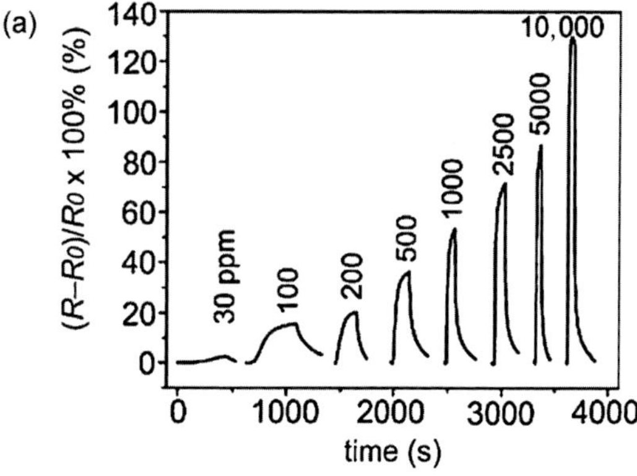

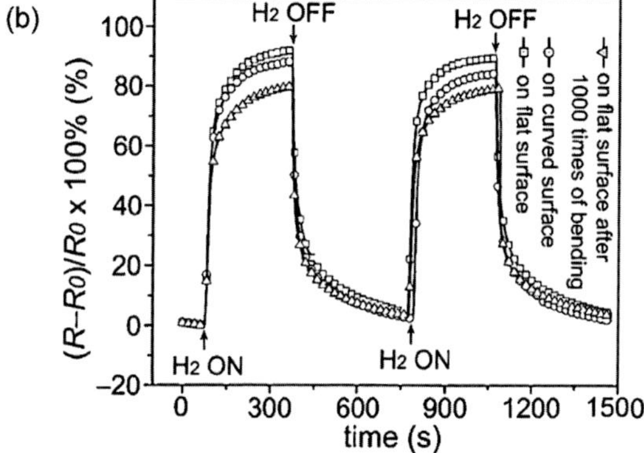

FIGURE 9.9 *(a) Resistance response over H_2 down to 30 ppm. (b) Sensor response after bending 1000 times. (Reprinted with permission from [46], © 2007 Wiley-VCH Verlag GmbH & Co. KGaA, Weinheim.)*

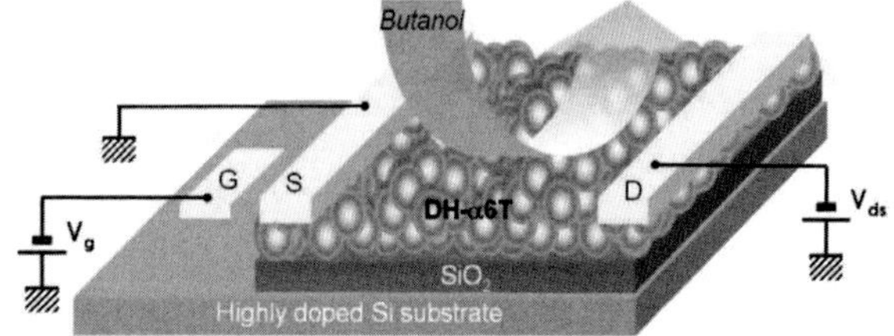

FIGURE 9.10 *Schematic drawing showing a three-terminal chemical field-effect transistor design. (Reprinted with permission from [51], © 2008 Elsevier B.V.)*

Chemocapacitance sensors

Capacitance measurement is a convenient method to monitor chemical adsorption/desorption quantitatively. Humidity is commonly measured through capacitance change with a hydrophilic porous polymer film placed

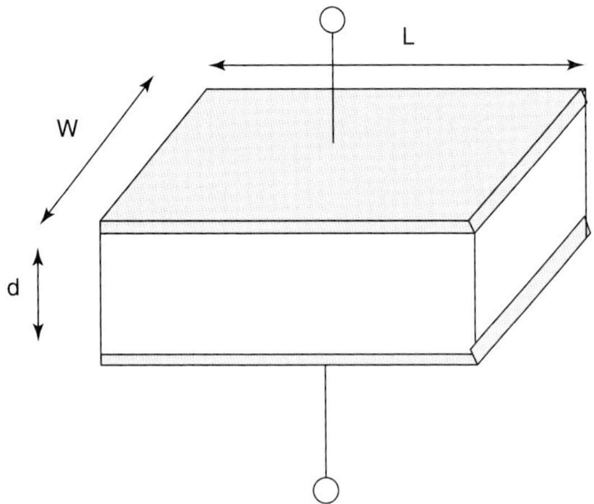

FIGURE 9.11 *Simple capacitor design with dielectric media between a pair of electrodes.*

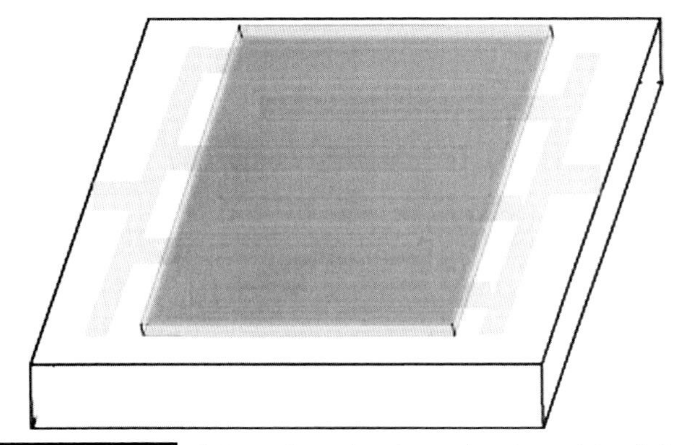

FIGURE 9.12 *Schematic drawing showing interdigitated electrodes over a substrate with a sensing layer on top.*

between two electrodes. There are two basic designs. The simplest one is to have a dielectric medium between two opposite electrodes (Figure 9.11). The capacitance (C, in farads, F) is calculated as:

$$C = \varepsilon_{\mathrm{r}}\varepsilon_0 \frac{WL}{d} \qquad (9.3)$$

where ε_{r} is relative permittivity of the dielectric media, ε_0 is relative permittivity of free space (or $\varepsilon_0 = 8.854 \times 10^{-12}$ F/m), W and L are sensor width and length (m), and d is the separation between the two plate electrodes. When the plate electrodes are made movable, the capacitance measurements lead to many other types of sensors such as mechanical sensors for vibration, change in acceleration, etc. A better chemical sensor design is to use a pair of interdigitated electrodes over a fixed substrate with sensing media deposited over the electrode area (Figure 9.12). For capacitors in parallel configuration, each has the same applied voltage, and the total capacitance is the summation of all individual capacitors. The chemical-sensitive capacitor is activated with an applied AC voltage. The voltage-dependent capacitance is measured by an impedance bridge and the sensing data are extracted from the shift of the capacitance–voltage curve [54]. This shift has the same meaning as the shift of the gate voltage (see Chemical field-effect transistors, above) [55]. For flexible capacitance sensing, polyethylene naphthalate (PEN) and polyimide (PI) films with 30–100 μm thickness have been used as the substrates, Cu thin film as the electrode and polycellulose acetate (CA), polycelluloseacetate butyrate (CAB), PMMA and polyvinylpyrrolidone (PVP) have been applied as the sensing layers [54]. Linear capacitance shift as a function of relative humidity between 0 and 100% was observed and the chemocapacitance humidity sensor was considered suitable to be incorporated into radiofrequency identification (RFID) devices.

Optical sensors

Chemical sensing based on optical detection has a long history. The absorption and fluorescent emission, changes in absorption and fluorescence upon interaction with analytes are commonly used for quantitative identification. Chemical sensors based on fluorescent conductive polymers [56], amplifying fluorescent conductive polymers [57], sol-gel, mesoporous materials, surfactant aggregates, quantum dots, etc., have been thoroughly reviewed [58] and will not be discussed here. Conductive polymers, when made into nanofibers and cast into nanofiber films, respond more rapidly to gas analytes compared to conventional thin films. The improvement is attributed to their high surface area, porosity and small diameters that enhance the diffusion of molecules and dopants into the nanofibers [59]. The use of single-polymer nanofiber to improve sensor response time and sensitivity has recently been demonstrated in optical sensing [60]. Polymers such as polyacrylamide (PAM), polyaniline (PANI) and PMMA doped with fluorescent dye have been drawn into nanofibers with diameters less than 300 nm. The diameters of these nanofibers are significantly less than the wavelength of visible light (380–750 nm). Application of a low-loss optical waveguide with submicrometer diameter is challenging. Optical waveguiding in a single nanowire is implemented by an evanescent coupling method [61], where a fiber taper drawn from a single-mode fiber with distal end about 500 nm in diameter is placed in parallel and close contact with one end of a polymer nanowire supported by a low-index substrate. Owing to the strong evanescent coupling between the nanowire and the fiber taper, light can be efficiently launched into and picked up from the nanowire within a few micrometers' overlap [62]. A schematic drawing showing the evanescent coupling between fiber taper and the single polymer nanofiber is shown in Figure 9.13. The inset indicates a PAM nanofiber (410 nm in diameter) with a 532 nm light launched from the left. The PAM nanofiber serves as a waveguide as well as a humidity sensor. The transmittance intensity decreases as humidity increases owing to moisture diffusing into the poly(acrylamide) nanofiber. Because of the small size and high surface area, the nanofiber sensor response time was reduced to 30 ms.

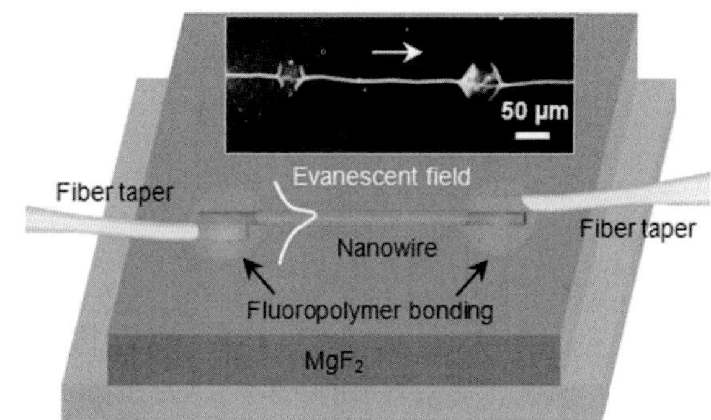

FIGURE 9.13 *A 410 nm single PAM nanofiber over MgF$_2$ support with both inlet and outlet coupled to fiber tapers. The inset indicates 532 nm laser launched from the left.*
(Reprinted with permission from [60], © 2008 American Chemical Society.)

Similar single nanofibers of PANI and doped PMMA were reported to be highly sensitive NO_2 and NH_3 sensors with detection limits of 0.1 and 3 ppm, respectively. Here, when a nanofiber is exposed to NO_2, the increase in the oxidation degree of PANI results in spectral absorption at the wavelength of the probing light (532 nm), and the absorbance is proportional to the degree of the oxidation which increases with the concentration of NO_2. Therefore, a single nanofiber of a conductive polymer or polymer with optical active dopant can be used as a waveguide and the increase or decrease in the absorbance may serve as the sensing signature. The nanofiber is intrinsically flexible and the substrate is certainly not limited to rigid materials. Further development is expected in the future.

A new class of nanostructured plasmonic chemical sensors has been developed in recent years [63]. Gold and silver nanoparticles are frequently studied because they exhibit a strong surface plasmon resonance (SPR) in visible wavelength. The aggregation of these nanoparticles will result in a significant color change that serves as the basis of sensitive chemical detection. For example, the surfaces of Au nanoparticles may be derivatized with oligonucleotide and in the presence of complementary target DNA, Au nanoparticles will aggregate and change their color from red to blue [64]. A similar approach was applied to trinitrotoluene (TNT) detection where Au nanoparticles prepared from trisodium citrate were first treated with cysteamine ($SHCH_2CH_2NH_2$). Upon addition of TNT solution, the Au nanoparticles solution turned from red to violet blue owing to TNT-induced Au nanoparticle aggregation (Figure 9.14) [65]. Although these chemical sensing techniques were demonstrated in solution, it is likely that solid-state thin-film methods will be developed in the future.

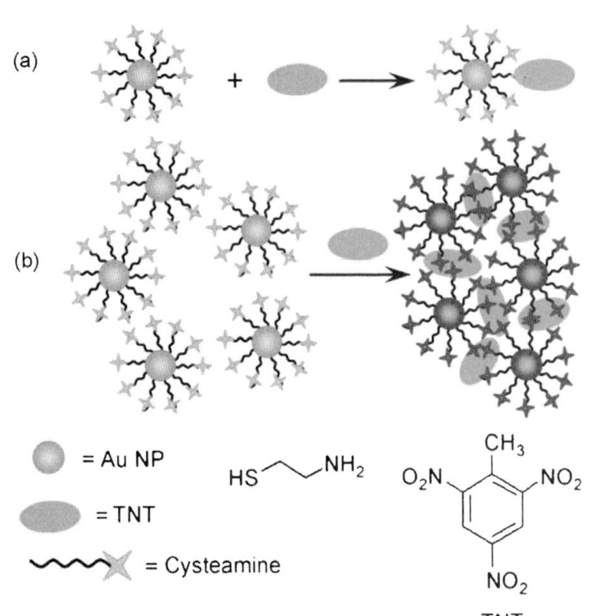

FIGURE 9.14 *Schematic drawing showing the working principle of nanoparticle-based chemical sensors.*
(Reprinted with permission from [65], © 2008 Wiley-VCH Verlag GmbH & Co. KGaA, Weinheim.)

INTEGRATION

Assembly and alignment: nanowires and carbon nanotubes

Placing nanowires and nanotubes in the desired location and proper configuration is an important step in the preparation of chemical sensors. For sensors based on percolation networks, a simple assembly procedure is drop-, dip- or spin-cast from a solution with well-dispersed nanowires or nanotubes. Substrates may be Si wafer with an insulating SiO_2 layer, glass, or for flexible applications polymer substrates such as PET (trade name Mylar) and polyimide (e.g. Kapton). Solution concentration and/or volume may be used to control the deposition density. For SWNT sensors, the deposition density should be low enough to avoid extensive contacts and metallic conductivity region. Around 10 nanotubes/μm^2 is sufficient to establish a percolation network (Figure 9.15). The nanowires/nanotubes solution is simply drop- or spin-cast onto the desired substrate. The dip- and spin-cast motion also applies a weak shear force to the nanowires and nanotubes. The randomly aligned nanotubes from drop-cast deposition can be improved by applying a unidirectional air flow to the liquid–solid substrate interface that creates a laminar flow of the suspension and therefore deposits aligned SWNTs [66].

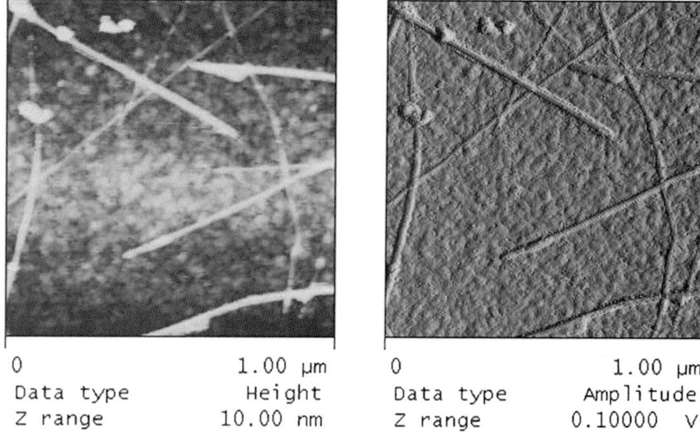

FIGURE 9.15 *AFM images of single-walled and multiwalled carbon nanotubes over Si_3N_4 substrate. Left: Height; and right: amplitude image, ~10 carbon nanotubes/μm^2.*

The substrate surface may be chemically modified with amine-terminated or phenyl-terminated silanes. A recent article reports that with amine-terminated substrate together with spin-cast deposition, SWNT-based FETs with enriched semiconductive SWNTs have been built with an on/off ratio as high as 900 000 [67]. This process may lead to highly sensitive chemical sensors. In addition, with phenyl-terminated substrate, collection of enriched metallic SWNTs has been observed.

Another effective assembly technique is the PDMS stamp dry-transfer technique [68–70]. High-quality SWNTs are typically prepared under CVD conditions at high temperature. The dry-transfer technique allows SWNTs over SiO_2/Si wafer to be first transferred to a PDMS elastomer stamp through a Cr/Au layer. The SWNTs/Cr/Au on PDMS are then laminated onto a plastic substrate with a thin layer of epoxy. Upon etching to remove the Cr/Au layer, PDMS can be released [68]. A complete process from CVD-prepared SWNTs to a decorated SWNTs/Pd/PET is shown in Figure 9.16. The key steps are: (1) grow SWNTs through CVD; (2) evaporate a Cr/Au layer followed by

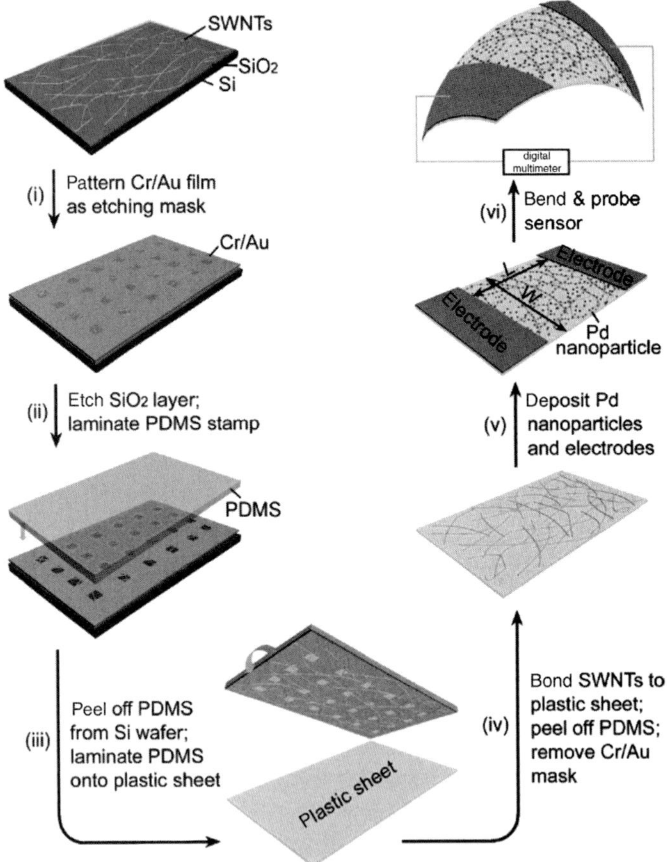

FIGURE 9.16 *Poly(dimethylsiloxane) (PDMS) stamp dry transfer technique. The schematic drawing includes single-walled carbon nanotube (SWNT) synthesis, transfer to plastic substrate and sensor fabrication.*
(Reprinted with permission from [70], © 2008 American Chemical Society.)

photolithography to develop etching pores; (3) remove the SiO_2 layer below SWNTs/Cr/Au with HF etching; (4) apply a PDMS stamp to remove the SWNTs/Cr/Au layer; (5) laminate the PDMS to a flexible PET substrate; (6) release PDMS through Cr/Au etching; and (7) evaporate the Pd thin film over the SWNT network to complete the flexible sensor fabrication. Highly aligned SWNTs grown on quartz single crystal and multiple layers of aligned SWNTs have been reported recently [71, 72]. These highly aligned SWNTs can be dry-transferred to a polyamide/ITO/PET plastic substrate to form a flexible thin-film transistor. Mobilities as high as $240 \ cm^2/V{\cdot}s$ have been measured [73]. The rapid development of high-performance flexible electronics will likely lead to high-performance chemical sensors in the future.

Since CVD nanowires are typically grown vertically to the substrate, in order to translate the vertical alignment into lateral alignment, a different technique based on differential roll printing (DRP) has been developed. Here, a cylindrical glass or quartz tube was first immersed in poly-L-lysine then gold colloid solutions to develop a catalyst bed. Vapor–liquid–solid growth of nanowires leads to a dense nanowire lawn on the outer surface of the substrate cylinder. The cylindrical nanowire growth substrate is then rolled over a tension-loaded receiving substrate, which may be a silica/Si wafer or a flexible Kapton sheet

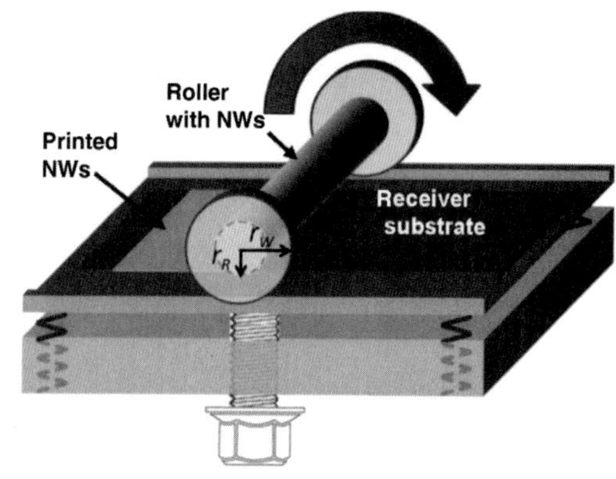

FIGURE 9.17 Differential roll printing of nanowires (NWs) onto a receiving substrate.
(Reprinted with permission from [74], © 2007 American Institute of Physics.)

(Figure 9.17). Octane is used as a lubricant to minimize the mechanical friction between nanowires. About 90% of the nanowires can be transferred to the substrate with a spread angle less than 5 degrees [74].

An ingenious general nanowire alignment technique is to use microfluidic channels fabricated with PDMS. A PDMS mold was brought into contact with a flat substrate layer. Solutions of well-suspended nanowires such as Si, InP, GaP, etc., were flowing through these channels with a controlled flow rate for a fixed period. Nanowires can be deposited preferentially along the flow direction and their angular spread decreased with increasing flow rate. Together with surface functionalization with a amine group as well as patterning, the specific deposition location may be controlled [75]. This flow-directed alignment technique was applied to deposit Si nanowires over a 4-inch wafer. The same technique was also applicable to plastic substrates. High-performance Si and CdS nanowire thin-film transistors were prepared with mobility $\sim$300 cm^2/V·s [76]. Another interesting nanowire and nanotube thin-film deposition is via blown bubble, where a bubble is expanded from a homogeneous suspension of these nanoscale materials and brought into contact with a substrate to carry out the deposition. Fairly large-area, $\sim$200 mm in diameter, flexible polymer sheet and highly curved surfaces can all be covered through this process. Integrated circuits such as arrays of nanowire transistors have been demonstrated [77].

A highly precise and somewhat elaborate process has been developed for nanowire deposition. The superlattice nanowire pattern transfer or SNAP process starts from molecular beam epitaxy (MBE) growth of a GaAs/AlGaAs superlattice. The AlGaAs layer is selectively etched down by 20–30 nm. Various metals such as Au, Cr, Pt, Al and Ti are evaporated onto the GaAs edges from a tilted 36 degree angle. The evaporated metal nanowires are then transferred to an Si- or plastic-substrate coated with a thin layer of epoxy. Upon additional etching to remove GaAs oxide, the highly aligned metal nanowires are transferred onto the substrates [43, 78, 79]. The method also combines the nanowire synthesis and assembly together in an integrated process.

Integrated package

Since the sensor response and/or the interpretation of the sensor response may depend on many environmental parameters, such as temperature, humidity, pressure and location, an integrated smart sensor package is desirable. A few environmental factors such as temperature and humidity sensors can be developed as an independent module. Along this line, a flexible microtemperature and humidity sensor on parylene thin film was designed and fabricated using a microelectromechanical system (MEMS) process. Based on the principles of the thermistor and the ability of a polymer to absorb moisture, the sensing device comprised of a chemiresistive gold wire and a polyimide chemocapacitor film [80]. As a proof-of-concept demonstration, a temperature, humidity and gas sensor (alcohols, hexane, ammonia, etc.) integrated on a plastic polyimide foil based on two capacitive chemical sensors and a resistance thermometer has been reported [81]. For the food monitoring industry during the food transportation logistic chain, a flexible microlaboratory that integrates physical and chemical sensors has been developed. The concept is based on RFID wireless data exchange that monitors time, date, location and gas sensing to ensure the quality of the food during transportation, storage and vending [82]. One more recent development is the integration of heterogeneous nanowire arrays to build multifunctional all-nanowire circuitry. In the report, CdSe and Ge/Si nanowire arrays are deterministically positioned on substrates with the DRP technique, and configured as photodiodes and transistors, respectively. In this large-scale integration, parallel arrays of nanowires are incorporated to enable circuit operation involving five orders of magnitude of current amplification of the signal from the nanosensor [83]. Although the report is based on SiO_2/Si substrate, the DRP methodology is also applicable to flexible plastic substrates.

CONCLUSION

Flexible chemical sensors are not a new concept. Traditional chemical sensors based on conductive polymers over polymer substrates are generally flexible. However, with the advent of high-quality nanowires and nanotubes, and the rapid development of flexible electronics which involves many breakthroughs in the synthesis and manipulation of these nanoscale materials, significant progress has been made in modern flexible chemical sensors. In traditional conductive polymer chemical sensors, self-organization in many solution-processed, semiconducting conjugated polymers results in complex microstructures, in which ordered microcrystalline domains are embedded in an amorphous matrix. This has important consequences for the electrical properties of these materials, i.e. charge transport is usually limited by the most difficult hopping processes and is therefore dominated by the disordered matrix, resulting in low charge carrier mobilities ($<10^{-5}$ cm^2/V·s) [84]. In this chapter, SWNTs, graphene and high-quality nanowires and nanotubes as the sensing materials have been reviewed. Mobility as high as 10^4 cm^2/V·s in SWNT has been reported [9]. Owing to the superior quality of SWNTs, high sensitivity, low detection limit and fast response time have been observed in these nanowire- and nanotube-based chemical sensors. For highly crystalline metal oxide nanowires, a few sensing reactions occurred at room temperature instead of at high temperature. Ingenious nanowire and nanotube assembly and alignment open up many possibilities for high-performance multifunctional flexible chemical sensor development. Large integrated sensors with built-in intelligence are also on the horizon. The future of flexible chemical sensors is indeed very promising.

ACKNOWLEDGMENT

Argonne National Laboratory's work was supported by the US Department of Energy, Office of Science, Office of Basic Energy Sciences, under contract DE-AC02-06CH11357.

REFERENCES

[1] Fraden J. AIP handbook of modern sensors: physics, designs and applications. New York: American Institute of Physics; 1993.
[2] Mallouk TE, Harrison DJ, editors. Interfacial design and chemical sensing, vol. 561. Washington DC: American Chemical Society; 1993.
[3] Harsanyi G. Polymer films in sensor applications: technology, materials, devices and their characteristics. Basel: Technomic Publishing; 1995.
[4] Rogers KR, Mulchandani A, Zhou W, editors. Biosensor and chemical sensor technology, vol. 613. Washington DC: American Chemical Society; 1995.
[5] Iijima S. Helical microtubules of graphitic carbon. Nature 1991;354:56–8.
[6] Dresselhaus MS, Dresselhaus G, Eklund PC. Science of Fullerenes and carbon Nanotubes. San Diego: Academic Press; 1996.

[7] Kong J, Franklin NR, Zhou C, Chapline MG, Peng S, Cho K, Dai H. Nanotube molecular wires as chemical sensors. Science 2000;287:622–5.

[8] Collins PG, Bradley K, Ishigami M, Zettl A. Extreme oxygen sensitivity of electronic properties of carbon nanotubes. Science 2000;287:1801–4.

[9] Zhou X, Park J-Y, Huang S, Liu J, McEuen PL. Band structure, phonon scattering, and their performance limit of single-walled carbon nanotbue transistor. Phys Rev Lett 2005;95:146805.

[10] Baddour CE, Briens C. Carbon nanotube synthesis: a review. Int J Chem React Eng 2005;3:R3.

[11] Bandow S, Asaka S, Zhao X, Ando Y. Purification and magnetic properties of carbon nanotubes. Appl Phys A 1998;67:23–7.

[12] Bonard J-M, Stora T, Salvetat J-P, Maier F, Stöckli T, Duschl C, et al. Purification and size-selection of carbon nanotubes. Adv Mater 1997;9:827–31.

[13] Chen J, Hamon MA, Hu H, Chen Y, Rao AM, Eklund PC, Haddon RC. Solution properties of single-walled carbon nanotubes. Science 1998;282:95–8.

[14] Thien-Nga L, Hernadi K, Ljubovic E, Garaj S, Forro LS. Mechanical purification of single-walled carbon nanotube bundles from catalytic particles. Nano Lett 2002;2:1349–52.

[15] Hersam MC. Progress toward monodisperse single-walled carbon nanotubes. Nat Nanotechnol 2008;3:387–94.

[16] Moshammer K, Hennrich F, Kappes MM. Selective suspension in aqueous sodium dodecyl sulfate according to electronic structure type allows simple separation of metallic from semiconducting single-walled carbon nanotubes. Nano Res 2009;2:599–606.

[17] Novoselov KS, Geim AK, Morozov SV, Jiang D, Zhang Y, Dubonos SV, et al. Electric field effect in atomically thin carbon films. Science 2004;306:666–9.

[18] Ezawa M. Peculiar width dependence of the electronic properties of carbon nanoribbons. Phys Rev B 2006;73:045432.

[19] Brey L, Fertig HA. Electronic strates of graphene nanoribbons studied with the Dirac equation. Phys Rev B 2006;73:235411.

[20] Son Y-W, Cohen ML, Louie SG. Energy gaps in graphene nanoribbons. Phys Rev Lett 2006;97:216803.

[21] Han MY, Özyilmaz B, Zhang Y, Kim P. Energy band-gap engineering of graphene nanoribbons. Phys Rev Lett 2007;98:206805.

[22] Li X, Wang X, Zhang L, Lee S, Dai H. Chemically derived, ultrasmooth graphene nanoribbon semiconductors. Science 2008;319:1229–32.

[23] Kosynkin DV, Higginbotham AL, Sinitskii A, Lomeda JR, Dimiev A, Price BK, Tour JM. Longitudinal unzipping of carbon nanotubes to form graphene nanotubes. Nature 2009;458:872–6.

[24] Jiao L, Zhang L, Wang X, Diankov G, Dai H. Narrow graphene nanoribbons from carbon nanotubes. Nature 2009;458:877–80.

[25] Lee JH, Shin DW, Makotchenko VG, Nazarov AS, Fedorov VE, Kim YH, et al. One-step exfoliation synthesis of easily soluble graphite and transparent conducting graphene sheets. Adv Mater 2009;21:4383–7.

[26] Pang S, Tsao HN, Feng XL, Müllen K. Patterned graphene electrodes from solution-processed graphite oxide films for organic field-effect transistors. Adv Mater 2009;21:3488–91.

[27] Becerril HA, Mao J, Liu Z, Stoltenberg RM, Bao Z, Chen Y. Evaluation of solution-processed reduced graphene oxide films as transparent conductors. ACS Nano 2008;2:463–70.

[28] Cote LJ, Cruz-Silva R, Huang J. Flash reduction and patterning of graphite oxide and its polymer composites. J Am Chem Soc 2009;131:11027–32.

[29] Fowler JD, Allen MJ, Tung VC, Yang Y, Kaner RB, Weiller BH. Practical chemical sensors from chemically derived graphene. ACS Nano 2009;3: 301–6.

[30] Robinson JT, Perkins FK, Snow ES, Wei Z, Sheehan PE. Reduced graphene oxide molecular sensors. Nano Lett 2008;8:3137–40.

[31] Zhang Y-H, Chen Y-B, Zhou K-G, Liu C-H, Zeng J, Zhang H-L, Peng Y. Improving gas sensing properties of graphene by introducing dopants and defects: a first-principles study. Nanotechnology 2009;20:185504.

[32] Kim KS, Zhao Y, Jang H, Lee SY, Kim JM, Kim KS, et al. Large-scale pattern growth of graphene films from stretchable transparent electrodes. Nature 2009;457:706–10.

[33] Huang MH, Mao S, Feick H, Yan H, Wu Y, Kind H, et al. Room-temperature ultraviolet nanowire nanolasers. Science 2001;292:1897–9.

[34] Kind H, Yan H, Messer B, Law M, Yang P. Nanowire ultraviolet photodetectors and optical switches. Adv Mater 2002;14:158–60.

[35] Law M, Kind H, Messer B, Kim F, Yang P. Photochemcial sensing of NO_2 with SnO_2 nanoribbon nanosensors at room temperature. Angew Chem Int Ed 2002;41:2405–8.

[36] Li C, Zhang D, Han S, Liu X, Tang T, Zhou C. Diameter-controlled growth of single-crystalline In_2O_3 nanowires and their electronic properties. Adv Mater 2003;15:143–6.

[37] Zhang D, Liu Z, Li C, Tang T, Liu X, Han S, et al. Detection of NO_2 down to ppb level using individual and multiple In_2O_3 nanowire devices. C. Nano Lett 2004;4:1919–24.

[38] Shen G, Chen P-C, Ryu K, Zhou C. Devices and chemical sensing applications of metal oxide nanowires. J Mater Chem 2009;19:828–39.

[39] Dan Y, Evoy S, Charlie Johnson AT. In: Nanowire research progress, Nova Science Publishers; 2008.

[40] Huang X-J, Choi Y-K. Chemical sensors based on nanostructured materials. Sens Actuat B 2007;122:659–71.

[41] Graf M, Gurlo A, Barsan N, Weimar U, Hierlemann A. Microfabricated gas sensor systems with sensitive nanocrystalline metal-oxide films. J Nanopart Res 2006;8:823–39.

[42] Kolmakov A, Moskovits M. Chemical sensing and catalysis by one-dimensional metal-oxide nanostructure. Annu Rev Mater Res 2004;34:151–80.

[43] McAlpine MC, Ahmad H, Wang D, Heath JR. Highly ordered nanowire arrays on plastic substrates for ultrasensitive flexible chemical sensors. Nat Mater 2007;6:379–84.

[44] Fan Z, Ho JC, Jacobson ZA, Yerushalmi R, Alley RL, Razavi H, Javey A. Wafer-scale assembly of highly oredered semicondcutor nanowire arrays by contact printing. Nano Lett 2008;8:20–5.

[45] Laidler KJ. Chemical kinetics. New York: McGraw-Hill; 1965.

[46] Sun Y, Wang HH. High-performance, flexible hydrogen sensors that use carbon nanotubes decorated with palladium nanoparticles. Adv Mater 2007;19: 2818–23.

[47] Favier F, Walter EC, Zach MP, Benter T, Penner RM. Hydrogen sensors and switches from electrodeposited palladium mesowire arrays. Science 2001; 293:2227–31.

[48] Xu T, Zach MP, Xiao ZL, Rosenmann D, Welp U, Kwok WK, Crabtree GW. Self-assembled monolayer-enhanced hydrogen sensing with ultrathin palladium films. Appl Phys Lett 2005;86:203104.

[49] Sun Y, Wang HH. Electrodeposition of Pd nanoparticles on single-walled carbon nanotubes for flexible hydrogen sensors. Appl Phys Lett 2007;90: 213107.

[50] Star A, Joshi V, Skarupo S, Thomas D, Gabriel J-CP. Gas sensor array based on metal-decorated carbon nanotubes. J Phys Chem B 2006;110:21014–20.

[51] Torsi L, Marinelli F, Angione MD, Dell'Aquila A, Cioffi N, De Giglio E, Sabbatini L. Contact effects in organic thin-film transistor sensors. Org Electron 2009; 10:233–9.

[52] Roberts ME, Mannsfeld SCB, Stoltenberg RM, Bao Z. Flexible, plastic transistor-based chemical sensors. Org Electron 2009;10:377–83.

[53] Roberts ME, Mannsfeld SCB, Queraltó N, Reese C, Locklin J, Knoll W, Bao Z. Water-stable organic transistors and their application in chemical and biological sensors. Proc Natl Acad Sci USA 2008;105:12134–9.

[54] Oprea A, Barsan N, Weimar U, Bauersfeld M-L, Ebling D, Wollenstein JU. Capacitive humidity sensors on flexible RFID labels. Sens Actuat B 2008;132: 404–10.

[55] Janata J, Josowicz M. Conducting polymers in electronic chemical sensors. Nat Mater 2003;2:19–24.

[56] McQuade DT, Pullen AE, Swager TM. Conjugated polymer-based chemical sensors. Chem Rev 2000;100:2537–74.

[57] Thomas SW, Joly GD, Swager TM. Chemical sensors based on amplifying fluorescent conjugated polymers. Chem Rev 2007;107:1339–86.

[58] Basabe-Desmonts L, Reinhoudt DN, Crego-Calama M. Design of fluorescent materials for chemical sensing. Chem Soc Rev 2007;36:993–1017.

[59] Virji S, Huang J, Kaner RB, Weiller BH. Polyaniline nanofibers gas sensors: examination of response mechanisms. Nano Lett 2004;4:491–6.

[60] Gu F, Zhang L, Yin X, Tong L. Polymer single-nanowire optical sensors. Nano Lett 2008;8:2757–61.

[61] Tong L, Gattass RR, Ashcom JB, He S, Lou J, Shen M, et al. Subwavelength-diameter silica wires for low-loss optical wave guiding. Nature 2003;426: 816–9.

[62] Tong L, Lou J, Gattass RR, He S, Chen X, Liu L, Mazur E. Assembly of silica nanowires on silica aerogels for microphotonic devices. Nano Lett 2005;5: 259–62.

[63] Stewart ME, Anderton CR, Thompson LB, Maria J, Gray SK, Rogers JA, Nuzzo RG. Nanostructured plasmonic sensors. Chem Rev 2008;108:494–521.

[64] Rosi NL, Mirkin CA. Nanostructures in biodiagnostics. Chem Rev 2005; 105:1547–62.

[65] Jiang Y, Zhao H, Zhu N, Lin Y, Yu P, Mao L. A simple assay for direct colorimetric visualization of trinitrotoluene at picomolar levels using gold nanoparticle. Angew Chem Int Ed 2008;47:8601–4.

[66] Vichchulada P, Zhang Q, Lay MD. Recent progress in chemical detection with single-walled carbon nanotube networks. Analyst 2007;132:719–23.

[67] LeMieux MC, Roberts M, Barman S, Jin YW, Kim JM, Bao Z. Self-sorted, aligned nanotube networks for thin-film transistors. Science 2008;321:101–4.

[68] Hur S-H, Park OO, Rogers JA. Extreme bendability of single-walled carbon nanotube networks transferred from high-temperature growth substrates to plastic and their use in thin-film transistors. Appl Phys Lett 2005;86:243502.

[69] Cao Q, Hur S-H, Zhu Z-T, Sun Y, Wang C, Meitl MA, et al. Highly bendable, transparent thin-film transistors that use carbon-nanotube-based conductors and semiconductors with elastomeric dielectrics. Adv Mater 2006;18: 304–9.

[70] Sun Y, Wang HH, Xia M. Single-walled carbon nanotubes modified with Pd nanoparticles: unique building blocks for high-performance, flexible hydrogen sensors. J Phys Chem C 2008;112:1250–9.

[71] Kocabas C, Shim M, Rogers JA. Spatially selective guided growth of high-coverage arrays and random networks of single-walled carbon nanotubes and their integration into electronic devices. J Am Chem Soc 2006;128:4540–1.

[72] Kang SJ, Kocabas C, Kim H-S, Cao Q, Meitl MA, Khang D-Y, Rogers JA. Printed multilayer superstructures of aligned single-walled carbon nanotubes for electronic applications. Nano Lett 2007;7:3343–8.

[73] Kocabas C, Kang SJ, Ozel T, Shim M, Rogers JA. Improved synthesis of aligned arrays of single-walled carbon nanotubes and their implementation in thin film type transistors. J Phys Chem C 2007;111:17879–86.

[74] Yerushalmi R, Jacobson ZA, Ho JC, Fan Z, Javey A. Large scale, highly ordered assembly of nanowire parallel arrays by differential roll printing. Appl Phys Lett 2007;91:203104.

[75] Huang Y, Duan X, Wei Q, Lieber CM. Directed assembly of one-dimensional nanostructures into functional networks. Science 2001;291:630–3.

[76] Duan X, Niu C, Sahi V, Chen J, Parce JW, Empedocles S, Goldman JL. High-performance thin-film transistors using semiconductor nanowires and nano-ribbons. Nature 2003;425:274–8.

[77] Yu G, Cao A, Lieber CM. Large-area blown bubble films of aligned nanowires and carbon nanotubes. Nat Nanotechnol 2007;2:372–7.

[78] Melosh NA, Boukai A, Diana F, Gerardot B, Badolato A, Petroff PM, Heath JR. Ultrahigh-density nanowire lattices and circuits. Science 2003;300:112–5.

[79] Heath JR. Superlattice nanowire pattern transfer (SNAP). Acc Chem Res 2008;41:1609–71.

[80] Lee C-Y, Wu G-W, Hsieh W-J. Fabrication of microsensors on a flexible substrate. Sens Actuat A 2008;147:173–6.

[81] Oprea A, Courbat J, Bârsan N, Briand D, de Rooij NF, Weimar U. Temperature, humidity and gas sensors integrated on plastic foil for low power application. Sens Actuat B 2009;140:227–32.

[82] Abad E, Zampolli S, Marco S, Scorzoni A, Mazzolai B, Juarros A, et al. Flexible tag microlab development: gas sensors integration in RFID flexible tags for food logistics. Sens Actuat B 2007;127:2–7.

[83] Fan Z, Ho JC, Jacobson ZA, Razavi H, Javey A. Large-scale, heterogeneous integration of nanowire arrays for image sensor circuitry. Proc Natl Acad Sci USA 2008;105:11066–70.

[84] Sirringhaus H, Brown PJ, Friend RH, Nielsen MM, Bechgaard K, Langeveld-Voss BMW, et al. Two-dimensional charge transport in self-organized, high-mobility conjugated polymers. Nature 1999;401:685–8.

[85] Li G, Martinez C, Semancik S. Controlled electrophoretic patterning of polyaniline from a collodial suspension. J Am Chem Soc 2005;127:4903–9.

[86] Su P-G, Wang C-P. Flexible humidity sensor based on TiO_2 nanoparticles-polypyrrole-poly[3-(methacrylamino)propyl] trimethyl ammonium chloride composite materials. Sens Actuat B 2008;129:538–43.

[87] Mubeen S, Zhang T, Yoo B, Deshusses MA, Myung NV. Palladium nanoparticles decorated single-walled carbon nanotube hydrogen sensor. J Phys Chem C 2007;111:6321–7.

[88] Paulose M, Varghese OK, Mor GK, Grimes CA, Ong KG. Unprecedented ultra-high hydrogen gas sensitivity in undoped titania nanotube. Nanotechnology 2006;17:398–402.

[89] Dunk RM, Peltzer ET, Walz PM, Brewer PG. Seeing a deep ocean CO_2 enrichment experiment in a new light: laser Raman detection of dissolved CO_2 in seawater. Environ Sci Technol 2005;39:9630–6.

[90] Zribi A, Knobloch A, Rao R. CO_2 detection using carbon nanotube networks and micromachined resonant transducers. Appl Phys Lett 2005;86:203112.

[91] Dixit V, Misra SCK, Sharma BS. Carbon monoxide sensitivity of vacuum deposited polyaniline semiconducting thin films. Sens Actuat B 2005;104:90–3.

[92] Sadek AZ, Wlodarski W, Shin K, Kaner RB, Kalantar-zadeh K. A layered surface acoustic wave gas sensor based on a polyaniline/In_2O_3 nanofibre composite. Nanotechnology 2006;17:4488–92.

[93] Fu D, Lim H, Shi Y, Dong X, Mhaisalkar SG, Chen Y, et al. Differentiation of gas molecules using flexible and all-carbon nanotube devices. J Phys Chem C 2008;112:650–3.

[94] Joshi RK, Kruis FE. Size-selected SnO1.8:Ag mixed nanoparticle films for ethanol, CO, and CH_4 detection. J Nanomater 2007; Article ID 67072.

[95] Baratto C, Comini E, Faglia G, Sberveglieri G, Zhab M, Zappettini A. Metal oxide nanocrystal for gas sensing. Sens Actuat B 2005;109:2–6.

[96] Hernández-Ramírez F, Tarancón A, Casals O, Arbiol J, Romano-Rodríguez A, Morante JR. High response and stability in CO and humidity measures using a single SnO_2 nanowire. Sens Actuat B 2007;121:3–17.

[97] Kuzmych O, Allen BL, Star A. Carbon nanotube sensors for exhaled breath components. Nanotechnology 2007;18:375502.

[98] Xie D, Jiang Y, Pan W, Li D, Wu Z, Li Y. Fabrication and characterization of polyaniline-based gas sensor by ultra-thin film technology. Sens Actuat B 2002;81:158–64.

[99] Qi P, Vermesh O, Grecu M, Javey A, Wang Q, Dai H, et al. Toward large arrays of multiplex functionalized carbon nanotube sensors for highly sensitive and selective molecular detection. Nano Lett 2003;3:347–51.

[100] Su P-G, Lee C-T, Chou C-Y, Cheng K-H, Chuang Y-S. Fabrication of flexible NO_2 sensors by layer-by-layer self-assembly of multi-walled carbon nanotubes and their gas sensing properties. Sens Actuat B 2009;139:488–93.

[101] Li J, Lu Y, Ye Q, Cinke M, Han J, Meyyappan M. Carbon nanotube sensors for gas and organic vapor detection. Nano Lett 2003;3:929–33.

[102] Zhang T, Nix MB, Yoo B-Y, Deshusses MA, Myung NV. Electrochemically functionalized single-walled carbon nanotube gas sensors. Electroanalysis 2006; 18:1153–8.

[103] Bekyarova E, Davis M, Burch T, Itkis ME, Zhao B, Sunshine S, Haddon RC. Chemically functionalized single-walled carbon nanotubes as ammonia sensors. J Phys Chem B 2004;108:19717–20.

[104] Dan Y, Lu Y, Kybert NJ, Luo Z, Johnson ATC. Intrinsic response of graphene vapor sensors. Nano Lett 2009;9:1472–5.

[105] Li C, Zhang D, Lei B, Han S, Liu X, Zhou C. Surface treatment and doping dependence of In_2O_3 nanowires as ammonia sensors. J Phys Chem B 2003;107:12451–5.

[106] Virji S, Fowler JD, Baker CO, Huang J, Kaner RB, Weiller BH. Polyaniline nanofiber composites with metal salts: chemical sensors for hydrogen sulfide. Small 2005;1:624–7.

[107] Liu Z, Fan T, Zhang D, Gong X, Xu J. Hierarchically porous ZnO with high sensitivity and selectivity to H_2S derived from biotemplates. Sens Actuat B 2009;136:499–509.

[108] Sun Z, Yuan H, Liu Z, Han B, Zhang X. A highly efficient chemical sensor material for H_2S:α-Fe_2O_3 nanotubes fabricated using carbon nanotube templates. Adv Mater 2005;17:2993–7.

[109] Suehiro J, Zhou G, Hara M. Detection of partial discharge in SF_6 gas using a carbon naotube-based gas sensor. Sens Actuat B 2005;105:164–9.

[110] Lu Y, Li J, Han J, Ng H-T, Binder C, Partridge C, Meyyappan M. Room temperature methane detection using palladium loaded single-walled carbon nanotube sensors. Chem Phys Lett 2004;391:344–8.

[111] Athawale AA, Bhagwat SV, Katre PP. Nanocomposite of Pd-polyaniline as a selective methanol sensor. Sens Actuat B 2006;114:263–7.

[112] Vieira SMC, Beecher P, Haneef I, Udrea F, Milne WI, Namboothiry MAG, et al. Use of nanocomposite to increase electrical 'gain' in chemical sensors. Appl Phys Lett 2007;91:203111.

[113] Tamaki J, Naruo C, Yamamoto Y, Matsuoka M. Sensing properties to dilute chlorine gas of indium oxide based thin film sensors prepared by electron beam evaporation. Sens Actuat B 2002;83:190–4.

[114] Bender F, Kim C, Mlsna T, Vetelino JF. Characterization of a WO_3 thin film chlorine sensor. Sens Actuat B 2001;77:281–6.

[115] Rose AE, Zhu Z, Madigan CF, Swager TM, Bulovic V. Sensitivity gains in chemosensing by lasing action in organic polymers. Nature 2005;434:876–9.

[116] Hrapovic S, Majid E, Liu Y, Male K, Luong JHT. Metallic nanoparticle–carbon nanotube composite for electrochemical determination of explosive nitroaromatic compounds. Anal Chem 2006;78:5504–12.

[117] Cattanach K, Kulkarni RD, Kozlov M, Manohar SK. Flexible carbon nanotube sensors for nerve agent stimulants. Nanotechnology 2006;17:4123–8.

[118] Kwon J-H, Lee K-S, Lee Y-H, Ju B-K. Single-walled carbon nanotube-based pH sensor fabricated by the spray method. Electrochem Solid State Lett 2006;9: H85–7.

CHAPTER 10

Mechanics of Stiff Thin Films of Controlled Wavy Geometry on Compliant Substrates for Stretchable Electronics

Jianliang Xiao[1], Hanqing Jiang[2], Yonggang Huang[1,3] and John A. Rogers[4]

[1]*Department of Mechanical Engineering, Northwestern University, Evanston, Illinois, USA*

[2]*Department of Mechanical and Aerospace Engineering, Arizona State University, Tempe, Arizona, USA*

[3]*Department of Civil and Environmental Engineering, Northwestern University, Evanston, Illinois, USA*

[4]*Department of Materials Science and Engineering, Beckman Institute, and Seitz Materials Research Laboratory, University of Illinois at Urbana–Champaign, Illinois, USA*

INTRODUCTION

Stretchable electronics have important applications in many emerging technologies, such as paper-like displays [1, 2], electronic eyes [3], bending actuators [4], synthetic sensitive skins [5, 6] and structural health monitoring devices [7]. There are two main approaches to achieve stretchability in electronics. One design uses isolated, rigid islands which are fabricated with semiconductor components (e.g. transistors) on top and linked by stretchable interconnects of conducting materials [8–14]. Another different, but complementary, method is to directly produce stretchable electronic devices based on well-defined sinusoidal distributions of surface relief through non-linear buckling processes [15]. Field-effect transistors, p–n diodes and other devices for electronic circuits can be directly integrated into semiconductor materials (e.g. silicon) to yield fully stretchable components. The latter approach can be applied not only in stretchable electronics but also in precision metrology [16, 17], and thus has attracted attention in many fields of research [18–25]. It has also been extended to two-dimensional buckling of nanomembranes [23–29] and molecular scale (e.g. carbon nanotubes) buckling [30, 31] to realize two-dimensional stretchable electronics and stretchable nanoelectronics, respectively.

CONTENTS

Semiconductor Nanomaterials for Flexible Technologies

275

In spite of many attractive features, owing to the intrinsic buckling process, the second approach has some limitations. First, the wavy semiconductor ribbons are formed spontaneously from buckling with fixed amplitudes and wavelengths determined by material properties (e.g. moduli and thickness), without any direct control over the geometries or the phases of the waves. Second, although already improved significantly from that of silicon (Si) itself (~1%), the maximum strains that can be accommodated are in the range of 20–30%, limited by the non-optimal wavy geometries that result from this process. Third, buckling introduces initial compressive strains into the brittle semiconductor materials, which yields high stretchability at the expense of reduced compressibility.

This chapter reviews two alternative approaches, including both experimental and theoretical studies, which overcome some or all of the limitations described above and thus have the potential for extensive applications in stretchable electronics and other emerging technologies.

CONTROLLED BUCKLING OF SEMICONDUCTOR NANORIBBONS

To control the buckling geometries and thereby to improve the stretchability, Sun et al. [32] demonstrated a mechanical strategy for nanoribbons of GaAs and Si that combines lithographically patterned surface bonding chemistry and a buckling process similar to that originally reported by Khang et al. [15]. Periodic or aperiodic geometries of semiconductor nanoribbons are possible in this procedure, for any selected set of individual nanoribbons, in large-scale organized arrays of such structures [32]. Specialized geometries designed for stretchable electronics enable strain ranges of up to nearly 150% [32], even in brittle materials such as Si and GaAs, which is as much as 10 times larger than previously reported results [15].

Fabrication of buckled nanoribbons

Figure 10.1 shows the process for controlled buckling in nanoribbons [32]. First, a mask is prepared for patterning surface chemical adhesion sites on an elastomeric substrate of poly(dimethylsiloxane) (PDMS) [32]. The patterning involves passing deep ultraviolet (UV) light (240–260 nm) through an amplitude photomask (fabricated in step i), referred to as a UVO mask [34], in conformal contact with the PDMS. This mask supports recessed relief in the transparent regions; exposure to UV light creates patterned areas of ozone in these areas, next to the surface of the PDMS [34]. The ozone converts the native hydrophobic surface, dominated by

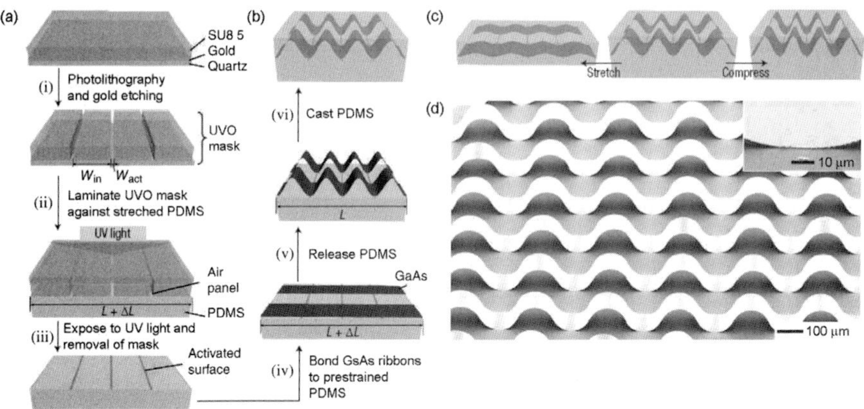

FIGURE 10.1 *Processing steps for engineering 3D buckled shapes in semiconductor nanoribbons. (a) Fabrication of a UVO mask and its use in patterning the surface chemistry on a poly(dimethylsiloxane) (PDMS) substrate. (b) Formation of buckled GaAs ribbons and the process of embedding them in PDMS. (c) Response of buckled GaAs ribbons to stretching and compressing. (d) SEM image of a sample formed using the procedures in (a) and (b). The prestrain used for generating this sample was 60%, with $W_{act} = 10$ µm and $W_{in} = 400$ µm. The inset shows an SEM image of a bonded region.*
(Reprinted with permission from [32] © 2006 Nature Publishing Group.)

$-OSi(CH_3)_2O-$ groups, to a highly polar and reactive surface (activated surface) that is terminated with $-O_nSi(OH)_{4-n}$ functionalities. This surface chemistry allows reactions with various inorganic materials to form strong chemical bonds [35]. The unexposed areas are unmodified by this process (inactivated surface). They interact with other surfaces only through weak van der Waals' forces [32, 36]. The procedures for controlled buckling involve such exposures on PDMS substrates (thickness ~4 mm) that are under large, uniaxial prestrains ($\varepsilon_{pre} = \Delta L/L$ for lengths changed from L to $L + \Delta L$) (step ii) [32]. For masks with patterns of lines and spaces, the widths of the activated (purple in the bottom frame of Figure 10.1a) and inactivated stripes are denoted by W_{act} and W_{in} (step iii). The activated areas can bond irreversibly to other materials (e.g. GaAs ribbons covered with SiO_2 films or Si ribbons with oxide layers) that have $-Si-OH$ groups on their surfaces due to strong siloxane linkages ($-O-Si-O$) that form through condensation reactions [35, 37]. The resulting adhesion sites can be exploited to create well-defined buckling geometries in nanoribbons, as described in the following [32].

The nanoribbons consisted of single-crystalline Si or GaAs [32]. The Si ribbons were prepared from silicon-on-insulator (SOI) wafers [15]. The GaAs ribbons were formed from multilayers of Si-doped n-type GaAs (120 nm;

carrier concentration of $4 \times 10^{17}/cm^3$), semi-insulating GaAs (SI-GaAs; 150 nm) and AlAs (200 nm) deposited on a (100) SI-GaAs wafer by molecular-beam epitaxy (MBE) [38]. Chemically etching through the epilayers and then removing the AlAs by hydrofluoric acid (HF) etching yielded the ribbons [32]. In the final step, a thin layer of SiO_2 (~30 nm) was deposited to provide the –Si–OH surface chemistry for bonding to PDMS [32].

By laminating the processed Si or GaAs ribbons, while still on the corresponding source wafers, against a UVO-treated prestretched PDMS substrate (ribbons along the direction of prestrain), baking in an oven at 90°C for 5 min and removing the wafer, all of the ribbons were transferred to the surface of the PDMS (step iv) [32]. Heating facilitated conformal contact and the formation of strong siloxane linkages in the activated areas, while weak van der Waals' interactions held the ribbons to the PDMS in the inactivated regions. Relaxing the strain in the PDMS caused the formation of buckles through the physical separation of the ribbons from the inactivated regions (step v) [32]. The ribbons remained bonded to the PDMS in the activated regions owing to the strong chemical interactions. The resulting pattern of buckles depends on the prestrain and the patterns of surface activation. For the case of a simple line pattern, W_{in} and the prestrain determine the widths and amplitudes. Sinusoidal waves with wavelengths and amplitudes much smaller than the buckles also formed in the same ribbons when W_{act} was >100 μm, owing to mechanical instabilities of the type that generate wavy ribbons of Si [15]. As a final step in the fabrication, the three-dimensional (3D) ribbon structures can be encapsulated in PDMS by casting and curing a liquid prepolymer (step vi). In this case, the liquid flows and fills the gaps formed between the ribbons and the substrate [32].

Figure 10.1(d) shows a scanning electron microscope (SEM) image of buckled GaAs ribbons on PDMS, where $\varepsilon_{pre} = 60\%$, $W_{act} = 10$ μm and $W_{in} = 400$ μm [32]. The image shows buckles with common geometries and spatially coherent phases for all ribbons in the array. The anchoring points coincide precisely with the lithographically defined adhesion sites. The inset shows an image of a bonded region; the width is ~10 μm, consistent with W_{act}. The results also reveal that the PDMS is approximately flat, even at the bonding sites [32]. This behavior is different from the strongly coupled wavy structures reported previously [15]; it suggests that the PDMS induces the displacements but is not intimately involved in the buckling process (that is, its modulus does not affect the geometries of the ribbons). In this sense, the PDMS represents a soft, non-destructive tool for manipulating the ribbons through forces applied at the adhesion sites [32].

Mechanics of controlled buckling of nanoribbons on elastomeric substrate

The procedures mentioned above provide a means to precisely control the buckling geometry and significantly improve the stretchability. Underlying mechanics analysis establishes the key relationships between the geometries of interfacial patterns and the buckling profiles, which yields design tools of prediction and optimization.

Methods of energy minimization are used to determine the buckling geometry [33]. The thin film is modeled as an elastic non-linear von Karman beam since the film thickness ($\sim 0.1\ \mu m$) is much smaller than any characteristic length in the film plane such as the buckling wavelength ($\sim 200\ \mu m$) and film width and length and the strain in the thin film is negligibly small [39]. The first requirement is always satisfied and the second one, to be shown later, also holds for this type of thin film/substrate system, which ensures that the von Karman approximation is always valid. The PDMS substrate is modeled as a semi-infinite elastic media because its thickness (a few millimeters) is four orders of magnitude larger than the film thickness ($\sim 0.1\ \mu m$) [33]. The total energy has three parts, the bending energy U_b due to thin film buckling, membrane energy U_m in the thin film, and energy U_s in the substrate.

Relaxing the prestrain in the patterned PDMS imposes compressive strains in the thin film, which induces buckling over the inactivated (weak bonding) region shown in Figure 10.1. The buckling profile can be expressed as [33]:

$$w = \begin{cases} w_1 = \dfrac{1}{2}A\left(1 + \cos\dfrac{\pi x}{L_1}\right), & -L_1 < x < L_1, \\ w_2 = 0, & L_1 < |x| < L_2, \end{cases} \tag{10.1}$$

where A is the buckling amplitude to be determined, x is the ribbon direction, $2L_1 = W_{in}/(1 + \varepsilon_{pre})$ is the buckling wavelength, $2L_2 = W_{in}/(1 + \varepsilon_{pre}) + W_{act}$ is the sum of activated and inactivated regions after relaxation (Figure 10.1b), and $-\varepsilon_{pre}$ is the compressive strain due to the relaxation of prestrain ε_{pre} in the substrate. The bending energy U_b in the thin film can be obtained from Eqn (10.1) as [33]:

$$U_b = \int_{-L_2}^{L_2} \frac{1}{2}\frac{\bar{E}_f h^3}{12}\left(\frac{d^2 w}{dx^2}\right)^2 dx = \frac{\bar{E}_f h^3 \pi^4 A^2}{96 L_1^3}, \tag{10.2}$$

where the bending rigidity is $\bar{E}_f h^3/12$, h is the film thickness, $\bar{E}_f = E_f/(1 - \nu_f^2)$, E_f and ν_f are the plane-strain modulus, Young's modulus and Poisson's ratio, respectively.

The membrane energy in the thin film is determined by the membrane strain ε_m, which is related to the displacement w in Eqn (10.1) and the in-plane displacement u by:

$$\varepsilon_m = du/dx + 1/2(dw/dx)^2, \tag{10.3}$$

The shear stress at the thin film/substrate interface is negligibly small since Young's modulus of PDMS substrate ($\sim$1 MPa) is five orders of magnitude smaller than the Si thin film ($\sim$100 GPa) [27], which leads to a constant membrane force and an in-plane displacement [33]:

$$u = \begin{cases} \dfrac{A^2\pi^2}{16L_1}(L_2 - L_1)x + \dfrac{A^2\pi}{32L_1}\sin\dfrac{2\pi x}{L_1} - \varepsilon_{pre}x, & -L_1 < x < L_1, \\[2ex] \left(\dfrac{A^2\pi^2}{16L_1L_2} - \varepsilon_{pre}\right)x - \dfrac{A^2\pi^2}{16L_1}\text{sign}(x), & L_1 < |x| < L_2. \end{cases} \tag{10.4}$$

The membrane strain then becomes a constant:

$$\varepsilon_m = \frac{A^2\pi^2}{16L_1L_2} - \varepsilon_{pre} \tag{10.5}$$

The membrane energy in the thin film is given by [33]:

$$U_b = \int_{-L_2}^{L_2} \frac{1}{2}\bar{E}_f h\varepsilon_m^2 dx = \bar{E}_f h L_2\varepsilon_m^2 \tag{10.6}$$

As shown in the SEM image of the GaAs thin film/PDMS substrate interface in the inset of Figure 10.1(d), the PDMS surface remains flat after relaxation for both activated and inactivated regions. Since there is vanishing displacement and stress traction at the PDMS substrate, the relaxed PDMS has zero energy $U_S = 0$.

Minimization of total energy $\partial U_{total}/\partial A = 0$ with respect to the buckling amplitude A gives [33]:

$$A = \frac{4}{\pi}\sqrt{L_1L_2(\varepsilon_{pre} - \varepsilon_c)} \quad \text{for } \varepsilon_{pre} > \varepsilon_c \tag{10.7}$$

where $\varepsilon_c = h^2\pi^2/12L_1^2$ is the critical strain for buckling (or the minimal strain to trigger the buckling), which is identical to the Euler buckling strain for a doubly clamped beam with thickness h, length $2L_1$ and modulus E_f. Once ε_{pre} exceeds ε_c, the membrane strain obtained from Eqn (10.5) is $-\varepsilon_c$, i.e. the thin film buckles and adjusts its buckling amplitude A such that the (compressive) membrane strain remains at ε_c.

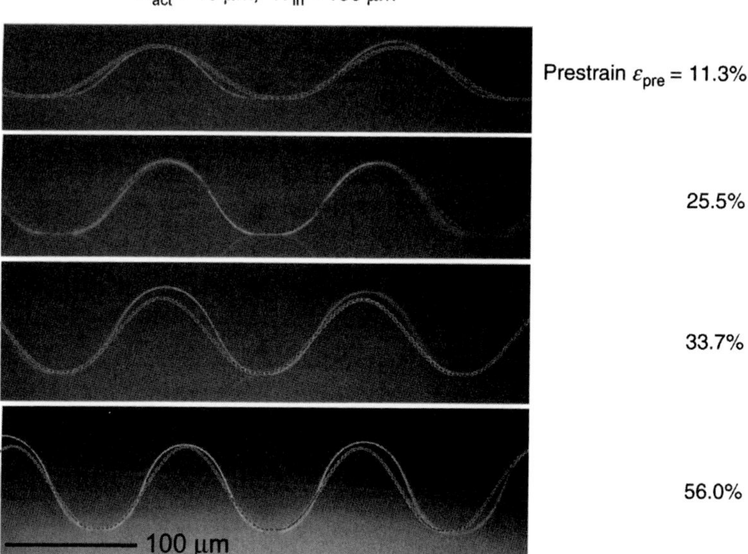

FIGURE 10.2 *Buckled GaAs thin films on patterned poly(dimethylsiloxane) (PDMS) substrate with $W_{act} = 10\ \mu m$ and $W_{in} = 190\ \mu m$ for different prestrain levels, 11.3%, 25.5%, 33.7% and 56.0% (from top to bottom). The red lines are the profiles of the buckled GaAs thin film predicted by the analytical solution.*
(Reprinted with permission from [33] © 2007 American Institute of Physics.)

This analytical solution agrees very well with experiments, as shown in Figure 10.2. The red lines are the profiles of the buckled GaAs thin film given by Eqns (10.1) and (10.7) for different prestrain levels with the same layout of interfacial patterns $W_{act} = 10\ \mu m$ and $W_{in} = 190\ \mu m$. Good agreement between analytical solutions and experiments is observed for both amplitude and wavelength, except for low prestrain (e.g. $\varepsilon_{pre} = 11.3\%$). This discrepancy is due to the assumption that the thin film buckles over the entire inactivated region, which may not hold at low prestrain [33].

Since the critical strain $\varepsilon_c = h^2\pi^2/12L_1^2$ is in the order of 10^{-6} for the buckling wavelength $2L_1 \sim 200\ \mu m$ and thin film thickness $h \sim 0.1\ \mu m$ [33], the buckling amplitude A in Eqn (10.7) is approximated as

$$A \approx \frac{4}{\pi}\sqrt{L_1 L_2 \varepsilon_{pre}} = \frac{2}{\pi(1 + \varepsilon_{pre})}\sqrt{W_{in}[W_{in} + W_{act}(1 + \varepsilon_{pre})]\varepsilon_{pre}} \qquad (10.8)$$

which does not depend on the thin film properties (e.g. thickness, Young's modulus) and is solely determined by the layout of interfacial patterns (W_{in} and W_{act}) and the prestrain [33]. This generic conclusion suggests a broader application of this approach: thin films made of any materials will form into almost the same buckled geometries (shown in Figure 10.1d) for the same interfacial patterns [33]. This approach was used to obtain very similar buckled geometries for Si and GaAs.

One of the most important issues in stretchable electronics is to reduce the strain level in electronics. Since the membrane strain ε_m is negligibly small (10^{-6}), the maximum strain in the thin film is the bending strain due to the local curvature d^2w/dx^2. This gives [33]:

$$\varepsilon_{\max} = \frac{h}{2}\max\left(\frac{d^2w}{dx^2}\right) = \frac{h\pi}{L_1^2}\sqrt{L_1 L_2 \varepsilon_{\mathrm{pre}}} \tag{10.9}$$

For the activated region W_{act} much smaller than the inactivated region W_{in}, Eqn (10.9) can be approximated by $\varepsilon_{\max} \approx (h\pi/L_1)\sqrt{\varepsilon_{\mathrm{pre}}}$ [33]. It is much smaller than the prestrain for a thin film if thickness h is much smaller than wavelength $2L_1$. For example, the maximum strain is only 0.6% for a 0.3 μm thin GaAs film buckled on a patterned PDMS substrate with $W_{\mathrm{act}} = 10$ μm, $W_{\mathrm{in}} = 400$ μm, and $\varepsilon_{\mathrm{pre}} = 60\%$, which is two orders of magnitude smaller than the prestrain. Also, this infinitesimal strain (0.6%) ensures that the von Karman beam assumption is always valid in this study. The precisely controlled buckling can significantly reduce the maximum strain in thin film, and improve the system stretchability.

STRETCHABLE AND COMPRESSIBLE STIFF THIN FILMS ON COMPLIANT WAVY SUBSTRATES

The approach introduced in the preceding section provides precise control over the buckling geometries of semiconductor nanoribbons and achieves very high stretchability, but the buckling process unavoidably produces initial compressive strain in the semiconductor nanoribbons, which significantly reduces the compressibility. Xiao et al. [40, 41] presented an alternative design that involves stiff thin films conformally integrated with a compliant substrate with prefabricated wavy structures of relief on its surface, thereby avoiding any initial film strain and thus achieving both high stretchability and compressibility. Furthermore, this approach also provides precise control over the wavy geometries.

Fabrication of stiff thin films on compliant wavy substrates

Fabrication of contoured compliant substrates involved a series of molding and imprinting steps from a rigid Si template [40]. In the first phase,

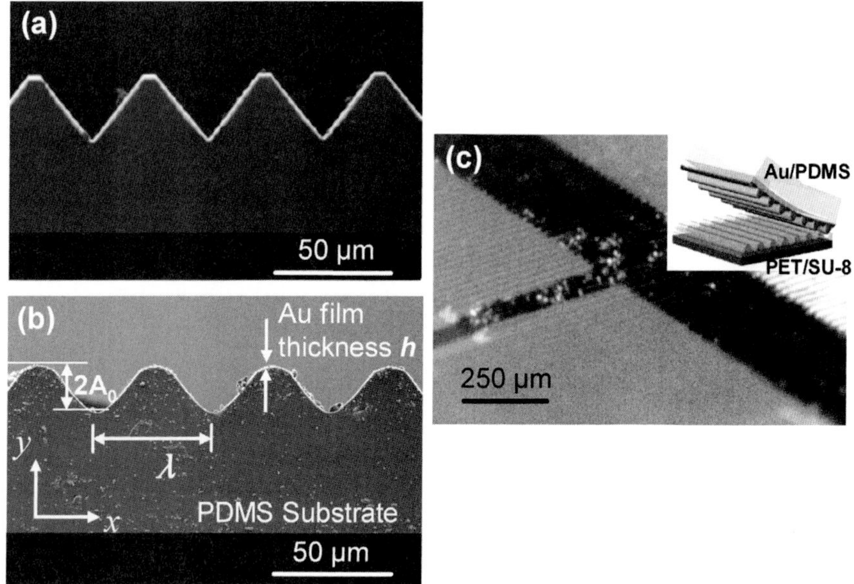

FIGURE 10.3 *Profile of the wavy compliant substrate and stiff thin-film system. Anisotropic etching of silicon produced a saw-tooth pattern (a), which formed the underlying sinusoidal structure for the wavy profile. After a sequential process of smoothing, molding and metallization, a final poly(dimethylsiloxane) (PDMS) substrate with a thin (300 nm) gold film was obtained (b). Here, $2A_0$ is the peak to valley initial amplitude, λ the wavelength, and h the thickness of the stiff thin film. (c) Optical image of an isolated wavy gold film created in this manner. The inset shows delamination of the contoured Au/PDMS substrate from the polyethylene terephthalate (PET)/SU-8 mold. (Reprinted with permission from [40] © 2008 American Institute of Physics.)*

plasma-enhanced chemical vapor deposition formed a thin (200 nm) silicon nitride film on a Si (100) wafer (SQI, Inc.); photolithography and etching defined patterns of lines and spaces in the nitride. After removing the photoresist, anisotropic etching of the Si (100) in a hot (90°C) isopropyl alcohol-buffered 25% KOH solution for 35 min produced a saw-tooth structure (Figure 10.3a) [40]. Removing the nitride mask with concentrated (49%) HF and then spin-casting a thin layer of photoresist (MicroChem S1805) at 3500 rpm for 90 s partially planarized the saw-tooth relief to yield a roughly sinusoidal shape. Replication, by two consecutive casting and curing steps, converted this structure into PDMS [40].

This process represents the first step of the second phase of the fabrication; it reproduces the surface profile of the Si/photoresist structure in a thin (300 µm) layer of PDMS on a glass substrate [40]. Pressing this PDMS/glass element against a layer (75 μm) of negative tone photoresist (SU-8 50; MicroChem Corp.) on a thin (100 μm) plastic substrate (PET), and then flood exposing the system to UV light ($\lambda = 365$ nm) for 30–40 s from both the top and bottom followed by a 5 min bake at 65°C cured the resist. Carefully removing the PDMS/glass yielded a molded SU-8 substrate with the inverted saw-tooth profile of the Si wafer. Spin-casting a layer of SU-8 2 (MicroChem Corp.) smoothed the relief valleys in the molded SU-8. Electron-beam deposition (Temescal BJD1800) of Au through a shadow mask produced patterned films (~300 nm thick), conformally coating the contoured SU-8 surface. Treating this substrate with mercaptopropyltrimethoxysilane (MPTMS) prepared the sample for a final PDMS molding step. After casting and curing the PDMS prepolymer, fast removal of the stamp transferred the metal films to the PDMS, resulting in the wavy Au-coated profile seen in Figure 10.3(b) [40]. Figure 10.3(c) presents an optical image of an isolated thin film (300 nm) on a wavy substrate. The mechanics of these structures were examined by surface profilometry (Dektak 3030) as a function of uniaxial strain [40].

Mechanics of stiff thin films on compliant wavy substrates

Energy minimization, conceptually similar to that described for the buckled system introduced previously, is used to calculate the change in profile amplitude under applied strain and further the maximum strain supported by the stiff thin film [40, 41]. The total system energy is composed of three parts, the bending energy U_b due to curvature change of the thin film, the membrane energy of thin film U_m, and substrate energy U_S [40, 41]. Figure 10.3(b) shows a stiff thin film of thickness h on a compliant substrate of a wavy surface profile $y = A_0 \cos kx$, where A_0 is the initial amplitude, $k = 2\pi/\lambda$, and λ is the wavelength. For small initial amplitudes, $A_0 \ll \lambda$, the curvature of the film is [40]:

$$\kappa_0 = \mathrm{d}^2 y/\mathrm{d}x^2 = -A_0 k^2 \cos kx. \qquad (10.10)$$

After application of a small uniaxial strain ε_a to the film/substrate system, the curvature change is [40]:

$$\Delta\kappa = \mathrm{d}^2 w/\mathrm{d}x^2 = (A_0 - A)k^2 \cos kx, \qquad (10.11)$$

where A is the amplitude after deformation and $w = (A - A_0) \cos kx$ is the displacement in the y direction. The bending energy per unit length of the film can then be obtained as [40]:

$$U_{\mathrm{b}} = \frac{1}{\lambda} \int_0^{\lambda} \frac{1}{2} \bar{E}_{\mathrm{f}} I (\Delta \kappa)^2 \mathrm{d}x = \frac{1}{48} \bar{E}_{\mathrm{f}} h^3 (A_0 - A)^2 k^4 \tag{10.12}$$

where $\bar{E}_{\mathrm{f}} = E_{\mathrm{f}}/(1 - \nu_{\mathrm{f}}^2)$ is the plane-strain modulus of the film and $\bar{E}_{\mathrm{f}} I = \bar{E}_{\mathrm{f}} h^3/12$ is the bending stiffness.

The membrane strain ε_{m} in the thin film depends on both the displacements in x and y direction, u and w. For an infinitesimal segment of the film with initial length $\mathrm{d}S = \sqrt{(\mathrm{d}x)^2 + (\mathrm{d}y)^2}$, it is elongated to $\mathrm{d}s = \sqrt{(\mathrm{d}x + \mathrm{d}u)^2 + (\mathrm{d}y + \mathrm{d}w)^2}$ after deformation. The membrane strain is $\varepsilon_{\mathrm{m}} = (\mathrm{d}s - \mathrm{d}S)/\mathrm{d}S$ and can be simplified to [40]:

$$\varepsilon_{\mathrm{m}} = \frac{\mathrm{d}u}{\mathrm{d}x} + \frac{1}{2} \left(\frac{\mathrm{d}w}{\mathrm{d}x} \right)^2 + \frac{\mathrm{d}y}{\mathrm{d}x} \frac{\mathrm{d}w}{\mathrm{d}x} \tag{10.13}$$

for small amplitudes $A_0 \ll \lambda$ and small strains, but finite rotation. Since the elastic modulus of the PDMS substrate ($\sim$2 MPa) is several orders of magnitude smaller than that of Au film ($\sim$70 GPa), the effect of interface shear is negligible [24]. This conclusion is confirmed independently by the finite element analysis presented later. Force equilibrium of the film requires a constant membrane strain for film thickness $h = \lambda$, giving the displacement [40]:

$$u = \varepsilon_{\mathrm{a}} x - \frac{1}{8} (A_0^2 - A^2) k \sin 2kx, \tag{10.14}$$

where ε_{a} is the applied strain in the x direction, and the condition $\int_0^{\lambda} u' \mathrm{d}x = \varepsilon_{\mathrm{a}} \lambda$ has been imposed to ensure the compatibility of deformation in the film and substrate. The membrane strain in the film is then obtained as [40]:

$$\varepsilon_{\mathrm{m}} = \varepsilon_{\mathrm{a}} - \frac{1}{4} (A_0^2 - A^2) k^2. \tag{10.15}$$

The membrane energy per unit length of the film is given by integration [40]:

$$U_{\mathrm{m}} = \frac{1}{\lambda} \int_0^{\lambda} \frac{1}{2} \bar{E}_{\mathrm{f}} h \varepsilon_{\mathrm{m}}^2 \mathrm{d}x = \frac{1}{2} \bar{E}_{\mathrm{f}} h \left[\varepsilon_{\mathrm{a}} - \frac{1}{4} (A_0^2 - A^2) k^2 \right]^2. \tag{10.16}$$

The substrate is modeled as a semi-infinite solid since its thickness is much larger than the wavelength λ. The wavy surface profile, $y = A_0 \cos kx$, can be transformed to a straight line in a new coordinate system (x, η), where

$\eta = y - A_0 \cos kx$ [40, 41]. Strains are then obtained from the displacements u_x and u_y as [41]:

$$\varepsilon_{xx} = \frac{\partial u_x}{\partial x} + \frac{\partial u_x}{\partial \eta} A_0 k \sin kx,$$

$$\varepsilon_{yy} = \frac{\partial u_y}{\partial \eta}, \tag{10.17}$$

$$2\varepsilon_{xy} = \frac{\partial u_x}{\partial \eta} + \frac{\partial u_y}{\partial x} + \frac{\partial u_y}{\partial \eta} A_0 k \sin kx.$$

The substrate stress is determined via the linear elastic relation (Young's modulus E_S and Poisson's ratio ν_S) as $\sigma_{xx} = \frac{E_S}{(1+\nu_S)(1-2\nu_S)}[(1-\nu_S)\varepsilon_{xx} + \nu_S\varepsilon_{yy}]$, $\sigma_{yy} = \frac{E_S}{(1+\nu_S)(1-2\nu_S)}[(1-\nu_S)\varepsilon_{yy} + \nu_S\varepsilon_{xx}]$ and $\sigma_{xy} = \frac{E_S}{1+\nu_S}\varepsilon_{xy}$ [41]. The equilibrium equations $\sigma_{xx,x} + A_0 k \sin kx \sigma_{xx,\eta} + \sigma_{xy,\eta} = 0$ and $\sigma_{xy,x} + A_0 k \sin kx \sigma_{xy,\eta} + \sigma_{yy,\eta} = 0$ provide two partial differential equations for u_x and u_y [41]:

$$(1-2\nu)\left(\frac{\partial^2 u_x}{\partial x^2} + \frac{\partial^2 u_x}{\partial \eta^2}\right) + \frac{\partial}{\partial x}\left(\frac{\partial u_x}{\partial x} + \frac{\partial u_y}{\partial \eta}\right) + A_0 k \sin kx\left[4(1-\nu)\frac{\partial^2 u_x}{\partial x \partial \eta} + \frac{\partial^2 u_y}{\partial \eta^2}\right]$$

$$+ 2(1-\nu)A_0 k^2 \cos kx \frac{\partial u_x}{\partial \eta} + 2(1-\nu)A_0^2 k^2 \sin^2 kx \frac{\partial^2 u_x}{\partial \eta^2} = 0,$$

$$(1-2\nu)\left(\frac{\partial^2 u_y}{\partial x^2} + \frac{\partial^2 u_y}{\partial \eta^2}\right) + \frac{\partial}{\partial \eta}\left(\frac{\partial u_x}{\partial x} + \frac{\partial u_y}{\partial \eta}\right) + A_0 k \sin kx\left[2(1-2\nu)\frac{\partial^2 u_y}{\partial x \partial \eta} + \frac{\partial^2 u_x}{\partial \eta^2}\right]$$

$$+ (1-2\nu)A_0 k^2 \cos kx \frac{\partial u_y}{\partial \eta} + (1-2\nu)A_0^2 k^2 \sin^2 kx \frac{\partial^2 u_y}{\partial \eta^2} = 0. \tag{10.18}$$

Boundary conditions for the top surface, $\eta = 0$, are $u_y = (A - A_0)\cos kx$ and a vanishing shear traction, $\tau_{nt} = 0$, due to negligible interface shear. The remote boundary conditions require $\varepsilon_{xx}|_{\eta \to -\infty} = \varepsilon_a$ due to the applied strain and vanishing tractions. For small amplitude, $A_0 \ll \lambda$, the displacements can be written as [41]:

$$u_x = u_x^{(0)} + (A_0 k)u_x^{(1)} + (A_0 k)^2 u_x^{(2)} + \ldots$$

$$u_y = u_y^{(0)} + (A_0 k)u_y^{(1)} + (A_0 k)^2 u_y^{(2)} + \ldots \tag{10.19}$$

Substitution into the equilibrium equations (10.18) and boundary conditions, together with the perturbation method, gives the analytical solutions [40, 41]:

$$u_x^{(0)} = \varepsilon_a x, \quad u_y^{(0)} = -\frac{\nu_S}{1-\nu_S}\varepsilon_a \eta, \tag{10.20}$$

$$u_x^{(1)} = \left[(1 - 2\nu_S + k\eta)\frac{A - A_0}{A_0} - (3 - 2\nu_S + k\eta)\varepsilon_a\right]\frac{e^{k\eta}\sin kx}{2k(1 - \nu_S)},$$

$$u_y^{(1)} = \left\{-2\nu_S\varepsilon_a + \left[(2 - 2\nu_S - k\eta)\frac{A - A_0}{A_0} + (2\nu_S + k\eta)\varepsilon_a\right]e^{k\eta}\right\}\frac{\cos kx}{2k(1 - \nu_S)},$$

$$\tag{10.21}$$

$$u_x^{(2)} = \frac{e^{k\eta}\sin 2kx}{16k(1 - \nu_S)^2}\left\{\begin{array}{l} 4(1 - \nu_S)\left[\dfrac{A - A_0}{A_0}(2 - 2\nu_S + k\eta) - \varepsilon_a(4 - 2\nu_S + k\eta)\right] \\[2ex] -e^{k\eta}\left\{\begin{array}{l}\dfrac{A - A_0}{A_0}[5 - 12\nu_S + 8\nu_S^2 + 2(3 - 4\nu_S)k\eta] \\[1.5ex] -\varepsilon_a[19 - 28\nu_S + 8\nu_S^2 + 2(5 - 4\nu_S)k\eta]\end{array}\right\} \end{array}\right\},$$

$$u_y^{(2)} = \frac{e^{k\eta}\cos 2kx}{16k(1 - \nu_S)^2}\left\{\begin{array}{l} 4(1 - \nu_S)\left[\dfrac{A - A_0}{A_0}(1 - 2\nu_S - k\eta) + \varepsilon_a(1 + 2\nu_S + k\eta)\right] \\[2ex] -2e^{k\eta}\left\{\begin{array}{l}\dfrac{A - A_0}{A_0}[2(1 - 3\nu_S + 2\nu_S^2) - (3 - 4\nu_S)k\eta] \\[1.5ex] +\varepsilon_a[2(1 + \nu_S - 2\nu_S^2) + (5 - 4\nu_S)k\eta]\end{array}\right\} \end{array}\right\}$$

$$+ \frac{1}{4k(1 - \nu_S)}\left\{\begin{array}{l} e^{k\eta}\left[\dfrac{A - A_0}{A_0}(1 - 2\nu_S - k\eta) + \varepsilon_a(1 + 2\nu_S + k\eta)\right] \\[2ex] -\left[\dfrac{A - A_0}{A_0}(1 - 2\nu_S) + \varepsilon_a(1 + 2\nu_S)\right] \end{array}\right\}$$

$$\tag{10.22}$$

The strain energy per unit length of the substrate can be obtained by integrating $\sigma_{ij}\varepsilon_{ij}/2$ over the substrate. The derivative with respect to A is [41]:

$$\frac{dU_S}{dA} = \frac{1}{4}\bar{E}_S k(A - A_0) + \frac{1}{4}\bar{E}_S\varepsilon_a k A_0 \tag{10.23}$$

where $\bar{E}_S = E_S/(1 - \nu_S^2)$ is the plane-strain modulus.

Minimization of total energy with respect to the profile amplitude, $\partial U_{\mathrm{TOT}}/\partial A = 0$, yields a cubic equation for A as [40]:

$$(A/A_0)^3 + p(A/A_0) - q = 0 \tag{10.24}$$

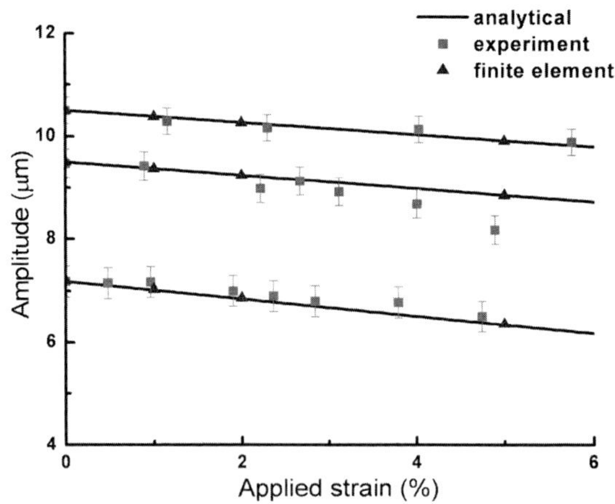

FIGURE 10.4 *Amplitude A versus the applied strain ε_a for a 0.3 μm thick Au film ($E_f = 78$ GPa, $\nu_f = 0.42$) on a wavy poly(dimethylsiloxane) (PDMS) substrate ($E_S = 2$ MPa, $\nu_S = 0.48$) with wavelength $\lambda = 49$ μm and several initial amplitudes $A_0 = 7.2$, 9.5 and 10.5 μm.*

(Reprinted with permission from [40] © 2008 American Institute of Physics.)

where $p = 4\varepsilon_a/(kA_0)^2 + h^2/(3A_0^2) + 2\bar{E}_S/(\bar{E}_f k^3 h A_0^2) - 1$ and $q = h^2/(3A_0^2) + 2(1 - \varepsilon_a)\bar{E}_S/(\bar{E}_f k^3 h A_0^2)$. For small applied strains, this equation can be solved analytically for the amplitude [40]:

$$A = A_0 \left[\frac{1 - 6\varepsilon_a(\bar{E}_S + 2\bar{E}_f kh)}{6\bar{E}_S + \bar{E}_f kh(k^2 h^2 + 6k^2 A_0^2)} \right] \tag{10.25}$$

As is evident from Eqn (10.25), the amplitude decreases linearly with the applied strain. Figure 10.4 compares simulated and measured amplitude changes for a 0.3 μm thick Au film ($E_f = 78$ GPa, $\nu_f = 0.42$) on a wavy PDMS substrate ($E_S = 2$ MPa, $\nu_S = 0.48$) under small uniaxial tension. The profile wavelength was held constant at $\lambda = 49$ μm for each of the three initial amplitudes $A_0 = 7.2$, 9.5 and 10.5 μm examined. The experimental measurements in Figure 10.4 (denoted by squares) and finite element method results (marked by triangles) agree remarkably well with the analytical solution of Eqn (10.25) without parameter fitting [40].

The maximum strain ε_{max} supported by the thin film can be determined by the sum of the membrane strain ε_m and the bending strain, $\varepsilon_b = \Delta\kappa h/2$ as [40]:

$$\varepsilon_{max} = \frac{\varepsilon_a 3\bar{E}_S(2 - k^2 A_0^2 + k^2 A_0 h) + \bar{E}_f k^3 h^2(6A_0 + h)}{6\bar{E}_S + \bar{E}_f k^3 h(6A_0^2 + h^2)} \tag{10.26}$$

Figure 10.5 shows the ratio of applied strain to the maximum film strain, $\varepsilon_a/\varepsilon_{max}$, versus the ratio of initial amplitude to wavelength, A_0/λ, for $A_0/h = 10$, 50 and 100 [40]. Each curve in Figure 10.5 has a plateau at large initial amplitude (e.g. $A_0 > \lambda/5$), indicating the maximum ratio of $\varepsilon_a/\varepsilon_{max}$. For very stiff ($\bar{E}_f \gg \bar{E}_S$) and thin films ($h \ll A_0$), the plateau is given simply by A_0/h such that the system stretchability (maximum strain before failure) is $A_0\varepsilon_c/h$, where ε_c is the failure strain of the thin film [40]. For Au thin film with a maximum elastic strain of $\varepsilon_c = 0.46\%$ [42] (beyond which plastic yielding occurs), the system stretchability is 4.6%, 23% and 46% for $A_0/h = 10$, 50 and 100, respectively [40]. Therefore, for a given film thickness h, a large initial amplitude and/or small wavelength, i.e. a wavier film (and substrate), reduces film strain, making the system more stretchable.

The above analyses and results also hold for compression. For example, the system compressibility is also given by $A_0\varepsilon_c/h$ for very stiff and thin films. This provides the contoured wavy film/substrate with an additional range of motion not captured by other buckling systems [40].

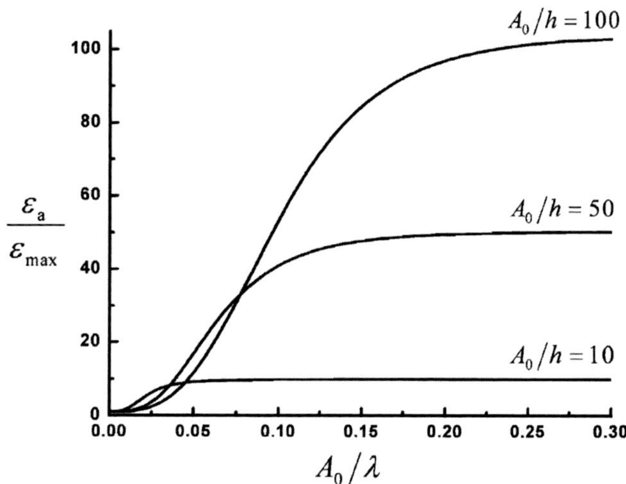

FIGURE 10.5 *Ratio of applied strain to the maximum film strain, $\varepsilon_a/\varepsilon_{max}$, versus the ratio of initial amplitude to wavelength, A_0/λ, obtained by the analytical solution for $A_0/h = 10$, 50 and 100, where h is the film thickness.*
(Reprinted with permission from [40] © 2008 American Institute of Physics.)

CONCLUSION

This chapter provides a review of two conceptually different, but related, approaches to achieve reversible, elastic response to large strain deformations in inorganic films that can be used for electronics. Recent work shows that these methods can be scaled to active devices (e.g. transistors, photodetectors, diodes and resistors) [14], circuits (e.g. logic gates, ring oscillators and differential amplifiers) [14] and full integrated systems (e.g. electronic eye cameras) [3]. Further optimization of the approaches and development of new application possibilities appear to represent promising directions for future work.

REFERENCES

[1] Rogers JA, Bao Z, Baldwin K, Dodabalapur A, Crone B, Raju VR, et al. Paper-like electronic displays: large-area rubber-stamped plastic sheets of electronics and microencapsulated electrophoretic inks. Proc Natl Acad Sci USA 2001;98: 4835–40.
[2] Jacobs HO, Tao AR, Schwartz A, Gracias DH, Whitesides GM. Fabrication of a cylindrical display by patterned assembly. Science 2002;296:323–5.
[3] Ko HC, Stoykovich MP, Song J, Malyarchuk V, Choi WM, Yu C-J, et al. A hemispherical electronic eye camera based on compressible silicon optoelectronics. Nature 2008;454:748–53.
[4] Watanabe M, Shirai H, Hirai T. Wrinkled polypyrrole electrode for electroactive polymer actuators. J Appl Phys 2002;92:4631–7.
[5] Lumelsky V, Shur MS, Wagner S. Sensitive skin. IEEE Sens J 2001;1:41–51.

[6] Someya T, Sekitani T, Iba S, Kato Y, Kawaguchi H, Sakurai T. A large-area, flexible pressure sensor matrix with organic field-effect transistors for artificial skin applications. Proc Natl Acad Sci USA 2004;101:9966–70.

[7] Nathan A, Park B, Sazonov A, Tao S, Chan I, Servati P, et al. Amorphous silicon detector and thin film transistor technology for large-area imaging of x-rays. Microelectron J 2000;31:883–91.

[8] Faez R, Gazotti WA, De Paoli MA. Amorphous silicon detector and thin film transistor technology for large-area imaging of x-rays. Polymer 1999;40:5497–503.

[9] Lacour SP, Wagner S, Huang ZY, Suo Z. Stretchable gold conductors on elastomeric substrates. Appl Phys Lett 2003;82:2404–6.

[10] Gray DS, Tien J, Chen CS. High-conductivity elastomeric electronics. Adv Mater 2004;16:393–9.

[11] Marquette CA, Blum LJ. Conducting elastomer surface texturing: a path to electrode spotting-application to the biochip production. Biosens Bioelectron 2004;20:197–203.

[12] Lacour SP, Jones J, Wagner S, Li T, Suo Z. Stretchable interconnects for elastic electronic surfaces. Proc IEEE 2005;93:1459–67.

[13] Someya T, Kato Y, Sekitani T, Iba S, Noguchi Y, Murase Y, et al. Conformable, flexible, large-area networks of pressure and thermal sensors with organic transistor active matrixes. Proc Natl Acad Sci USA 2005;102:12321–5.

[14] Kim D-H, Song J, Choi WM, Kim H-S, Kim R-H, Liu Z, et al. Materials and noncoplanar mesh designs for integrated circuits with linear elastic responses to extreme mechanical deformations. Proc Natl Acad Sci USA 2008;105:18675–80.

[15] Khang DY, Jiang HQ, Huang Y, Rogers JA. A stretchable form of single-crystal silicon for high-performance electronics on rubber substrates. Science 2006;311:208–12.

[16] Stafford CM, Harrison C, Beers KL, Karim A, Amis EJ, Vanlandingham MR, et al. A buckling-based metrology for measuring the elastic moduli of polymeric thin films. Nat Mater 2004;3:545–50.

[17] Nolte AJ, Rubner MF, Cohen RE. Determining the Young's modulus of polyelectrolyte multilayer films via stress-induced mechanical buckling instabilities. Macromolecules 2005;38:5367–70.

[18] Bowden N, Brittain S, Evans AG, Hutchinson JW, Whitesides GM. Spontaneous formation of ordered structures in thin films of metals supported on an elastomeric polymer. Nature 1998;393:146–9.

[19] Jiang H, Khang D-Y, Song J, Sun Y, Huang Y, Rogers JA. Finite deformation mechanics in buckled thin films on compliant supports. Proc Natl Acad Sci USA 2007;104:15607–12.

[20] Kim J, Lee HH. Wave formation by heating in thin metal film on an elastomer. J Polym Sci B 2001;39:1122–8.

[21] Lacour SP, Wagner S, Huang Z, Suo Z. Stretchable gold conductors on elastomeric substrates. Appl Phys Lett 2003;82:2404–6.

[22] Jiang H, Khang D-Y, Fei H, Kim H, Huang Y, Xiao J, Rogers JA. Finite width effect of thin-films buckling on compliant substrate: experimental and theoretical studies. J Mech Phys Solids 2008;56:2585–98.

[23] Chen X, Hutchinson JW. Herringbone buckling patterns of compressed thin films on compliant substrates. J Appl Mech 2004;71:597–603.

[24] Huang ZY, Hong W, Suo Z. Nonlinear analyses of wrinkles in a film bonded to a compliant substrate. J Mech Phys Solids 2005;53:2101–18.

[25] Huang R, Suo Z. Wrinkling of a compressed elastic films on a viscous layer. J Appl Phys 2002;91:1135–42.

[26] Chen X, Hutchinson JW. A family of herringbone patterns in thin films. Scr Mater 2004;50:797–801.

[27] Huang ZY, Hong W, Suo Z. Evolution of wrinkles in hard films on soft substrates. Phys Rev E 2004;70:030601.

[28] Peterson RL, Hobart KD, Kub FJ, Yin H, Sturm JC. Reduced buckling in one dimension versus two dimensions of a compressively strained film on a compliant substrate. Appl Phys Lett 2006;88:201913.

[29] Huang R, Im HS. Dynamics of wrinkle growth and coarsening in stressed thin films. Phys Rev E 2006;74:026214.

[30] Khang D-Y, Xiao J, Kocabas C, Maclaren S, Banks T, Jiang H, et al. Molecular scale buckling mechanics on individual aligned single-wall carbon nanotubes on elastomeric substrates. Nano Lett 2008;8:124–30.

[31] Xiao J, Jiang H, Khang D-Y, Wu J, Huang Y, Rogers JA. Mechanics of buckled carbon nanotubes on elastomeric substrates. J Appl Phys 2008;104:033543.

[32] Sun Y, Choi WM, Jiang H, Huang YY, Rogers JA. Controlled buckling of semiconductor nanoribbons for stretchable electronics. Nat Nanotechnol 2006; 1:201–7.

[33] Jiang H, Sun Y, Rogers JA, Huang Y. Mechanics of precisely controlled thin film buckling on elastomeric substrate. Appl Phys Lett 2007;90:133119.

[34] Childs WR, Motala MJ, Lee KJ, Nuzzo RG. Masterless soft lithography: patterning UV/ozone-induced adhesion on poly(dimethylsiloxane) surfaces. Langmuir 2005;21:10096–105.

[35] Duffy DC, McDonald JC, Schueller OJA, Whitesides GM. Rapid protyping of microfluidic systems in poly(dimethylsiloxane). Anal Chem 1998;70:4974–8.

[36] Huang YY, Zhou WX, Hsia KJ, Menard E, Park JU, Rogers JA, Alleyne AG. Stamp collapse in soft lithography. Langmuir 2005;21:8058–6.

[37] Sun Y, Rogers JA. Fabricating semiconductor nano/microwires and transfer printing ordered arrays of them onto plastic substrates. Nano Lett 2004;4:1953–9.

[38] Sun Y, Kumar V, Adesida I, Rogers JA. Buckled and wavy ribbons of GaAs for high-performance electronics on elastomeric substrates. Adv Mater 2006;18:2857–62.

[39] Timoshenko SP, Gere JM. Theory of elastic stability. New York: McGraw-Hill; 1961.

[40] Xiao J, Carlson A, Liu ZJ, Huang Y, Jiang H, Rogers JA. Stretchable and compressible thin films of stiff materials on compliant wavy substrates. Appl Phys Lett 2008;93:013109.

[41] Xiao J, Carlson A, Liu ZJ, Huang Y, Jiang H, Rogers JA. Analytical and experimental studies of the mechanics of deformation in a solid with a wavy surface profile. J Appl Mech-Trans ASME 2010;77: 011003.

[42] Weihs TP, Hong S, Bravman JC, Nix WD. Mechanical deflection of cantilever microbeams—a new technique for testing the mechanical properties of thin films. J Mater Res 1998;3:931–42.

Index

S

Samsung, 129, 143
scattering parameters/S parameters, 100, 121
Schottky barrier, 42, 43, 44, 76, 216, 220, 223
Schottky barrier height, 74, 76
Schottky contact(s), 74, 218, 220
screen printing, 119, 143, 166
secondary ion mass spectrum/SIMS, 78, 79, 177
seed-mediated hydrothermal reaction, 188
self-aligned source/drain formation method, 82
self-assembled block copolymer nanostructures, 206
self-assembly, 17, 117
self-assembly monolayer(s)/SAM(s), 137, 138, 139, 201, 205
self-sustaining electrolyte, 234
semiconducting carbon nanotube networks/s-CNNs, 107, 113, 115, 120, 121
semiconductor/electrolyte interfaces, 163, 164, 192
semi-infinite elastic media, 279
shadow masks, 144
sharp-bending microwave loss, 99
shear traction, 286
short circuit current (or photocurrent), 162, 177, 186, 191, 221
short circuiting, 99
short-open-load-through (SLOT) calibration, 84
shunt diode, 97
Si nanomembranes/SiNMs, 68, 77, 78, 79, 80, 82, 90, 91, 92, 93, 94, 95, 96
SiNM PIN diode, 95, 96
Si nanoribbons/microribbons, 169
Si nanowire transistors, 42
Si nanowires, 40, 42, 182, 189, 191, 240, 256, 265
p-Si nanowires, 45, 50
Si/PtSi interfaces, 42
Si/SiGe/Si sandwich strips, 72
SiGe alloy, 69, 70, 72, 73, 76
SiGe/Si/SiGe sandwich structure, 70
signal-ground-signal microwave probes, 121

silica nanoparticle, 107
silicon monoxide/SiO, 83, 98, 99, 101
silicon-on-insulator/SOI, 51, 72, 77, 78
SOI MOSFET, 43, 58
SOI source, 71, 93
SOI substrate(s), 81, 82, 85, 86, 88, 94, 95
SOI template layer, 77, 79
SOI wafers, 69, 91, 277
siloxane linkages, 277, 278
simulated AM, 1.5 illumination, 177
single mode optical waveguide, 53
single nanowire cavities, 47
single nanowire field-effect transistor /NW-FET, 40
single nanowire laser(s), 47, 53
single nanowire (optical) waveguides, 47, 51
single-crystal organic nanowires, 17
single-crystal semiconductor (or Si) nanomembranes, 76, 77, 97
single-crystalline silicon (Si), 105, 121
single-pole single-throw (SPST) switch, 95, 96
single-tube inverter, 118
single-walled carbon nanotubes/ SWNTs, 105, 106, 107, 108, 109, 110, 114, 115, 118, 121, 124, 135, 138, 139, 248, 251, 256, 258, 259, 267
aligned SWNTs, 108, 109, 124, 264
arrays of isolated SWNTs, 105
metallic SWNTs/m-SWNTs, 105, 106, 107, 108, 113, 115, 121, 263
m-SWNT pathways, 113
semiconducting SWNTs/s-SWNTs, 105, 107, 108, 115, 120, 121, 124
SWNT array transistors, 122
SWNT bundles, 105, 138
SWNT chips, 123
SWNT circuit, 119
SWNT field emitters, 138

SWNT film(s), 107, 110, 113, 115, 117, 124
transparent conductive SWNT films, 115
SWNT networks, 111, 264
SWNT paper, 235, 236
SWNT radio, 123
SWNT sensors, 263
SWNT thin film(s), 106, 107, 108, 112, 124
SWNT transistor radio, 122
SWNT-SWNT junctions, 108, 110, 115, 120
site-selective growth, 201
small signal models, 122
small-molecule (organic) semiconductors, 11, 17
small-signal equivalent circuit, 86, 87
small-signal scattering (S-) parameters, 84, 96
smart cards, 67, 228, 232, 240
Sn-based anode battery, 236
SnO_2 nanobelts, 254, 255
sodium dodecyl sulfate/ SDS, 251
solar battery, 160
solar cell(s), 27, 159, 160, 161, 162, 163, 170, 178, 189, 198
flexible solar cells, 159, 160, 163, 191, 192
solar electricity, 159
solar energy, 159, 164, 182
solar fuel, 159
solar panels, 160
solar thermal, 159
sol-gel process, 166
solid-state reaction diffusion, 34
solid-state reactions, 34, 42, 239
solid-vacuum potential barrier, 129
Sony, 228, 236
space charge region, 162
spin-on-glass, 143
Spint tip emitters, 130
spiral inductors, 98, 99, 100
spray-coating method, 20
SPST PIN diode shunt-series diode RF switch, 97
stand-alone optical waveguides, 52
standard test conditions/STC, 162
sticking coefficient, 32